Engineering Analysis Methods

Engineering Analysis Methods

I.S. Habib
University of Michigan
Dearborn

Lexington Books
D.C. Heath and Company
Lexington, Massachusetts
Toronto London

Library of Congress Cataloging in Publication Data

Habib, I. S.

Engineering analysis methods.

Bibliography: p.

Includes index.

1. Differential equations. 2. Differential equations, Partial. 3. Integral equations. I. Title.

TA347.D45H3 515'.35'02462 75-3786

ISBN 0-669-99671-8

Published simultaneously in Canada

Printed in the United States of America

International Standard Book Number: 0-669-99671-8

Library of Congress Catalog Card Number: 75-3786

To Nuha, Dina and Tania

Contents

List of Figures

Preface

In view of the increasing use of mathematical analysis for applications in engineering sciences, it has become evident that a notable improvement in mathematical preparations is required of those seeking careers in these areas. Furthermore, mathematics in physics and engineering in general has been presented in the literature as viewed by the mathematician, where the continuous emphasis on rigor has deterred students from developing interest in the subject, and has not left ample room for valuable and diversified physical applications.

In teaching applied mathematics to senior and graduate students at the University of Michigan—Dearborn, and in examining the nature and emphasis in engineering sciences at these levels in various institutions, it became clear to us that an existing and continuously increasing need confronted us and needed to be filled adequately.

The purpose of this text, therefore, is to provide students of science and engineering at the senior and first-year graduate level with the mathematical exposure that, from our experience, proved to be needed in the analysis stemming from the various diversified areas pertaining to students' interests. However, to retread the ground covered by others in the field of applied mathematics and with the same approach and emphasis would be wasteful. Therefore, in the treatment of the topics covered we emphasize the development of mathematical models for various physical problems. We draw applications from areas such as applied mechanics, fluid mechanics, heat transfer, electrical engineering, elasticity, chemical engineering, and related areas of physics.

It was decided in this text not to trade off the physical applications, the derivations of the mathematical models and statements, and the detailed physical interpretation of solutions, for an increasing mathematical rigor. It will be easier for an instructor to go to any extent of rigor; however, it is more difficult to present in the classroom diversified applications for mathematical models and to translate the mathematical forms into physical information more closely related to student experience in engineering sciences.

The philosophy of the treatment of the various subject stresses the following: (1) obtaining a mathematical model from the physical information encountered; (2) examining the model in light of its nature, whether it is a differential equation, integral equation, or any other mathematical expression; (3) treating the model by mathematical methods to yield the solution to the problem; and (4) interpreting the mathematical results as they apply to the original physical problem.

The bulk of the material in this text treats the areas of partial differential equations and integral equations. Because integral equations are appearing in the current literature with an increasing frequency, especially in connection with the practical computation of numerical solutions, this topic should take an important share of this text.

The first three chapters in the book treat materials considered to be a review and thus needed in preparing the student to analyze the material in the rest of the book. The first chapter covers basic methods of solving ordinary differential equations; the second chapter deals with series solution of ordinary differential equations and special functions; while the third treats the topic of Fourier series, including Sturm-Liouiville systems. Following this background material, the topic of classification of partial differential equations of the second order, which most commonly are encountered by students, is treated in chapter 4 in light of the transformation of such equations to canonical form for the hyperbolic, parabolic, and elliptic type equations. In chapter 5 partial differential equations of the hyperbolic type are treated with great detail. The wave equation is formulated for physical problems from various areas in engineering sciences, and methods of solution such as the method of separation of variables and the method of propagating waves are presented for problems that vary substantially in the complexities and applications. Chapter 6 treats partial differential equations of the parabolic type, such as the heat diffusion equation, while chapter 7 treats equations of the elliptic type. In chapters 8 and 9 the topic of linear integral equations is covered with emphasis on the methods that a student at the level of senior or first-year graduate is likely to use. One semester is considered adequate to cover the material in this text.

I would like to acknowledge the helpful comments presented by my students while taking my courses in engineering analysis. My sincere thanks goes to my friend and colleague T. Y. Na who read the manuscript and presented very constructive suggestions.

Suggestions and comments by the readers of this book will be gratefully received.

Engineering Analysis Methods

Linear Ordinary Differential Equations

Ordinary differential equations occupy a prominent place in applied mathematics, specifically in areas such as the study of response of elastic systems to external forces, the analysis of electric circuits and servomechanism, boundary value problems, thermal and mass diffusion, fluid mechanics, electromagnetic theory, and others. The most important and frequently occurring equations in these areas are linear. Therefore, it is important for the reader, prior to his exposure to the more advanced topics in the text, to be provided with a review of the basic analytical methods of solving linear ordinary differential equations.

1.1 Basic Concepts

A differential equation is an equation involving the derivatives or differentials of a certain number of variables. In this chapter we consider equations that involve one or several derivatives of a certain unspecified function u with respect to a certain variable, say x. The simplest of such equations can be written as

$$\frac{du}{dx} = g(x) \tag{1.1}$$

The term *ordinary* distinguishes the equation from a *partial differential equation*, which involves partial derivatives of an unknown function u of two or more variables, that is,

$$\frac{\partial^2 u}{\partial x^2} = \frac{\partial^2 u}{\partial t^2} \tag{1.2}$$

A general representation of linear ordinary differential equations can be written as

$$F(x, y, y', y'', \ldots, y^{(n)}) = 0 \tag{1.3}$$

where $y' = \dfrac{dy}{dx},$

$y'' = \dfrac{d^2y}{dx^2}, \ldots,$

$$y^{(n)} = \frac{d^n y}{dx^n}$$

The order of a differential equation is the highest derivative of the function u occurring in the equation. A first order linear equation can be written in the form

$$a_0(x)\frac{du}{dx} + a_1(x)u = g(x) \tag{1.4}$$

The *absence of products or nonlinear terms* of the function u and its derivatives makes equation (1.4) *linear*.

A *solution* of a differential equation is any functional relation not involving derivatives or integrals of the unknown function u so that when this functional relation for u is substituted in the differential equation, the equation becomes an identity.

1.2 Linear First Order Equations—Separable Equations

A large number of ordinary differential equations appear in the form

$$g(u)\frac{du}{dx} = f(x) \tag{1.5}$$

which can be separated into

$$g(u)\,du = f(x)\,dx \tag{1.6}$$

In Equation (1.6) one side involves the variable u only, while the other side involves the variable x only; thus, it is called separable. Integrating both sides of Equation (1.6) gives

$$\int g(u)\,du = \int f(x)\,dx + \text{constant} \tag{1.7}$$

When the integrals in Equation (1.7) are evaluated, the solution to Equation (1.5) is then obtained. As a simple application, let us find an expression for the size of a population $N(t)$, if the rate of growth of that population is proportional to its size at that time. Here, the mathematical expression is stated as

$$\frac{dN}{dt} = KN, \qquad K = \text{constant of proportionality}$$

Separating this equation and integrating yields

$$\int \frac{dN}{N} = \int K\,dt + C_1$$

or

$$N(t) = C\,e^{Kt} = N(0)\,e^{Kt}$$

where we invoked that the initial value for $N(t)$ at $t = 0$ is given as $N(0)$.

1.3 Exact First Order Equations

Certain first order equations take the form

$$P(x, y)\,dx + Q(x, y)\,dy = 0 \tag{1.8}$$

If the left-hand side of this equation is an exact or total differential, the equation is said to be exact. To explain the meaning of an exact differential, let us take the following equation

$$w(x, y) = C \qquad (C, \text{constant}) \tag{1.9}$$

Differentiating this equation yields

$$dw = \left(\frac{\partial w}{\partial x}\right) dx + \left(\frac{\partial w}{\partial y}\right) dy = 0 \tag{1.10}$$

or

$$P(x,y)\,dx + Q(x,y)\,dy = 0 \tag{1.11}$$

where we set

$$P(x,y) = \left(\frac{\partial w}{\partial x}\right) \tag{1.12}$$

and

$$Q(x,y) = \left(\frac{\partial w}{\partial y}\right) \tag{1.13}$$

Differentiating Equation (1.12) with respect to y and Equation (1.13) with respect to x we get

$$\frac{\partial P}{\partial y}(x, y) = \frac{\partial}{\partial y}\left(\frac{\partial w}{\partial x}\right) = \frac{\partial^2 w}{\partial y\,\partial x} \tag{1.14}$$

and

$$\frac{\partial Q}{\partial x}(x, y) = \frac{\partial}{\partial x}\left(\frac{\partial w}{\partial y}\right) = \frac{\partial^2 w}{\partial x\,\partial y} \tag{1.15}$$

From calculus, if P and Q have continuous first derivatives, the right-hand sides of Equations (1.14) and (1.15) are equal. Accordingly, we can write

$$\frac{\partial P}{\partial y}(x, y) = \frac{\partial Q}{\partial x}(x, y) \tag{1.16}$$

This condition is a necessary and a sufficient condition for $P(x,y)\,dx + Q(x,y)\,dy$ to be an exact differential. Hence, the solution to Equation (1.8) will be of the form

$$w(x,y) = C \tag{1.17}$$

Example 1.1 Solve the differential equation

$$(2y - 3x)\frac{dy}{dx} + (2x - 3y) = 0 \tag{1.18}$$

Solution: Rewriting Eq. (1.18) as

$$(2y - 3x)\,dy + (2x - 3y)\,dx = 0 \tag{1.19}$$

and setting

$$P(x,y) = (2x - 3y) \quad \text{and} \quad Q(x,y) = (2y - 3x)$$

we have then

$$\frac{\partial P}{\partial y} = -3, \qquad \frac{\partial Q}{\partial x} = -3$$

Therefore, Equation (1.18) is exact. According to Equations (1.12) and (1.13), we can write

$$\frac{\partial w}{\partial x} = 2x - 3y \tag{1.20}$$

and

$$\frac{\partial w}{\partial y} = 2y - 3x \tag{1.21}$$

Equation (1.20) when integrated gives

$$w(x,y) = x^2 - 3yx + \phi(y) = C \tag{1.22}$$

To determine $\phi(y)$, Equation (1.22) is substituted in Equation (1.21) to give

$$-3x + \phi'(y) = 2y - 3x \tag{1.23}$$

or

$$\phi'(y) = \frac{d\phi}{dy} = 2y \tag{1.24}$$

Hence,

$$\phi(y) = y^2 + C_1 \tag{1.25}$$

Introducing Equation (1.25) in Equation (1.22) yields

$$w(x,y) = C = x^2 - 3yx + y^2 + C_1 \tag{1.26}$$

Obviously, the general solution can be then written as

$$y^2 - 3yx + x^2 = \text{constant} \tag{1.27}$$

1.4 Integrating Factor

Sometimes a differential equation is written in the form

$$P(x,y)\,dx + Q(x,y)\,dy = 0 \tag{1.28}$$

but the condition for exactness is not satisfied. Equation (1.28), however, may be transformed into an exact differential by multiplying the equation by a suitable function $F(x, y)$ of the form $x^\alpha y^\beta$ to give

$$\underbrace{x^\alpha y^\beta P(x,y)\,dx}_{\overline{P}} + \underbrace{x^\alpha y^\beta Q(x,y)\,dy}_{\overline{Q}} = 0 \tag{1.29}$$

α and β are chosen such that

$$\frac{\partial \overline{P}}{\partial y}(x, y) = \frac{\partial \overline{Q}}{\partial x}(x, y) \tag{1.30}$$

Example 1.2 Solve the differential equation

$$(3y + 4xy)\,dx - 2x\,dy = 0 \tag{1.31}$$

Solution: Multiplying by $x^\alpha y^\beta$ and using Equation (1.30) yields

$$3x^\alpha(\beta + 1)y^\beta + 4x^{\alpha+1}(\beta + 1)y^\beta = 2y^\beta(\alpha + 1)x^\alpha \tag{1.32}$$

This equation yields $\beta = -1$ and $\alpha = -1$. Hence, an integrating factor is given by $F(x, y) = x^\alpha y^\beta = 1/xy$. Using this integrating factor in Equation (1.31) we obtain

$$\frac{(3y + 4xy)}{xy}dx - \frac{2x}{xy}dy = 0 \tag{1.33}$$

or

$$3\frac{dx}{x} + 4\,dx = 2\frac{dy}{y} \tag{1.34}$$

This equation when integrated yields

$$\frac{y^2}{x^3} = Ce^{4x} \tag{1.35}$$

The following are two important cases in which integrating factors can easily be determined:

1. When

$$\frac{\dfrac{\partial P}{\partial y} - \dfrac{\partial Q}{\partial x}}{Q} = f(x)$$

In this case, it can easily be verified that an integrating factor is given by

$$\exp\left[\int_a^x f(\xi)\, d\xi\right]$$

2. When

$$\frac{\dfrac{\partial P}{\partial y} - \dfrac{\partial Q}{\partial x}}{P} = f(y)$$

An integrating factor here is given by

$$\exp\left[-\int_a^y f(\eta)\, d\eta\right]$$

Any convenient value may be used for the lower integration limit a.

1.5 Nonhomogeneous Equation—First Order

Consider the nonhomogeneous equation

$$\frac{dy}{dx} + f(x)y = r(x) \tag{1.36}$$

and rewrite it as

$$(fy - r)\, dx + dy = 0 \tag{1.37}$$

Let an integrating factor $F(x)$ exist such that

$$F(x)\,(fy - r)\, dx + F(x)\, dy = 0 \tag{1.38}$$

is exact. Therefore, according to Equation (1.16) we should have

$$\frac{\partial}{\partial y}[F(fy - r)] = \frac{dF}{dx} \tag{1.39}$$

or

$$Ff = \frac{dF}{dx} \tag{1.40}$$

This equation when integrated gives the integrating factor

$$F(x) = e^{\int f(x)\,dx} \tag{1.41}$$

Multiplying Equation (1.36) by this factor yields

$$e^{\int f(x)\,dx}\left[\frac{dy}{dx} + fy\right] = e^{\int f(x)\,dx}\, r(x) \tag{1.42}$$

which may be written as

$$\frac{d}{dx}\left[y\, e^{\int f(x)\,dx}\right] = e^{\int f(x)\,dx}\, r(x) \tag{1.43}$$

Integrating both sides of Equation (1.43) gives the solution in the form

$$y\, e^{\int f(x)\,dx} = \int e^{\int f(x)\,dx}\, r(x)\,dx + C \tag{1.44}$$

or

$$y(x) = e^{-\int f(x)\,dx}\left[\int e^{\int f(x)\,dx}\, r(x)\,dx + C\right] \tag{1.45}$$

1.6 Variation of Parameters

Let us consider again the nonhomogeneous equation

$$\frac{dy}{dx} + f(x)\,y = r(x) \tag{1.46}$$

The homogeneous part of this equation

$$\frac{dy}{dx} + f(x)\,y = 0 \tag{1.47}$$

has the following solution as obtained by separating the variables

$$y_h = C\,e^{-\, f(x)\,dx} \tag{1.48}$$

In order to find the general solution to Equation (1.46), the method of variation of parameters suggests that the constant C in Equation (1.48) be replaced by a variable $u(x)$ to be determined. Accordingly, the general solution to Equation (1.46) is written as

$$y(x) = u(x)\,y_h(x) \tag{1.49}$$

Substituting this equation in Equation (1.46) yields

$$u y_h' + y_h u' + f u y_h = r \tag{1.50}$$

or

$$u' y_h + u [y_h' + f y_h] = r \tag{1.51}$$

Because y_h is the solution to the homogeneous equation, the quantity in the brackets of Equation (1.51) is zero. Hence, Equation (1.51) becomes

$$u' y_h = r \tag{1.52}$$

or

$$\frac{du}{dx} = \frac{r}{y_h} \tag{1.53}$$

Integrating Equation (1.53) gives

$$u(x) = \int \frac{r}{y_h} dx + C \tag{1.54}$$

Therefore, the solution to Equation (1.46) can now be written as

$$y(x) = u(x)\, y_h(x) = y_h(x) \int \frac{r}{y_h} dx + C y_h$$

or

$$y(x) = y_h(x) \left[\int \frac{r}{y_h} dx + C \right] \tag{1.55}$$

This is the same result obtained in Article 1.5 and represented by Equation (1.45).

1.7 Bernoulli Equation

Bernoulli equation is a nonlinear equation given as

$$\frac{dy}{dx} + P(x)y = Q(x)y^n \tag{1.56}$$

The equation clearly is linear for $n = 0$ or 1. n in Equation (1.56) need not be an integer. Rewriting Equation (1.56) in the form

$$y^{-n} \frac{dy}{dx} + P(x)\, y^{-n+1} = Q(x) \tag{1.57}$$

and defining a new variable

$$u = y^{-n+1} \tag{1.58}$$

there results

$$\frac{du}{dx} = \frac{du}{dy}\frac{dy}{dx} = (1 - n)y^{-n}\frac{dy}{dx} \tag{1.59}$$

Using Equation (1.59) in Equation (1.56) yields

$$\frac{du}{dx} + (1 - n)P(x)u = (1 - n)Q(x) \tag{1.60}$$

Equation (1.60) is a linear first order equation for $u(x)$ and can be solved by the methods of Article 1.5 or 1.6.

1.8 Ordinary Differential Equations of Higher Order

The terminology used for the first order equations regarding linearity and homogeneity applies to equations of higher orders.

Linear equations behave according to the superposition principle, which can be stated as follows: If a solution of a *homogeneous linear differential equation* over a certain interval is multiplied by a constant, the resulting function is also a solution. The sum of two or more solutions on a certain interval is also a solution. The functions

$$e^{x} \quad \text{and} \quad e^{-x}$$

are solutions to the differential equation

$$\frac{d^2y}{dx^2} - y = 0$$

The following functions are also solutions to this equation

$$C_1e^{x},\ C_2e^{-x},\ e^{x} + e^{-x},\ e^{x} - e^{-x}$$

This can be verified by substituting these functions into the differential equation.

The superposition theorem does not apply to nonhomogeneous equations and to nonlinear equations. For example, the functions $e^{x} - 1$ and $e^{-x} - 1$ are solutions to the nonhomogeneous equation

$$\frac{d^2y}{dx^2} - y = 1$$

However, the functions $2(e^{x} - 1)$, $2(e^{-x} - 1)$, and $(e^{x} - 1) + (e^{-x} - 1)$ are not solutions to this equation as can be verified by direct substitution. The function $\sqrt{2x}$ is a solution to the nonlinear differential equation

$$y\frac{dy}{dx} = 1$$

However, $C\sqrt{2x}$ is not a solution to this equation.

1.9 Linear Differential Equations with Constant Coefficients

In many engineering applications such as mechanical and electrical vibrations, heat transfer, and control theory, the second order linear ordinary differential equation plays an important role. The homogeneous form of the equation is

$$\frac{d^2y}{dx^2} + a\frac{dy}{dx} + by = 0 \tag{1.61}$$

The appearance of this equation suggests solutions of the form

$$y = e^{\lambda x} \tag{1.62}$$

where λ is a constant. This is so because all the derivatives of $e^{\lambda x}$ are constant multiples of the function itself, that is,

$$\frac{d^n}{dx^n}e^{\lambda x} = \lambda^n e^{\lambda x}$$

Introducing this form in Equation (1.61) yields

$$(\lambda^2 + a\lambda + b)e^{\lambda x} = 0 \quad \text{or} \quad (\lambda^2 + a\lambda + b) = 0 \tag{1.63}$$

with roots

$$\lambda_1 = \frac{1}{2}[-a + \sqrt{a^2 - 4b}], \quad \lambda_2 = \frac{1}{2}[-a - \sqrt{a^2 - 4b}]$$

The following three cases result from Equation (1.63):

1. Two distinct roots,

$$a^2 > 4b; \quad \lambda_1 \neq \lambda_2$$

In this case the solution to Equation (1.61) becomes

$$y(x) = A\,e^{\lambda_1 x} + B\,e^{\lambda_2 x} \tag{1.64}$$

2. Equal roots (or double root),

$$a^2 = 4b; \quad \lambda_1 = \lambda_2 = \lambda$$

Here, one solution is given by

$$y_1(x) = A e^{\lambda x} \tag{1.65}$$

To find another solution, the method of variation of parameters is used in which $y_2(x)$ takes the form

$$y_2(x) = u(x) e^{\lambda x} \tag{1.66}$$

Introducing this equation into the differential equation gives

$$u(x) = Cx + D \tag{1.67}$$

and

$$y_2(x) = (Cx + D)e^{\lambda x} \tag{1.68}$$

Actually, Equation (1.68) has two arbitrary constants and does represent the general solution of the differential equation. If we write

$$y(x) = y_1(x) + y_2(x) = A e^{\lambda x} + Cx e^{\lambda x} + D e^{\lambda x} \tag{1.69}$$

This equation, when simplified, yields

$$y(x) = (B + Cx)e^{\lambda x} \tag{1.70}$$

which is the form given by Equation (1.68).

3. Two complex conjugate roots,

$$a^2 < 4b; \quad \lambda_1 \neq \lambda_2$$

If we let

$$\lambda_1 = p + iq \quad \text{and} \quad \lambda_2 = p - iq$$

we can write

$$y(x) = A_1 e^{(p+iq)x} + B_1 e^{(p-iq)x} \tag{1.71}$$

Using Euler formula

$$e^{i\phi} = \cos \phi + i \sin \phi; \qquad e^{-i\phi} = \cos \phi - i \sin \phi$$

Equation (1.71) takes the form

$$y(x) = e^{px}[A \cos qx + B \sin qx] \tag{1.72}$$

Example 1.3. In the study of buckling of a uniform column under an axial loan P, the deflection $y(x)$ is obtained from the differential equation (see Figure 1-1)

$$\frac{d^2y}{dx^2} + \gamma^2 y = 0; \quad \gamma^2 = \frac{P}{EI} \tag{1.73}$$

where E is the modulus of elasticity and I is the moment of inertia. For such problems, there are various modes of buckling. To each mode there corresponds a critical load and an equilibrium position. Find an expression for the first critical load.

Solution: Equation (1.73) is obtained by equating the bending moment $(-EI(d^2y/dx^2))$ at any point x to the bending moment (Py) at any point x due to the force P.

The solution to Equation (1.73) is

$$y(x) = A \sin \gamma x + B \cos \gamma x \tag{1.74}$$

subject to:

1. $x = 0 \quad y = 0$,
2. $x = l \quad y = 0$

Condition 1. yields $B = 0$. Hence, the solution reduces to

$$y(x) = A \sin \gamma x \tag{1.75}$$

Applying condition 2., we get

$$0 = A \sin \gamma l, \qquad A \neq 0 \tag{1.76}$$

Hence, we must have $\sin \gamma l = 0$ or

$$\gamma l = n\pi \quad \text{and} \quad \gamma_n = n\pi/l, \qquad n = 1, 2, 3, \ldots \tag{1.77}$$

To each value of n, there corresponds a solution or a deflection representing a mode of buckling.[a] With $\gamma^2 = P/EI$ we can write

$$P_n = EI\,\gamma_n^2 = (n^2\pi^2/l^2)\,EI \tag{1.78}$$

When $n = 1$ we obtain the first critical load as

$$P_1 = \pi^2 EI/l^2 \tag{1.79}$$

1.10 Nonhomogeneous Equation—Second Order

There are many general methods of determining the particular solution to the nonhomogeneous equation

$$\frac{d^2y}{dx^2} + a\frac{dy}{dx} + by = r(x) \tag{1.80}$$

Two methods are presented here, and they are termed

1. The method of undetermined coefficients. This method is restricted to linear equations with constant coefficients.

[a] Such solutions are referred to as eigenfunctions, and the corresponding values of γ_n are called eigenvalues.

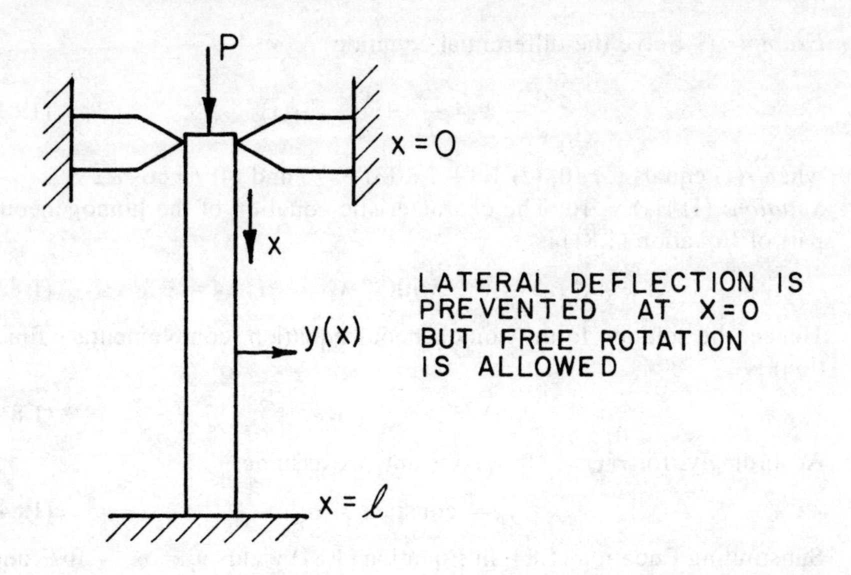

Figure 1-1. A Column Under an Axial Load in Example 1.1

2. The method of variation of parameters. This method has a more general applicability than method 1.

The Method of Undetermined Coefficients

This method suggests that a particular solution y_p for Equation (1.80) can be assumed of a form similar to $r(x)$ with unknown coefficients. The coefficients are subsequently determined by substituting the assumed solution in Equation (1.80). The following shows the forms of $r(x)$ and the assumed forms of the particular solution y_p:

$r(x)$:	$a_n x^n$ $(n = 0, 1, 2, \ldots)$	$A\,e^{px}$	$A \sin qx + B \cos qx$ (A or B may be zero)
Form of y_p:	$\alpha_n x^n$ $(n = 0, 1, 2, \ldots)$	$C\,e^{px}$	$C \sin qx + D \cos qx$
λ:	0	p	iq

provided that the characteristic equation $\lambda^2 + a\lambda + b = 0$ does not have the roots shown above. If the value of λ shown is a simple root of the characteristic equation, the choice of y_p should be multiplied by x; if it is a double root, it should be multiplied by x^2. The following example illustrates the procedure.

Example 1. Solve the differential equation

$$\frac{d^2y}{dx^2} + 4\frac{dy}{dx} + 3y = r(x) \tag{1.81}$$

when $r(x)$ equals (1) 10, (2) $10 + 2x$, (3) e^{-3x}, and (4) $F_o \cos \omega x$.
Solution: (1) $r(x) = 10$. The characteristic equation of the homogeneous part of Equation (1.81) is

$$\lambda^2 + 4\lambda + 3 = 0 \quad \text{with} \quad \lambda_1 = -1, \lambda_2 = -3 \tag{1.82}$$

Hence, the solution to the homogeneous equation (complementary function) is

$$y_h(x) = Ae^{-x} + Be^{-3x} \tag{1.83}$$

Accordingly, for $r(x) = 10 =$ constant, we assume

$$y_p = \text{constant} = \alpha_o \tag{1.84}$$

Substituting Equation (1.84) in Equation (1.81) yields $y_p = \alpha_o = 10/3$ and the general solution becomes

$$y = y_h + y_p = Ae^{-x} + Be^{-3x} + 10/3 \tag{1.85}$$

(2) $r(x) = 10 + 2x$. In this case we assume

$$y_p = \alpha_o + \alpha_1 x \tag{1.86}$$

Introducing this equation in Equation (1.81) yields

$$4\alpha_1 + 3\alpha_o + 3\alpha_1 x = 10 + 2x \tag{1.87}$$

This equation is an identity and, hence, the coefficients of like powers of x on either side should be equal. Therefore, $\alpha_1 = 2/3$; $\alpha_o = 22/9$; and the general solution becomes

$$y = A\, e^{-x} + B\, e^{-3x} + (2/3)\, x + 22/9 \tag{1.88}$$

(3) $r(x) = e^{-3x}$. In this case, e^{-3x} is a solution to the homogeneous equation, Equation (1.83). Hence, the assumed form of y_p should be

$$y_p = Cxe^{-3x} \tag{1.89}$$

Substituting this equation in Equation (1.81) yields $C = -1/2$, and the general solution becomes

$$y = Ae^{-x} + Be^{-3x} - (1/2)xe^{-3x} \tag{1.90}$$

(4) $r(x) = F_0 \cos \omega x$. Here, we let

$$y_p = C \cos \omega x + D \sin \omega x \tag{1.91}$$

When this equation is used in Equation (1.81) we get

$$C = F_o \frac{3 - \omega^2}{(3 - \omega^2)^2 + 16\omega^2}; \quad D = F_o \frac{4\omega}{(3 - \omega^2)^2 + 16\omega^2} \tag{1.92}$$

These constants are used in Equation (1.91) to yield y_p.

The Method of Variation of Parameters

This method is more general than the method of the undetermined coefficients in that it may be applied to linear equations with variable coefficients. It is to be noted, however, that for equations with constant coefficients, the method of undetermined coefficients is easier to apply.

In applying the method of variation of parameters, it is assumed that the solution to the homogeneous part of the equation is known a priori; that is, in Equation (1.80) the part given by

$$y'' + ay' + by = 0 \tag{1.93}$$

has the solution

$$y_h = A y_1(x) + B y_2(x) \tag{1.94}$$

The method consists of replacing A and B in Equation (1.94) by unknown functions $u(x)$ and $v(x)$ to be determined such that the equation

$$y_p(x) = u(x) y_1(x) + v(x) y_2(x) \tag{1.95}$$

satisfies Equation (1.80). Differentiating Equation (1.95) yields

$$y_p' = u y_1' + v y_2' + u' y_1 + v' y_2 \tag{1.96}$$

It is invoked that u and v can be determined such that

$$u' y_1 + v' y_2 = 0 \tag{1.97}$$

Accordingly, from Equation (1.96) we have

$$y_p' = u y_1' + v y_2' \tag{1.98}$$

Differentiating this equation again gives

$$y_p'' = u y_2'' + v y_2'' + u' y_1' + v' y_2' \tag{1.99}$$

Substituting Equations (1.95), (1.98) and (1.99) in Equation (1.80) yields

$$u' y_1' + v' y_2' = r \tag{1.100}$$

Equations (1.97) and (1.100) constitute two simultaneous differential equations for u and v. Solving these two equations using Cramer's rule, we obtain

$$u' = -\frac{r y_2}{|W|}, \qquad v' = \frac{r y_1}{|W|} \tag{1.101}$$

where W is the Wronskian of the homogeneous solution and is given by

$$W = y y_2' - y_1' y_2 \tag{1.102}$$

Integrating Equation (1.101) and substituting the result in Equation (1.95) we get the general solution in the form

$$y(x) = y_h(x) + y_p(x) \tag{1.103}$$

or

$$y(x) = y_1 \left[-\int \frac{r y_2}{|W|} dx + A \right] + y_2 \left[\int \frac{r y_1}{|W|} dx + B \right] \tag{1.104}$$

Example 1.5. Solve the equation

$$y'' + \lambda^2 y = \sin \lambda x \tag{1.105}$$

Solution: The homogeneous part of this equation yields

$$y_1(x) = \sin \lambda x, \qquad y_2(x) = \cos \lambda x \tag{1.106}$$

$$|W| = |\sin \lambda x\,(-\lambda \sin \lambda x) - \lambda \cos \lambda x \cos \lambda x| = \lambda \tag{1.107}$$

Applying Equation (1.104) we get

$$\begin{aligned} y(x) &= \sin \lambda x \left[-\int^x \frac{\sin \lambda x \cos \lambda x}{\lambda} dx + A \right] \\ &\quad + \cos \lambda x \left[\int^x \frac{\sin \lambda x \sin \lambda x}{\lambda} dx + B \right] \end{aligned} \tag{1.108}$$

or

$$\begin{aligned} y(x) &= A \sin \lambda x + B \cos \lambda x - \frac{1}{\lambda^2} \sin^3 \lambda x \\ &\quad + \frac{x \cos \lambda x}{2\lambda} - \frac{\sin 2\lambda x \cos \lambda x}{4\lambda^2} \end{aligned} \tag{1.109}$$

1.11 Cauchy or Euler Equation

The equation

$$x^2 y'' + a x y' + b y = 0 \quad (a, b \text{ are constants}) \tag{1.110}$$

is referred to as Cauchy or Euler Equation. This equation can be solved by assuming a solution of the form

$$y = x^m \tag{1.111}$$

Introducing Equation (1.111) in (1.110) yields

$$[m(m - 1) + am + b]x^m = 0, \; x^m \neq 0$$

or

$$m^2 + (a - 1)m + b = 0 \tag{1.112}$$

This equation yields two roots, m_1 and m_2. *If m_1 and m_2 are different*, the solution to Equation (1.110) becomes

$$y(x) = Ax^{m_1} + Bx^{m_2} \tag{1.113}$$

If Equation (1.112) has a double root, that is, $m_1 = m_2 = m$, one solution is given by $y = x^m$. Another solution is obtained by the method of the variation of parameters. The resulting form of $y_2(x)$ is

$$y_2(x) = x^m \ln x \tag{1.114}$$

Hence, in this case the general solution becomes

$$y(x) = (A + B \ln x) \, x^m \tag{1.115}$$

It is to be noted that by setting $x = e^t \; (x > 0)$ in Equation (1.110), the equation can be transformed into an equation with constant coefficients.

2 Series Solutions of Differential Equations

In this chapter solutions to the linear homogeneous differential equations in the form of power series are considered. It was found in chapter 1 that solutions to differential equations with constant coefficients were obtained in the form of elementary functions. The complexities associated with the equations having variable coefficients usually prevent such elementary functions. A very important group of differential equations belong to this class, such as Bessel's and Legendre's equations. These equations are of frequent occurrence in practice such that the series solution to these equations led to the definition of new mathematical functions referred to as special functions, for example, Bessel, Legendre, gamma, beta, and error functions.

2.1 Series Solution—Frobenius' Method

Among the most important types of differential equations is the second order homogeneous equation which has the standard form,

$$y'' + P(x)\,y' + Q(x)\,y = 0 \tag{2.1}$$

The solution to Equation (2.1) can be written as

$$y(x) = C_1\,y_1(x) + C_2\,y_2(x) \tag{2.2}$$

where $y_1(x)$ and $y_2(x)$ are the two linearly independent solutions of the equation. There are cases where the physical problem leads to a nonhomogeneous equation of the form

$$y'' + P(x)\,y' + Q(x)\,y = r(x) \tag{2.3}$$

In this case, as was discussed in chapter 1, another solution y_p is added to Equation (2.2) to account for the nonhomogeneous term, and the solution to Equation (2.3) becomes

$$y(x) = C_1y_1(x) + C_2y_2(x) + y_p(x) \tag{2.4}$$

Therefore, we concentrate now on the series solution of the homogeneous equation.

A differential equation of the form given by Equation (2.1) has at least one solution which can be expressed in the form

$$y(x) = x^c \sum_{n=0}^{\infty} a_n x^n$$
$$= x^c(a_0 + a_1 x + a_2 x^2 + a_3 x^3 + \ldots), \quad a_0 \neq 0 \qquad (2.5)$$

The exponent c is any real number (may be complex) and chosen so that $a_0 \neq 0$, and the series is convergent for $|x|$ less than some constant K. It is assumed (can be proved) that term by term differentiation of the series is permissible and that the series can be added together term by term and multiplied together term by term. To illustrate the method, we apply it first to the linear oscillator equation in the following example.

Example 2.1

$$\frac{d^2y}{dx^2} + \lambda^2 y = 0 \qquad (2.6)$$

It is well known that the solution to Equation (2.6) is

$$y(x) = A \cos \lambda x + B \sin \lambda x$$

Let

$$y(x) = \sum_{n=0}^{\infty} a_n x^{c+n} \quad a_0 \neq 0 \qquad (2.7)$$

Correspondingly

$$y'(x) = \sum_{n=0}^{\infty} (c + n) a_n x^{c+n-1}$$

$$y''(x) = \sum_{n=0}^{\infty} (c + n)(c + n - 1) a_n x^{c+n-2}$$

Substituting now these expressions in Equation (2.6) yields

$$\sum_{n=0}^{\infty} (c + n)(c + n - 1) a_n x^{c+n-2}$$
$$+ \lambda^2 \sum_{n=0}^{\infty} a_n x^{c+n} = 0 \qquad (2.8)$$

In Equation (2.8), the coefficients of each power of x must vanish individually.

The lowest power of x in Equation (2.8) is x^{c-2} appearing for $n = 0$ in the first summation. Setting the coefficient of x^{c-2} equal to zero yields

$$a_0 c(c - 1) = 0 \quad a_0 \neq 0 \qquad (2.9)$$

Equation (2.9) determines the value of c and is called the *indicial equation*.

It is the equation coming from the coefficient of the lowest power of x. Because $a_0 \neq 0$, Equation (2.9) gives

$$c(c - 1) = 0 \tag{2.10}$$

with the following two roots

$$c = 0, \qquad c = 1$$

We like to emphasize that the indicial equation and its roots are of critical importance in the present analysis.

Next, we require that the remaining coefficients in Equation (2.8) vanish, that is, for $n = 1$

$$a_1 c(c + 1) = 0 \tag{2.11}$$

and for $n \geq 2$

$$(c + n)(c + n - 1)a_n + \lambda^2 a_{n-2} = 0, \qquad n \geq 2 \tag{2.12}$$

or

$$a_n = -a_{n-2}\frac{\lambda^2}{(c + n)(c + n - 1)}, \qquad n \geq 2 \tag{2.13}$$

This equation is referred to as a two-term *recurrence relation*. If we have a_{n-2}, we can compute a_n, a_{n+2}, a_{n+4}, etc.

When $c = 0$, Equation (2.11) shows that a_1 is arbitrary, and Equation (2.13) becomes

$$a_n = -a_{n-2}\frac{\lambda^2}{n(n - 1)}, \qquad n \geq 2 \tag{2.14}$$

Correspondingly, for n even we have

$$a_2 = -a_0\frac{\lambda^2}{1\cdot 2} = -\frac{\lambda^2 a_0}{2!}$$

$$a_4 = -a_2\frac{\lambda^2}{3\cdot 4} = +\frac{\lambda^4 a_0}{4!}$$

$$a_6 = -a_4\frac{\lambda^2}{5\cdot 6} = -\frac{\lambda^6 a_0}{6!}, \text{etc.}$$

Similarly, for n odd we obtain

$$a_3 = -a_1\frac{\lambda^2}{2\cdot 3} = -\frac{\lambda^2 a_1}{3!}$$

$$a_5 = -a_3\frac{\lambda^2}{4\cdot 5} = +\frac{\lambda^4 a_1}{5!}$$

$$a_7 = -a_5 \frac{\lambda^2}{6 \cdot 7} = -\frac{\lambda^6 a_1}{7!}, \text{ etc.}$$

Using these coefficients in Equation (2.7) yields

$$y(x) = a_0 \left[1 - \frac{(\lambda x)^2}{2!} + \frac{(\lambda x)^4}{4!} - \frac{(\lambda x)^6}{6!} + \ldots \right]$$

$$+ \frac{a_1}{\lambda} \left[\lambda x - \frac{(\lambda x)^3}{3!} + \frac{(\lambda x)^5}{5!} - \frac{(\lambda x)^7}{7!} + \ldots \right] \qquad (2.15)$$

The series in the first bracket of Equation (2.15) is recognizable as cos λx while the series in the second bracket is sin λx. Hence, because a_0 and a_1 are still arbitrary, we can write

$$y(x) = A \cos \lambda x + B \sin \lambda x$$

When the other root of the indicial equation ($c = 1$) is used, a_1 is then zero as can readily be found from Equation (2.11). With $a_1 = a_3 = a_{\text{odd}} = 0$, the resulting series is also a cosine series, namely, $a_0 \cos \lambda x$.

Restrictions on Frobenius' Method

In order to apply the method of Frobenius to solve Equation (2.1) it is assumed that the coefficients $P(x)$ and $Q(x)$ are polynomials or can be expanded into such forms by a Taylor's series expansion. If the coefficients are regular (analytic) at $x = 0$, then $x = 0$ is called an ordinary point of the differential equation. If

$$\lim_{x \to 0} P(x) \approx \frac{1}{x} \quad \text{and} \quad \lim_{x \to 0} Q(x) \approx \frac{1}{x^2}$$

the point $x = 0$ is called a regular singular point. In this case $xP(x)$ and $x^2 Q(x)$ are regular at $x = 0$. Otherwise, the point is called an irregular singular point. Having these conditions in mind, the following rules can be stated:

1. If $P(x)$ and $Q(x)$ are regular everywhere (no singular points), the differential equation given by Equation (2.1) has two distinct solutions of the form $y(x) = \Sigma_{n=0}^{\infty} a_n x^n$.
2. If the differential equation possesses only regular singular points, then there will always be one solution of the form $y(x) = x^c \Sigma_{n=0}^{\infty} a_n x^n$. Another solution can be obtained by the method of variation of parameters.
3. When the differential equation has irregular singular points, then no

general method exists to obtain a solution. The equation, however, may or may not have a solution.

The schematic shown below summarizes the approach to the method:

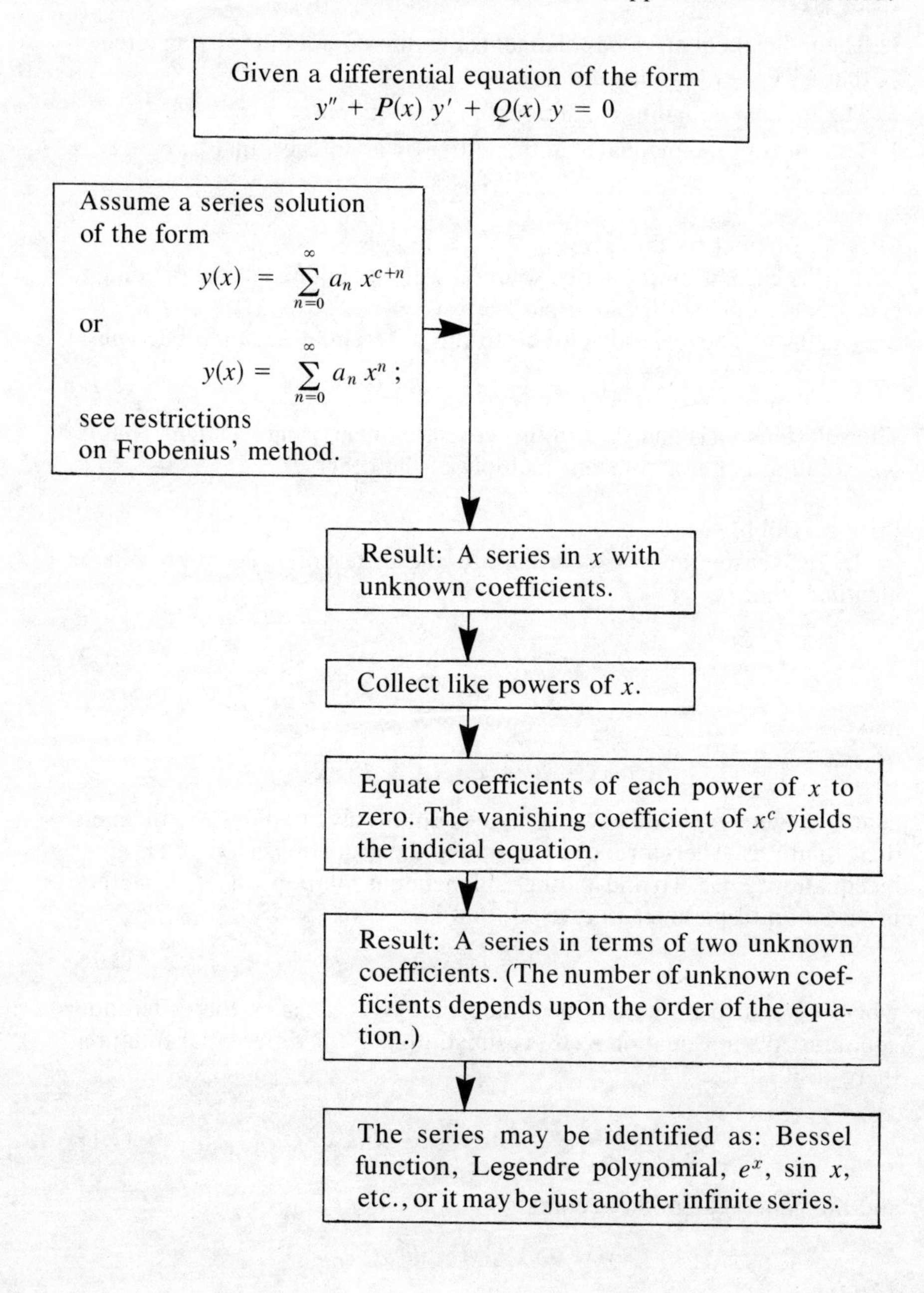

We found in Example 2.1 that the indicial equation had the roots 0 and 1. Usually, there will be three different possibilities corresponding to the following three cases, and they require slightly different treatment. These cases are:

1. The indicial equation has distinct roots that do not differ by an integer, that is, $c_1 - c_2 \neq$ integer or zero.
2. The indicial equation has a double root, that is, $c_1 = c_2$.
3. The roots of the indicial equation differ by an integer, that is, $c_1 - c_2 =$ integer.

Case 1. Distinct roots, c_1, c_2; $c_1 - c_2 \neq$ integer.

In this case, a simple series solution will be obtained corresponding to each root. Suppose the solutions were $y_1(x) = \Sigma_{n=0}^{\infty} a_n x^{c_1+n}$ and $y_2(x) = \Sigma_{n=0}^{\infty} a_n x^{c_2+n}$. The general solution to the differential equation becomes

$$y(x) = A\,y_1(x) + B\,y_2(x) \tag{2.16}$$

The solutions $y_1(x)$ and $y_2(x)$ in this case are linearly independent, that is, one solution is not a constant multiple of the other.

Case 2. Double roots, $c_1 = c_2 = c$.

In this case, the two solutions to the differential equation will be identical, that is,

$$y_1(x) = y_2(x) = \sum_{n=0}^{\infty} a_n x^{c+n} \tag{2.17}$$

and

$$A\,y_1(x) + B\,y_2(x) = (A + B)y(x)$$

is not a general solution because it does not involve two linearly independent solutions. Therefore, one solution to the problem is $y_1(x)$ and is given by equation (2.17). To find another independent solution, $y_2(x)$, the method of variation of parameters is used, that is,

$$y_2(x) = u(x)\,y_1(x) \tag{2.18}$$

where $u(x)$ is to be determined such that $y_2(x)$ satisfies the differential equation. When Equation (2.18) is substituted in the differential equation, there results

$$y_2(x) = y_1(x) \ln x + x^c \sum_{n=1}^{\infty} a_n x^n \tag{2.19}$$

and the general solution becomes

$$y(x) = A\,y_1(x) + B\,y_2(x)$$

or

$$y(x) = (A + B \ln x)y_1(x) + Bx^c \sum_{n=1}^{\infty} a_n x^n \tag{2.20}$$

Case 3. The roots of the indicial equation differ by an integer, $c_1 - c_2 =$ integer.

When the roots of the indicial equation differ by an integer, one solution is determined by

$$y_1(x) = \sum_{n=0}^{\infty} a_n x^{c_1+n} \tag{2.21}$$

where c_1 is the root with the larger real value. It may not be possible to find another solution by the method used in case 1 because the recurrence relation for the coefficients in the series corresponding to one of the roots leads to infinite coefficients. To avoid this difficulty, the method of variation of parameters is used, that is, we set

$$y_2(x) = u(x)\, y_1(x) \tag{2.22}$$

where $u(x)$ is determined by requiring that $y_2(x)$ satisfies the differential equation. The result is

$$y_2(x) = D\, y_1(x) \ln x + x^{c_2} \sum_{n=1}^{\infty} a_n x^n \tag{2.23}$$

where the coefficient D may or may not vanish depending upon the differential equation. Hence the general solution becomes

$$y(x) = A\, y_1(x) + B\, y_2(x)$$

or

$$y(x) = (A + C \ln x)\, y_1(x) + Bx^{c_2} \sum_{n=1}^{\infty} a_n x^n \tag{2.24}$$

We like to note that sometimes the smaller of the two roots c_1, c_2 gives a general solution of the form

$$y(x) = \sum_{n=0}^{\infty} a_n x^{c+n} \tag{2.25}$$

with two arbitrary constants.

2.2 Bessel Equation and Bessel Functions

Bessel equation arises in problems involving circular cylindrical boundaries as well as in applications in the areas of electric field theory, heat transfer, elasticity, vibration, fluid flow problems, and many others. The equation is one of the most important equations in mathematical physics, and it is of the form

$$x^2 y'' + x y' + (x^2 - \nu^2)y = 0 \tag{2.26}$$

where the parameter ν is any given real number. Equation (2.26) is called Bessel equation or order ν. The equation sometimes appears in the form

$$x^2 y'' + xy' + (\lambda^2 x^2 - \nu^2) y = 0 \tag{2.27}$$

where λ is another real number. Equation (2.27) can be reduced to the form of Equation (2.26) by a simple transformation such as $z = \lambda x$. When this transformation is used in Equation (2.27) the result becomes

$$z^2 y'' + zy' + (z^2 - \nu^2)y = 0 \qquad \left(' = \frac{d}{dz}\right) \tag{2.28}$$

which is of the same form as Equation (2.26). Equation (2.26) belongs to the general form given by Equation (2.1) and, hence, we will consider a series solution of the type

$$y = \sum_{n=0}^{\infty} a_n x^{c+n} \qquad a_0 \neq 0 \tag{2.29}$$

Using Equation (2.29) in Equation (2.26) yields

$$\sum_{n=0}^{\infty} (n + c)(n + c - 1)a_n x^{c+n} + \sum_{n=0}^{\infty} (n + c)a_n x^{c+n} + \sum_{n=0}^{\infty} a_n x^{c+n+2} - \nu^2 \sum_{n=0}^{\infty} a_n x^{c+n} = 0 \tag{2.30}$$

Equating the sum of the coefficients of x^{c+n} to zero results in

$$a_0[c(c - 1) + c - \nu^2] = 0, \quad \text{or} \quad a_0(c^2 - \nu^2) = 0 \tag{2.31}$$

$$a_1[(c + 1)c + (c + 1) - \nu^2] = 0, \quad \text{or} \quad a_1[(c + 1)^2 - \nu^2] = 0 \tag{2.32}$$

$$a_2[(c + 2)^2 - \nu^2] + a_0 = 0, \quad \text{or} \quad a_2[(c + 2)^2 - \nu^2] = -a_0 \tag{2.33}$$

$$\vdots$$

$$a_n[(c + n)^2 - \nu^2] + a_{n-2} = 0, \quad \text{or}$$

$$a_n[(c + n - \nu)(c + n + \nu)] + a_{n-2} = 0 \tag{2.34}$$

The indicial equation is obtained by setting the coefficient of a_0 equal to zero. Hence, Equation (2.31) gives

$$(c - \nu)(c + \nu) = 0, \quad \text{or} \quad c_1 = \nu, \qquad c_2 = -\nu \tag{2.35}$$

We select first the root $c_1 = \nu$ and obtain the corresponding solution. For this value of c_1, equation (2.32) yields $a_1 = 0$ and equation (2.34) yields the recurrence relation

$$a_n = -\frac{1}{n(2\nu + n)} a_{n-2} \tag{2.36}$$

Because $a_1 = 0$, Equation (2.36) yields $a_3 = 0$, $a_5 = 0$, $a_7 = 0$, . . . setting $n = 2m$ in Equation (2.36) there results

$$a_{2m} = -\frac{1}{2^2\, m(\nu + m)} a_{2m-2}, \qquad m = 1, 2, 3, \ldots \tag{2.37}$$

This equation gives a_2, a_4, a_6, . . ., a_{2m} in terms of a_0, that is

$$a_2 = -\frac{1}{2^2 1(\nu + 1)} a_0 = -\frac{\nu!}{2^2 1!(\nu + 1)!} a_0$$

$$a_4 = -\frac{1}{2^2\, 2(\nu + 2)} a_2 = +\frac{\nu!}{2^4\, 2!(\nu + 2)!} a_0$$

$$a_6 = -\frac{1}{2^2\, 3(\nu + 3)} a_4 = -\frac{\nu!}{2^5\, 3!(\nu + 3)!} a_0$$

and in general

$$a_{2m} = (-1)^m \frac{\nu!}{2^{2m}\, m!(\nu + m)!} a_0 \tag{2.38}$$

Inserting these coefficients in the assumed series solution (Eq. (2.29)), we have

$$y(x) = a_0 x^{\nu} \left[1 - \frac{\nu!\, x^2}{2^2 1!(\nu + 1)!} + \frac{\nu!\, x^4}{2^4 2\sim(\nu + 2)!} - \ldots \right] \tag{2.39}$$

In summation form we get

$$y(x) = \underbrace{a_0\, \nu!\, 2^{\nu}}_{A} \underbrace{\sum_{m=0}^{\infty} \frac{(-1)^m (x/2)^{2m+\nu}}{m!(\nu + m)!}}_{J_\nu(x)} \tag{2.40}$$

This summation is known as *Bessel function of the first kind of order* ν. When ν is replaced by $-\nu$ in Equation (2.40), we obtain

$$J_{-\nu}(x) = \sum_{m=0}^{\infty} \frac{(-1)^m (x/2)^{2m-\nu}}{m!(m - \nu)!} \tag{2.41}$$

If ν is not an integer, J_ν and $J_{-\nu}$ are linearly independent. In this case, the solution to the differential equation becomes

$$y(x) = A\,J_\nu(x) + B\,J_{-\nu}(x), \qquad \nu \neq \text{integer} \tag{2.42}$$

If ν is an integer, say n, the roots of the indicial equation are then n and $-n$. In this case, as can be verified from Equations (2.40) and (2.41), we have

$$J_{-n}(x) = (-1)^n\,J_n(x), \qquad n = \text{integer} \tag{2.43}$$

and hence, J_{-n} and J_n are linearly dependent. If we specify one solution $y_1(x)$ to be

$$y_1(x) = J_n(x) \tag{2.44}$$

another solution can then be obtained by the method of variation of parameters. According to Equation (2.23), the solution should be of the form

$$y_2(x) = J_n(x)\ln x + x^{c_2}\sum_{n=1}^{\infty} a_n x^n \tag{2.45}$$

When Equation (2.45) is substituted in Equation (2.26) and after simplification, there results the second independent solution of the form

$$\begin{aligned} y_2(x) \equiv Y_n(x) &= \frac{2}{\pi} J_n(x)\left(\ln\frac{x}{2} + \gamma\right) \\ &\quad + \frac{x^n}{\pi}\sum_{m=0}^{\infty}\frac{(-1)^{m-1}\,[\phi(K) + \phi(K+n)]}{2^{2m+n}\,m!(m+n)!}x^{2m} \\ &\quad - \frac{x^{-n}}{\pi}\sum_{m=0}^{n-1}\frac{(n-m-1)!}{2^{2m-1}\,m!}x^{2m} \end{aligned} \tag{2.46}$$

where

$$\gamma = \text{Euler constant} = 0.5772\ldots$$

$$\phi(K) = \sum_{m=1}^{K}\frac{1}{m} = 1 + \frac{1}{2} + \ldots + \frac{1}{K}, \qquad (K \geq 1) \tag{2.47}$$

$Y_n(x)$ is known as the Bessel function of the second kind of order n or Neuman's function of order n.

Consistent with Equation (2.46) $Y_n(x)$ is defined by the equation

$$Y_n(x) = \frac{(\cos n\pi)\,J_n(x) - J_{-n}(x)}{\sin n\pi} \tag{2.48}$$

From Equations (2.46) and (2.48), it can be seen that if $n = 0$, integer or not an integer, $Y_n(x)$ is linearly independent of $J_n(x)$. Hence, it can be stated that, *for all values of* ν the general solution of Bessel's equation given by equation (2.26) is

$$y(x) = A\, J_\nu(x) + B\, Y_\nu(x) \tag{2.49}$$

It is to be noted, however, that $Y_\nu(x)$ is not needed unless ν is 0 or an integer as Equation (2.42) applies for $\nu \neq$ integer.

The functions $J_n(x)$ and $Y_n(x)$ for $n = 0$ are shown in Figure 2-1.

2.3 Modified Bessel Functions

If we consider the equation

$$x^2 y'' + xy' - (x^2 + \nu^2)y = 0 \tag{2.50}$$

and write it in the form

$$x^2 y'' + xy' + (i^2 x^2 - \nu^2)\, y = 0 \quad (i = \sqrt{-1}) \tag{2.51}$$

we can readily see that it is of the form given by Equation (2.27). The solution can then be expressed as

$$y(x) = A\, J_\nu(ix) + B J_{-\nu}(ix) \qquad n \neq \text{integer} \tag{2.52}$$

Because the argument of $J_{\pm\nu}$ in Equation (2.52) is imaginary, the series for $J_\nu(ix)$ becomes (see Equation (2.40))

$$\begin{aligned} J_\nu(ix) &= \sum_{m=0}^{\infty} \frac{(-1)^m\, (ix/2)^{2m+\nu}}{m!(\nu + m)!} \\ &= i^\nu \underbrace{\sum_{m=0}^{\infty} \frac{(x/2)^{2m+\nu}}{m!(\nu + m)!}}_{I_\nu(x)} \end{aligned} \tag{2.53a}$$

In this case another Bessel function $I_\nu(x)$ with real argument is defined which satisfies the differential equation, that is,

$$i^{-\nu}\, J_\nu(ix) \equiv I_\nu(x) \tag{2.53b}$$

$I_\nu(x)$ is referred to as the modified Bessel function of the first kind of order ν.

When n is an integer, $I_\nu(x)$ and $I_{-\nu}(x)$ are equal. Another independent solution is then obtained by the method of variation of parameters in the form

$$K_\nu(x) = \frac{\pi}{2 \sin \nu\pi}[I_{-\nu}(x) - I_\nu(x)] \tag{2.54}$$

This function is referred to as the modified Bessel function of the second

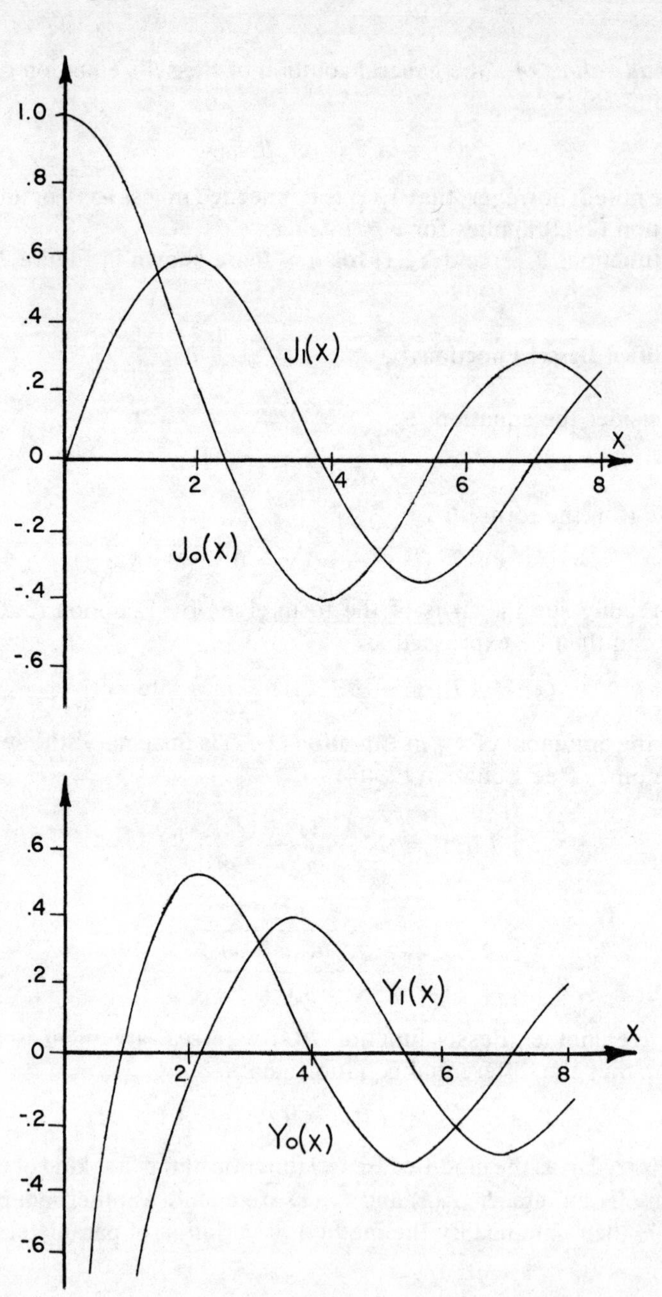

Figure 2-1. Bessel Functions of the First Kind ($J_n(x)$) and of the Second Kind ($Y_n\ (x)$)

kind of order ν. For all values of ν, K_ν is linearly independent of I_ν. Hence, the general form of the solution to Equation (2.50) can be written as

$$y(x) = A\,I_\nu(x) + B\,K_\nu(x) \tag{2.55}$$

The functions $I_\nu(x)$ and $K_\nu(x)$ are shown in Figure 2-2 for $\nu = 0$ and 1.

Recurrence Relations for Bessel Functions

The following recurrence relations for Bessel functions are presented without proof:

$$J_{\nu+1}(x) + J_{\nu-1}(x) = \frac{2\nu}{x} J_\nu(x) \tag{2.56}$$

$$J_{\nu-1}(x) - J_{\nu+1}(x) = 2\,J_\nu'(x) \tag{2.57}$$

$$\nu J_\nu(x) + x\,J_\nu'(x) = x\,J_{\nu-1}(x) \tag{2.58}$$

$$\nu J_\nu(x) - x\,J_\nu'(x) = x\,J_{\nu+1}(x) \tag{2.59}$$

$$\frac{d}{dx}\left[\frac{J_\nu(x)}{x^\nu}\right] = -\frac{J_{\nu+1}(x)}{x} \tag{2.60}$$

$$\frac{d}{dx}\left[x^\nu J_\nu(x)\right] = x^\nu J_{\nu-1}(x) \tag{2.61}$$

$$\frac{d}{dx}\left[x^{-\nu} J_\nu(x)\right] = -\,x^{-\nu} J_{\nu+1}(x) \tag{2.62}$$

Equations (2.56) to (2.62) apply as well for $Y_\nu(x)$.

Limiting Relations for Bessel Functions

For very small values of x $(x \to 0)$, the leading terms in the respective series yield the following approximations

$$J_0(x) \approx 1 - \frac{1}{2}\left(\frac{x}{2}\right)^2 \tag{2.63a}$$

$$J_n(x) \approx \frac{1}{\Gamma(n+1)}\left(\frac{x}{2}\right)^n \qquad n = 1, 2, 3, \ldots \tag{2.63b}$$

$$Y_0(x) \approx \frac{2}{\pi} \ln x \tag{2.64a}$$

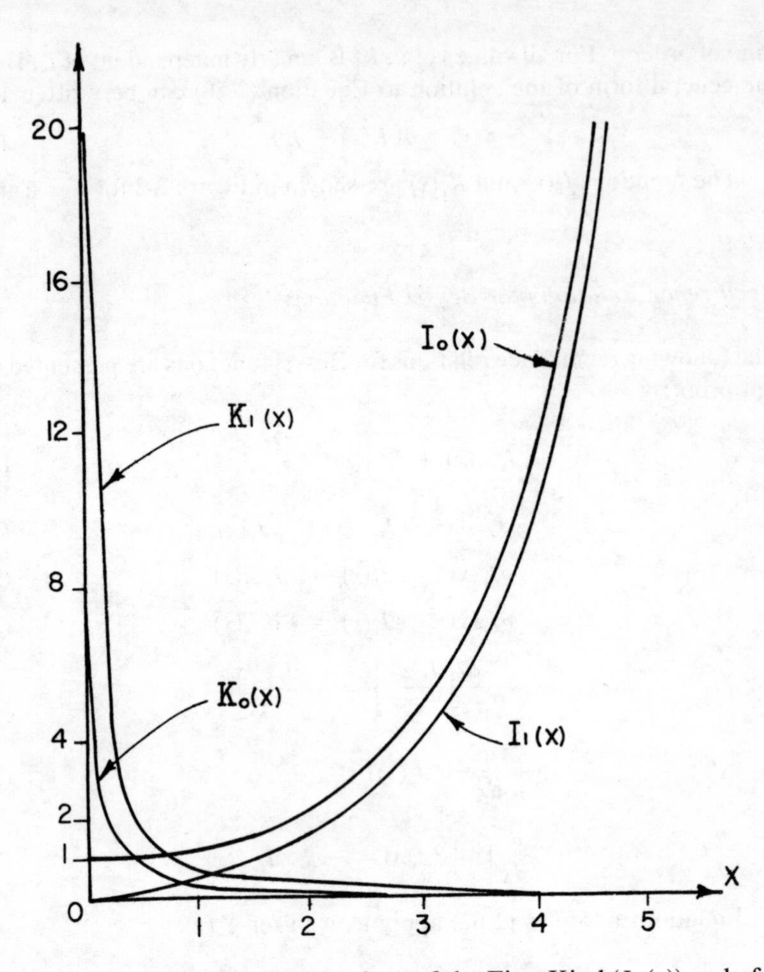

Figure 2-2. Modified Bessel Functions of the First Kind ($I_n(x)$) and of the Second Kind ($K_n(x)$)

$$Y_n(x) \approx -\left(\frac{2}{x}\right)^n \frac{(n-1)!}{\pi} \qquad n = 1, 2, 3, \ldots \tag{2.64b}$$

$$I_n(x) \approx \left(\frac{x}{2}\right)^n \frac{1}{n!} \;; \quad K_n(x) \approx 2^{n-1}(n-1)!x^{-n} \tag{2.65a,b}$$

$$K_0(x) \approx -\ln x \tag{2.66}$$

For large values of x ($x \to \infty$), the following limiting relations are valid

$$J_n(x) \approx \sqrt{\frac{2}{\pi x}} \cos\left(x - \frac{\pi}{4} - \frac{n\pi}{2}\right) \tag{2.67}$$

$$Y_n(x) \approx \sqrt{\frac{2}{\pi x}} \sin\left(x - \frac{\pi}{4} - \frac{n\pi}{2}\right) \tag{2.68}$$

$$I_n(x) \approx \frac{e^x}{\sqrt{2\pi x}}; \quad K_n(x) \approx \sqrt{\frac{\pi}{2x}} e^{-x} \tag{2.69a,b}$$

Bessel functions $J_\nu(x)$, $\nu = \pm 1/2, \pm 3/2, \pm 5/2, \ldots$, are elementary functions, that is,

$$J_{1/2}(x) = \sqrt{\frac{2}{\pi x}} \sin x; \quad J_{-1/2}(x) = \sqrt{\frac{2}{\pi x}} \cos x \tag{2.70a,b}$$

$$J_{3/2}(x) = \sqrt{\frac{2}{\pi x}}\left(\frac{\sin x}{x} - \cos x\right) \tag{2.71a}$$

$$J_{-(3/2)}(x) = -\sqrt{\frac{2}{\pi x}}\left(\frac{\cos x}{x} + \sin x\right) \tag{2.71 b}$$

$$I_{1/2}(x) = \sqrt{\frac{2}{\pi x}} \sinh x \; ; \; I_{-(1/2)}(x) = \sqrt{\frac{2}{\pi x}} \cosh x \tag{2.72 a,b}$$

Example 2.2 An electric current I flows through a solid wire of radius R, thermal conductivity k, and electric resistivity γ, which varies with temperature according to $\gamma = \gamma_0[1 + \beta(T - T_0)]$ where γ_0 is the electric resistivity at a known temperature T_0 and β is the temperature coefficient of resistance. If the surface temperature of the wire is maintained at a constant temperature T_∞, find the steady state radial temperature distribution in the wire.

Solution: Let us take a unit length of the wire and write an energy balance on a differential ring element as shown in Figure 2-3.

$$A_r q_r + g(2\pi r\Delta r) = A_{r+\Delta r} q_{r+\Delta r} \tag{2.73}$$

or

$$-(A_{r+\Delta r} q_{r+\Delta r} - A_r q_r) = -g(2\pi r\Delta r) \tag{2.74}$$

Dividing by Δr and having $\Delta r \rightarrow 0$ yields

$$-\frac{d}{dr}(A_r q_r) = -2\pi r g \tag{2.75}$$

With

$$q_r = -k\frac{dT}{dr} \quad \text{``Fourier's Conduction Law''} \tag{2.76}$$

and

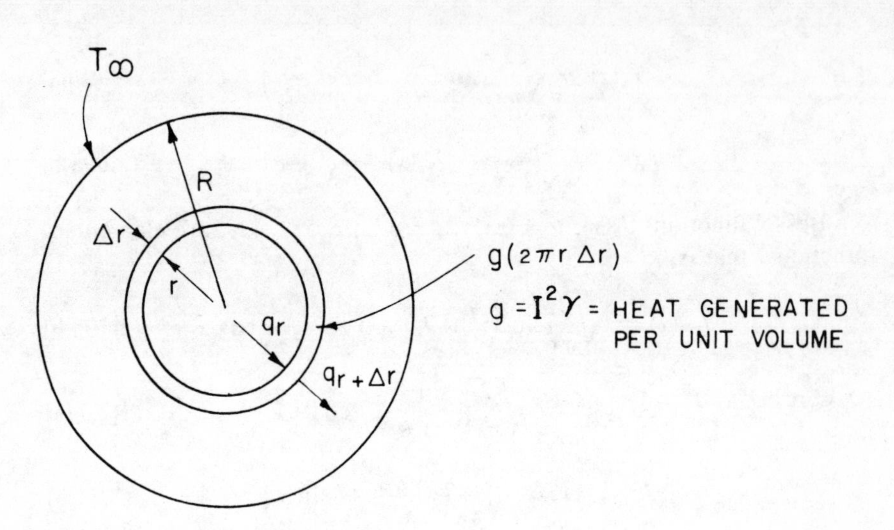

Figure 2-3. Heat Flow Through a Differential Ring Element in Example 2.2

$$g = 2\pi r\, I^2\, \gamma_0[1 + \beta(T - T_0)] \tag{2.77}$$

Equation (2.75), for a constant thermal conductivity k, becomes

$$\frac{d^2T}{dr^2} + \frac{1}{r}\frac{dT}{dr} + \sigma^2\, T = -A_0 \tag{2.78}$$

where

$$\sigma^2 = I^2\, \gamma_0 \frac{\beta}{k} \quad \text{and} \quad A_0 = \frac{I^2\, \gamma_0 \beta T_0 - I^2\, \gamma_0}{k}$$

Equation (2.78), when compared with Equation (2.27), is found to be a nonhomogeneous Bessel equation in which the parameter $\nu = 0$. Therefore, the solution to Equation (2.78) can be written as

$$T(r) = C_1 J_0(\sigma r) + C_2\, Y_0(\sigma r) - A_0/\sigma^2 \tag{2.79}$$

Two boundary conditions are needed to evaluate C_1 and C_2.

$$r = 0: \qquad T \text{ should be finite} \tag{2.80}$$

$$r = R: \qquad T = T_\infty \tag{2.81}$$

Because $Y_0(\sigma r)$ approaches infinity as r approaches zero, Equation (2.80) requires that C_2 should be zero. Hence, $T(r)$ becomes

$$T(r) = C_1\, J_0(\sigma r) - A_0/\sigma^2 \tag{2.82}$$

Applying Equation (2.81) yields

$$T_\infty = C_1 J_0(\sigma R) - A_0/\sigma^2 \tag{2.83}$$

or

$$C_1 = \left[T_\infty + \frac{A_0}{\sigma^2}\right]\frac{1}{J_0(\sigma R)} \tag{2.84}$$

Finally, using Equation (2.84) in Equation (2.82) yields the temperature distribution as

$$\left[\frac{T(r) + A_0/\sigma^2}{T_\infty + A_0/\sigma^2}\right] = \frac{J_0(\sigma r)}{J_0(\sigma R)} \tag{2.85}$$

2.4 Legendre Equation—Legendre Functions

In the solution of potential problems involving spherical geometries, whether in fluid flow, temperature distribution, electric and magnetic fields, Legendre functions play important roles. They appear in the expression for fluid velocities around a sphere in a uniform flow field, in the expressions for the components of electric and magnetic field vectors in conical horns, in the analysis of frequencies in spherical resonators, in the temperature distribution interior or exterior to spheres and many others. Legendre equation has the form

$$(1 - x^2)y'' - 2xy' + n(n + 1)y = 0 \tag{2.86}$$

From Equation (2.86) we can see that $x = 0$ is an ordinary point so that a series solution of the form

$$y = \sum_{k=0}^{\infty} a_k x^k \tag{2.87}$$

is applicable and convergent within a circle of unit radius. Introducing Equation (2.87) in Equation (2.86) yields

$$\sum_{k=2}^{\infty} k(k - 1)a_k x^{k-2} - \sum_{k=2}^{\infty} k(k - 1)a_k x^k - 2\sum_{k=1}^{\infty} k a_k x^k + n(n + 1)\sum_{k=0}^{\infty} a_k x^k = 0 \tag{2.88}$$

Equating the coefficients of like powers of x to zero gives the following recurrence relation

$$a_{k+2} = -\frac{(n - k)(n + k + 1)}{(k + 1)(k + 2)}a_k, \quad k \geq 0 \tag{2.89}$$

This recurrence relation gives $a_2, a_4, a_6, \ldots$ in terms of a_0, and $a_3, a_5, a_7, \ldots$ in terms of a_1. Because a_0 and a_1 are arbitrary, the solution results in two arbitrary constants and this is to be expected because of the order of the equation. Hence, Equation (2.87) will be a solution to Equation (2.86) if the coefficients of expansion are given by Equation (2.89). When these coefficients are used in the series, the solution can be written as

$$y(x) = a_0 \left[1 + \sum_{k=1}^{\infty} (-1)^k \, n(n-2) \ldots (n-2k+2)(n+1)(n+3) \ldots (n+2k-1) x^{2k} \right]$$

$$+ a_1 \left[x + \sum_{k=1}^{\infty} (-1)^k (n-1)(n-3) \ldots (n-2k+1)(n+2)(n+4) \ldots (n+2k) x^{2k+1} \right] \tag{2.90}$$

$$= a_0 u_n(x) + a_1 v_n(x) \tag{2.91}$$

The functions $u_n(x)$ and $v_n(x)$ are linearly independent.

From the recurrence relation Equation (2.89), we see that when $n = k$,

$$a_{n+2} = a_{n+4} = a_{n+6} = 0$$

Therefore,

For n even, the series for $u_n(x)$ terminates with x^n, and hence it becomes a polynomial of degree n; the series for $v_n(x)$ does not terminate.

For n odd, the series for $v_n(x)$ terminates with x^n and, also, it becomes a polynomial of degree n; the series for $u_n(x)$ does not terminate.

These polynomials can be normalized so they have the value of unity when $x = 1$. This can be done by dividing $u_n(x)$ and $v_n(x)$ by $u_n(1)$ and $v_n(1)$ respectively. The hew polynomial is called *Legendre polynomial* of degree n and is denoted by $P_n(x)$. Hence,

$$\textit{For n even,} \quad P_n(x) = \frac{u_n(x)}{u_n(1)} \tag{2.92}$$

$$\textit{For n odd,} \quad P_n(x) = \frac{v_n(x)}{v_n(1)} \tag{2.93}$$

The first few Legendre polynomials are

$$P_0(x) = 1,$$

$$P_1(x) = x,$$

$$P_2(x) = (1/2)(3x^2 1)$$

$$P_3(x) = 1/2\,(5x^3 - 3x),$$

$$P_4(x) = (1/8)(35x^4 - 30x^2 + 3)$$

and they are graphed in Figure 2-4.

When n is an integer, it was found that only one of the solutions $u_n(x)$ and $v_n(x)$ is a polynomial, while the other is an infinite series. The infinite series, when suitably normalized, is referred to as *Legendre function* of the second kind and it is denoted by $Q_n(x)$. The function is conventionally defined as

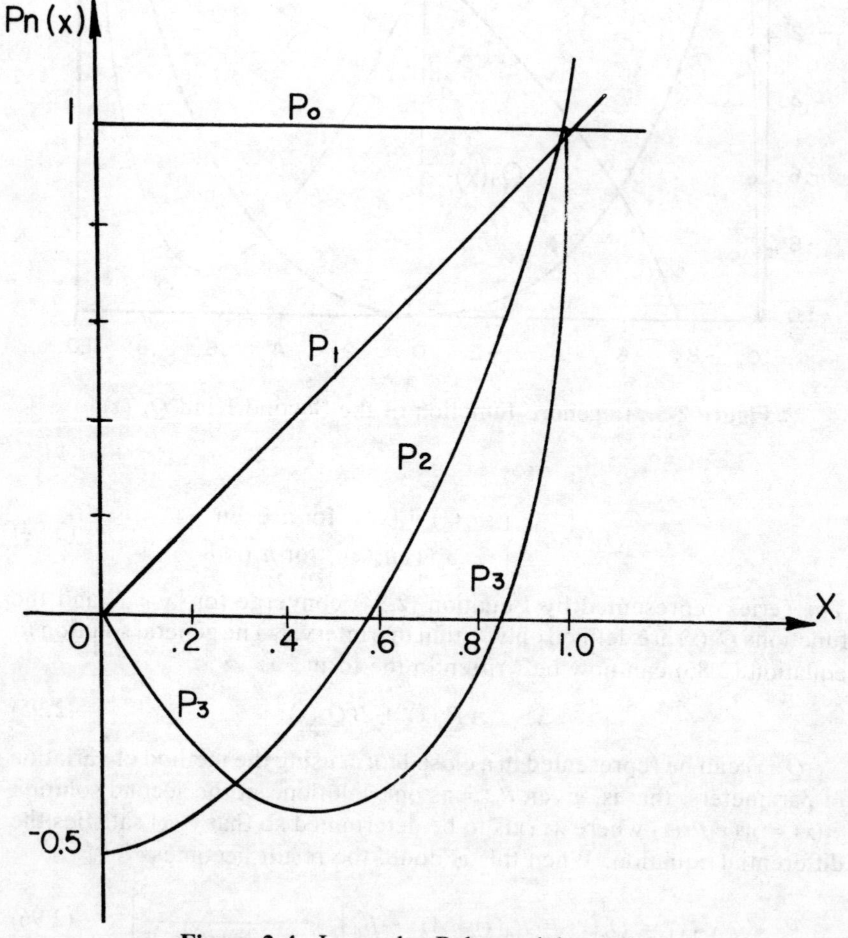

Figure 2-4. Legendre Polynomial $P_n(x)$

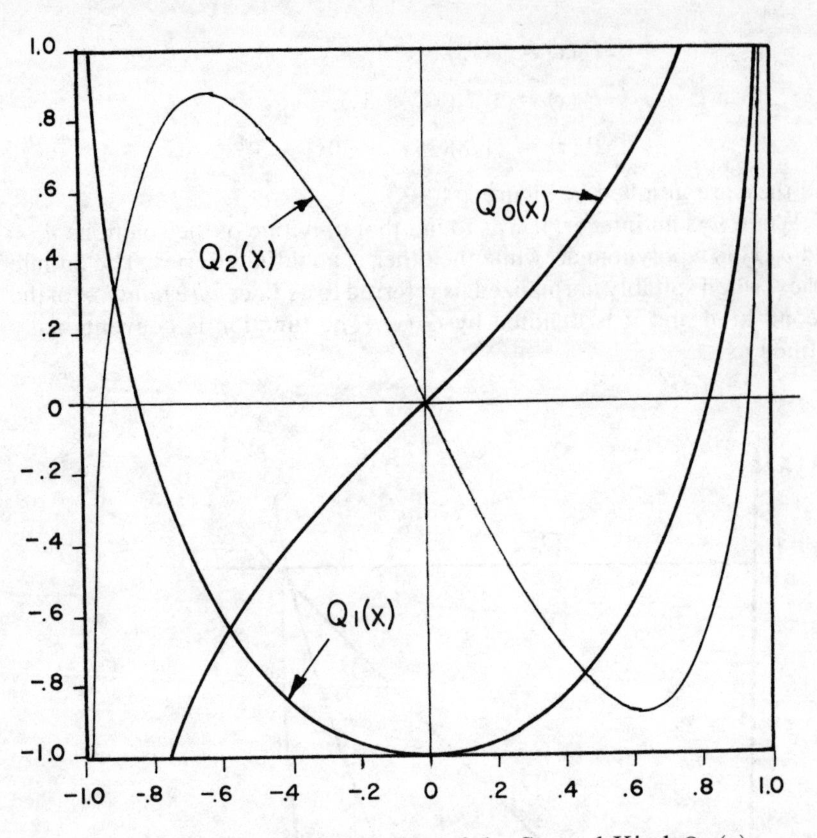

Figure 2-5. Legendre Function of the Second Kind $Q_n\ (x)$

$$Q_n(x) = \begin{cases} u_n(1)\ v_n(x) & \text{for } n \text{ even} \\ -\ v_n(1)\ u_n(x) & \text{for } n \text{ odd} \end{cases} \tag{2.94}$$

The series represented by Equation (2.94) converge for $|x|<1$, and the functions $Q_n(x)$ are defined only within this interval. The general solution to equation (2.86) can now be written in the form

$$y(x) = A\,P_n(x) + B\,Q_n(x) \tag{2.95}$$

$Q_n(x)$ can be represented in a closed form using the method of variation of parameters, that is, given $P_n(x)$ as one solution, let the second solution $y_2(x) = w(x)P_n(x)$ where $w(x)$ is to be determined so that $y_2(x)$ satisfies the differential equation. When this is done, the result becomes

$$y_2(x) = Q_n(x) = P_n(x)\left[A_n + B_n\int^x \frac{dx}{(1 - x^2)[P_n(x)]^2}\right] \tag{2.96}$$

The first few of these functions are

$$Q_0(x) = (1/2) \ln [(1 + x)/(1 - x)]$$

$$Q_1(x) = x\, Q_0(x) - 1$$

$$Q_2(x) = P_2(x)\, Q_0(x) - (3/2)\, x$$

$$Q_3(x) = P_3(x)\, Q_0(x) - (5/2) x^2 + 2/3$$

Figure 2-5 shows the graph of some of these functions.

Legendre polynomial can be expressed also in the following form, referred to as Rodrigues' formula

$$P_n(x) = \frac{1}{2^n\, n!} \frac{d^n}{dx^n} (x^2 - 1)^n \tag{2.97}$$

Important Recurrence Relations for Legendre Polynomials

$$x P_n'(x) - P_{n-1}'(x) = n P_n(x) \qquad n \geq 1 \tag{2.98}$$

$$P_{n+1}'(x) - P_{n-1}'(x) = (2n + 1) P_n(x) \qquad n \geq 1 \tag{2.99}$$

$$(n + 1) P_{n+1}(x) - (2n + 1) x P_n(x) + n P_{n-1}(x) = 0 \qquad n \geq 1 \tag{2.100}$$

$$P_{n+1}'(x) = x\, P_n'(x) + (n + 1) P_n(x) \qquad n \geq 1 \tag{2.101}$$

3 Fourier Series

3.1 Series Expansion of Functions

The use of series for the computation of functions is well known in introductory mathematics. For example, Maclaurin series can be used to compute the numerical value of a certain function for a given value of the argument, such as

$$\ln(1 + x) = x - \frac{x^2}{2} + \frac{x^3}{3} - \frac{x^4}{4} + \ldots \tag{3.1}$$

By specifying the value of x, the function $\ln(1 + x)$ can be evaluated by using the series in power of x, which is often easier to handle. The same holds true for $\sin x$ expressed as

$$\sin x = x - \frac{x^3}{3!} + \frac{x^5}{5!} - \frac{x^7}{7!} + \ldots \tag{3.2}$$

In certain problems, however, such as those dealing with oscillations, it is more convenient to use a series of periodic functions such as sines, cosines, or a combination of both.

We shall see that in dealing with many physical problems, such as the theory of sound, heat conduction, electromagnetic waves, electric circuits, and mechanical vibrations, it will be necessary to expand a certain function into a trigonometric series. As such a series is periodic, the functions to be expanded should also be periodic. Within each period, however, the function can be continuous or may be discontinuous. If it is discontinuous, the number of such discontinuities within the interval of expansion should be finite.

3.2 Periodic Functions

Let us consider the graphs of the following functions (Figure 3-1)

$$y = \sin 2x; \quad y = x \text{ when } -2 < x < +2; \quad y = \begin{cases} \dfrac{2hx}{l} & 0 < x < \dfrac{l}{2} \\ \dfrac{2h}{l}(l - x) & \dfrac{l}{2} < x < l \end{cases}$$

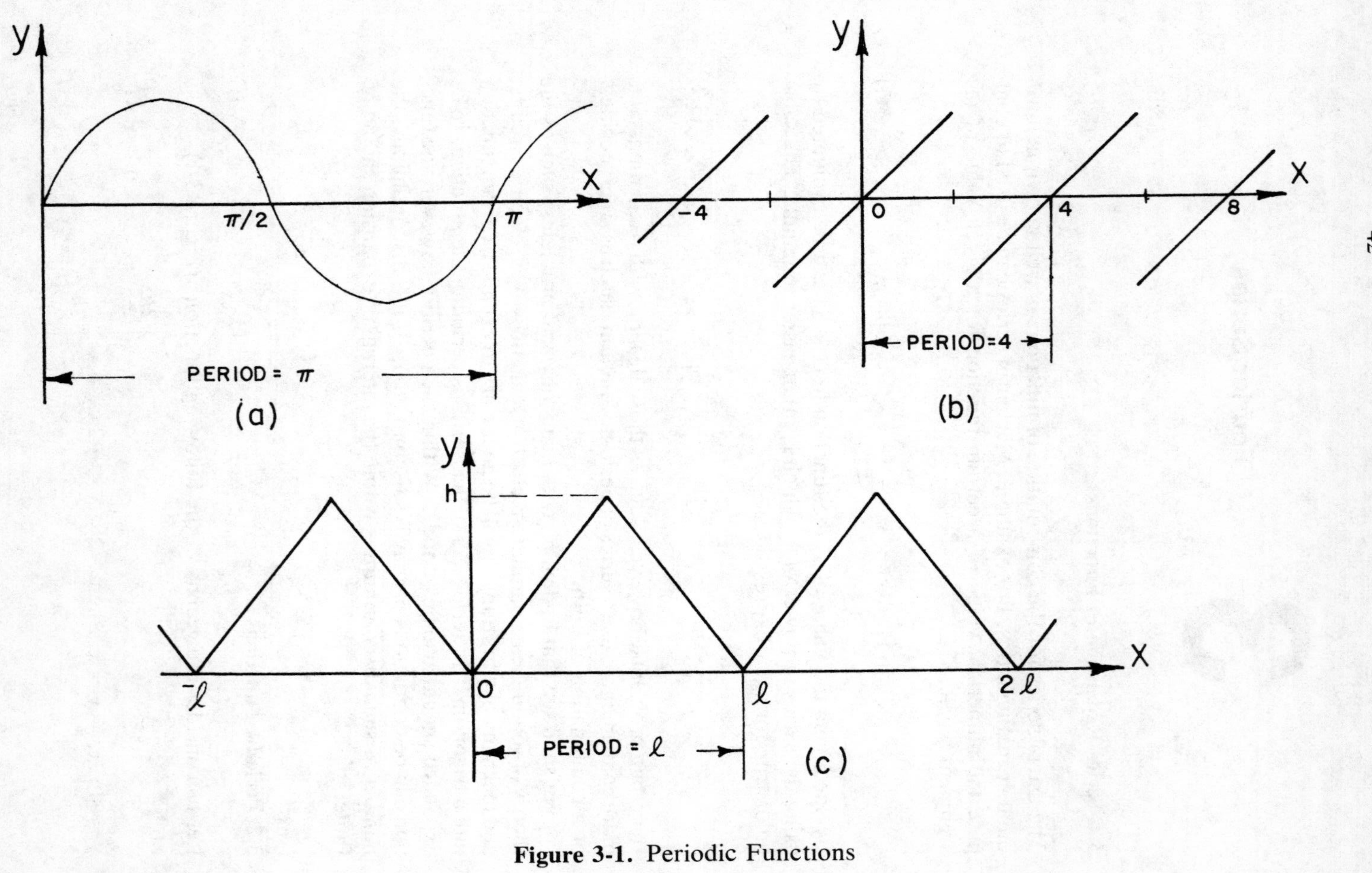

Figure 3-1. Periodic Functions

It can readily be seen that

$$\sin 2x = \sin 2(x + \pi) = \sin 2(x + n\pi) \text{ in (a)}$$

$$y(x) = y(x + n4) \text{ in (b)}$$

$$y(x) = y(x + nl) \text{ in (c)}$$

This holds true for all values of x. Therefore, if we have a function that is defined for all real x such that

$$f(x) = f(x + P)$$

the function is said then to be periodic with period P. If n is any integer, we can write as well

$$f(x) = f(x + nP) \tag{3.3}$$

When two functions $F_1(x)$ and $F_2(x)$ have the same period P, a function made up of their combination is also periodic with period P, that is,

$$H(x) = AF_1(x) + BF_2(x) \quad \text{with period } P \tag{3.4}$$

for example,

$$H(x) = A \sin(x + nP) + B \cos(x + nP)$$

3.3 Fourier Series

Suppose that a function $f(x)$ is periodic with period 2π and can be represented by the following trigonometric series

$$f(x) = a_0 + \sum_{n=1}^{\infty} (a_n \cos nx + b_n \sin nx) \tag{3.5}$$

It will be required then to determine the coefficients a_0, a_n, and b_n of this series.

Integrating both sides of Equation (3.5) from $-\pi$ to π yields

$$\int_{-\pi}^{\pi} f(x)\,dx = \int_{-\pi}^{\pi} a_0\,dx + \int_{-\pi}^{\pi} \left(\sum_{n=1}^{\infty} (a_n \cos nx + b_n \sin nx) \right) dx$$

Interchanging the summation and integration in the second integral on the right-hand side results in

$$\int_{-\pi}^{\pi} f(x)\,dx = 2\pi a_0 + \sum_{n=1}^{\infty} \Bigg(\underbrace{\int_{-\pi}^{\pi} a_n \cos nx\,dx}_{\text{I}} + \underbrace{\int_{-\pi}^{\pi} b_n \sin nx\,dx}_{\text{II}} \Bigg)$$

Integrals I and II are zeros as can be verified by direct integration. Hence,

$$a_0 = \frac{1}{2\pi}\int_{-\pi}^{\pi} f(x)\,dx \tag{3.6}$$

Equation (3.6) shows that a_0 is the average of $f(x)$ over the period of the function. To determine a_n, we multiply Equation (3.5) by cos mx and integrate from $-\pi$ to π and obtain

$$\begin{aligned}\int_{-\pi}^{\pi} f(x)\cos mx\,dx &= a_0\int_{-\pi}^{\pi}\cos mx\,dx \\ &+ \sum_{n=1}^{\infty}\left[a_n\int_{-\pi}^{\pi}\cos nx\cos mx\,dx\right. \\ &+ \left. b_n\int_{-\pi}^{\pi}\sin nx\cos mx\,dx\right]\end{aligned} \tag{3.7}$$

Now,

$$\int_{-\pi}^{\pi}\cos mx\,dx = 0, \qquad \int_{-\pi}^{\pi}\sin nx\cos mx\,dx = 0$$

and

$$\int_{-\pi}^{\pi}\cos nx\cos mx\,dx = \begin{cases}\pi & \text{if } m = n \\ 0 & \text{if } m \neq n\end{cases}$$

Therefore, Equation (3.7) yields

$$a_n = \frac{1}{\pi}\int_{-\pi}^{\pi} f(x)\cos mx\,dx \tag{3.8}$$

Similarly, to find b_n, we multiply Equation (3.5) by sin mx and integrate from $-\pi$ to π and obtain

$$\begin{aligned}\int_{-\pi}^{\pi} f(x)\sin mx\,dx &= a_0\int_{-\pi}^{\pi}\sin mx\,dx \\ &+ \sum_{n=1}^{\infty}\left[a_n\int_{-\pi}^{\pi}\sin mx\cos nx\,dx\right. \\ &+ \left. b_n\int_{-\pi}^{\pi}\sin mx\sin nx\,dx\right]\end{aligned} \tag{3.9}$$

Here also the following relations hold

$$\int_{-\pi}^{\pi}\sin mx\,dx = 0, \qquad \int_{-\pi}^{\pi}\sin mx\cos nx\,dx = 0$$

$$\int_{-\pi}^{\pi} \sin mx \sin nx \, dx = \begin{cases} \pi & \text{if } m = n \\ 0 & \text{if } m \neq n \end{cases}$$

Hence, Equation (3.9) yields

$$b_n = \frac{1}{\pi}\int_{-\pi}^{\pi} f(x) \sin mx \, dx \tag{3.10}$$

The series given by Equation (3.5) with a_0, a_n, and b_n given by Equations (3.6), (3.8), and (3.10), respectively, is called the Fourier series for $f(x)$. The conditions for convergence of this series require that $f(x)$ be bounded in the interval $-\pi \leq x \leq \pi$, and if $f(x)$ is discontinuous, the number of discontinuities should be finite within the interval.

3.4 Change of Scale or Interval

If the function $f(t)$ has a period $2l$, we may write

$$t = \frac{lx}{\pi} \quad \text{or} \quad x = \frac{t\pi}{l}$$

This means that when

$$x = \pm\pi, \quad t = \pm l$$

In this case, Equation (3.5) becomes

$$f(t) = a_0 + \sum_{n=1}^{\infty}\left[a_n \cos\frac{n\pi t}{l} + b_n \sin\frac{n\pi t}{l}\right] \tag{3.11}$$

and the coefficients are given by the following relations

$$a_0 = \frac{1}{2l}\int_{-l}^{l} f(t)\, dt \tag{3.12}$$

$$a_n = \frac{1}{l}\int_{-l}^{l} f(t) \cos\frac{n\pi t}{l}\, dt, \quad n = 1, 2, 3, \ldots \tag{3.13}$$

$$b_n = \frac{1}{l}\int_{-l}^{l} f(t) \sin\frac{n\pi t}{l}\, dt, \quad n = 1, 2, 3, \ldots \tag{3.14}$$

Example 3.1 In electronic circuits that are designed to handle pulses considered as step functions, the analysis of a square wave is quite common. Let us consider the wave shown in Figure 3-2 and defined by

$$f(x) = \begin{cases} K & \text{for } 0 < x < \pi \\ 0 & \text{for } -\pi < x < 0 \end{cases} \tag{3.15}$$

We want to find Fourier series expansion of the specified function.
Solution: Using Equations (3.6), (3.8), and (3.10) to find the coefficients a_0, a_n, and b_n, respectively, yields

$$a_0 = \frac{1}{2\pi}\left\{\int_{-\pi}^{0} f(x)\,dx + \int_{0}^{\pi} f(x)\,dx\right\}$$

$$= \frac{1}{2\pi}\left\{0 + \int_{0}^{\pi} K\,dx\right\} = \frac{K}{2} \tag{3.16}$$

$$a_n = \frac{1}{\pi}\int_{0}^{\pi} K \cos mx\,dx = 0, \quad m = 1, 2, 3, \ldots \tag{3.17}$$

$$b_n = \frac{1}{\pi}\int_{0}^{\pi} K \sin mx\,dx = \frac{k}{m\pi}[1 - \cos m\pi] \tag{3.18}$$

or

$$b_n = \frac{2K}{m\pi} \quad m \text{ odd}, \quad b_n = 0 \quad m \text{ even} \tag{3.19}$$

Therefore, the Fourier series expansion of Equation (3.15) becomes

$$f(x) = \frac{K}{2} + \frac{2K}{\pi}\left[\frac{\sin x}{1} + \frac{\sin 3x}{3} + \frac{\sin 5x}{5} + \ldots\right] \tag{3.20}$$

Let us graph now this series and compare the representation of the function $f(x)$ when the number of terms used in the series is varied (Figure 3-3)

1. $f(x) = \dfrac{K}{2}$

2. $f(x) = \dfrac{K}{2} + \dfrac{2K}{\pi}\sin x$

3. $f(x) = \dfrac{K}{2} + \dfrac{2K}{\pi}\sin x + \dfrac{2K}{3\pi}\sin 3x$

4. $f(x) = \dfrac{K}{2} + \dfrac{2K}{\pi}\sin x + \dfrac{2K}{3\pi}\sin 3x + \dfrac{2K}{5\pi}\sin 5x$

It can readily be seen that as the number of terms in the series increases, the series becomes a closer approximation to the function $f(x)$.

The function represented by Equation (3.20) and interpreted as a square wave can be used to represent a periodic loading of a beam of length nl as

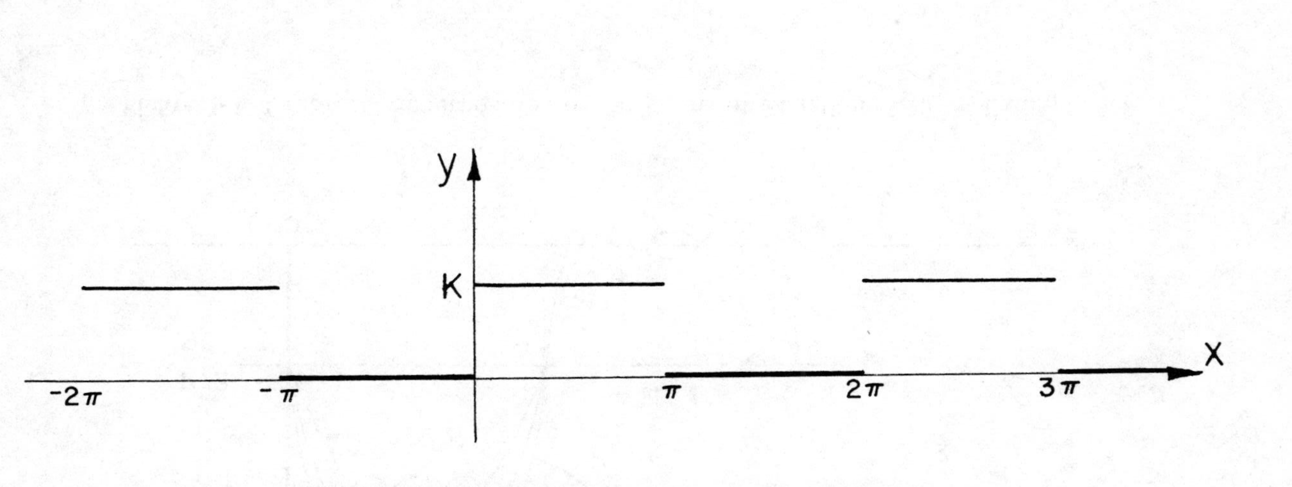

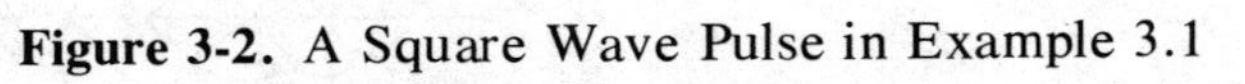

Figure 3-2. A Square Wave Pulse in Example 3.1

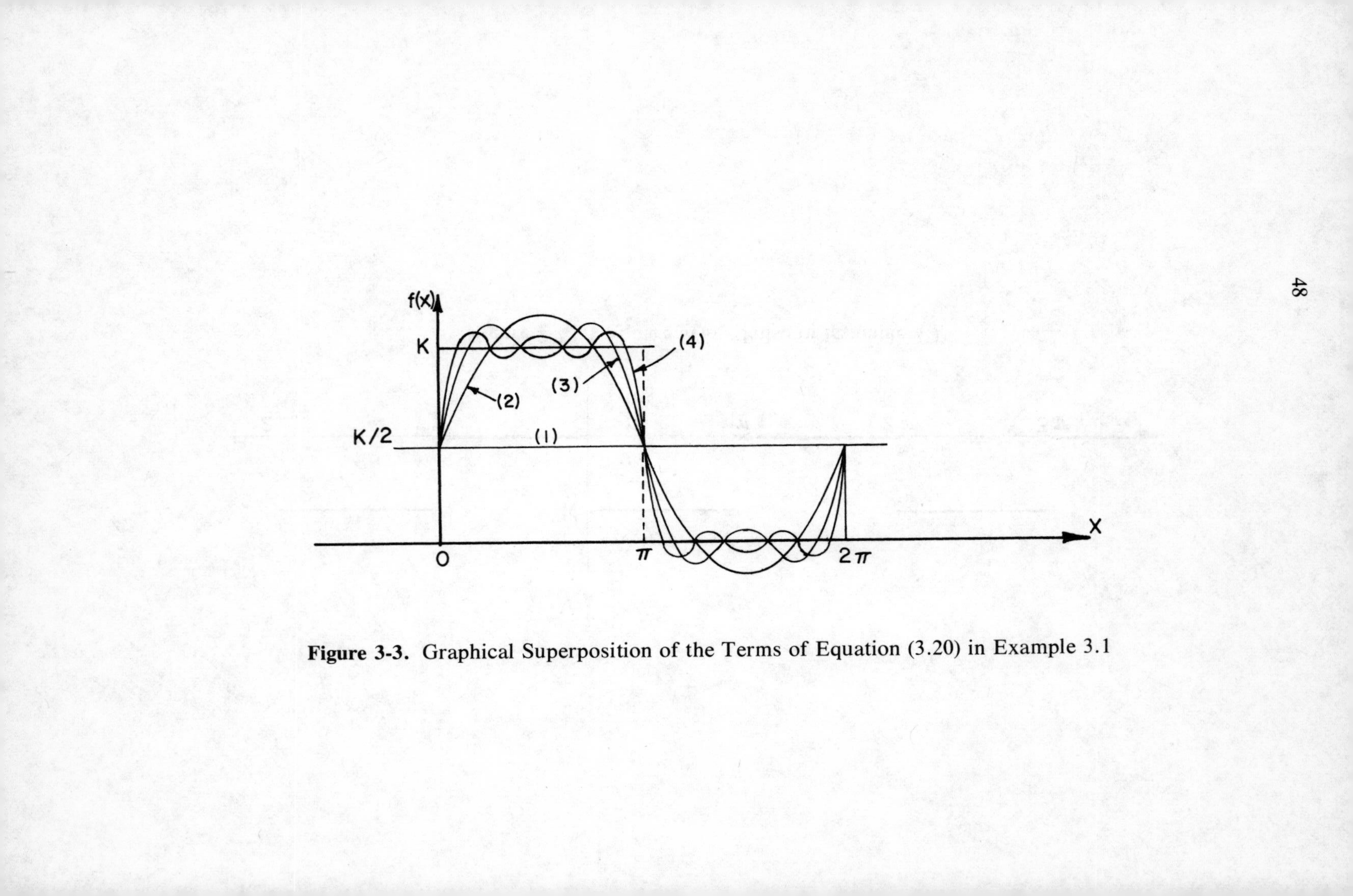

Figure 3-3. Graphical Superposition of the Terms of Equation (3.20) in Example 3.1

shown in Figure 3-4. Taking $n = 6$ so that the length of the beam becomes $6l$, Equation (3.20) gives

$$f(x) = \frac{P_0}{2} + \frac{2P_0}{\pi}\left[\sin\frac{\pi x}{l} + \frac{1}{3}\sin\frac{3\pi x}{l} + \frac{1}{5}\sin\frac{5\pi x}{l} + \ldots\right] \tag{3.21}$$

The meaningful interpretation of this series exists only in $0 \le x \le 6l$.

Example 3.2 Find the Fourier series representation of a full wave rectifier in which the negative portion of the sine wave is inverted as shown in Figure 3-5.

The function is defined by

$$f(t) = \begin{cases} \sin \omega t & 0 \le \omega t \le \pi \\ -\sin \omega t & -\pi \le \omega t \le 0 \end{cases} \tag{3.22}$$

Solution: Equation (3.5) in which x is ωt in this problem, becomes

$$f(x) = a_0 + \sum_{n=1}^{\infty} a_n \cos m\omega t + b_n \sin m\omega t \tag{3.23}$$

The coefficients of the series are given by

$$a_0 = \frac{1}{2\pi}\int_{-\pi}^{0} -\sin \omega t \, d(\omega t) + \frac{1}{2\pi}\int_{0}^{\pi} \sin \omega t \, d(\omega t)$$

or

$$a_0 = \frac{1}{\pi}\int_{0}^{\pi} \sin \omega t \, d(\omega t) = \frac{2}{\pi} \tag{3.24}$$

$$a_n = \frac{1}{\pi}\int_{-\pi}^{0} -\sin \omega t \cos m\omega t \, d(\omega t)$$

$$+ \frac{1}{\pi}\int_{0}^{\pi} \sin \omega t \cos m\omega t \, d(\omega t)$$

$$= \frac{2}{\pi}\int_{0}^{\pi} \sin \omega t \cos m\omega t \, d(\omega t)$$

or

$$a_n = \begin{cases} \left(\frac{2}{\pi}\right)\left(-\frac{2}{m^2 - 1}\right); & m \text{ even} \\ 0; & m \text{ odd} \end{cases} \tag{3.25}$$

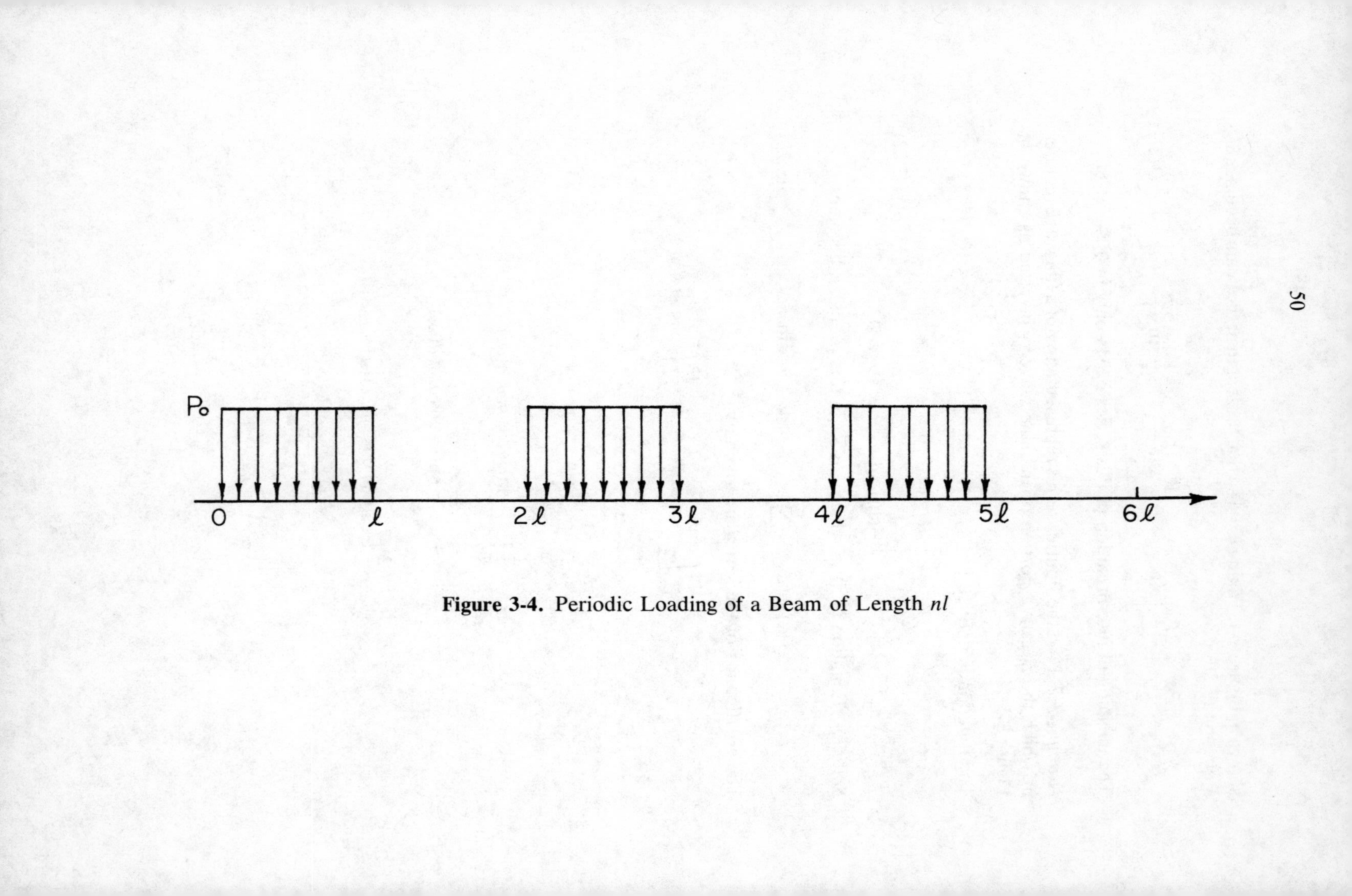

Figure 3-4. Periodic Loading of a Beam of Length *nl*

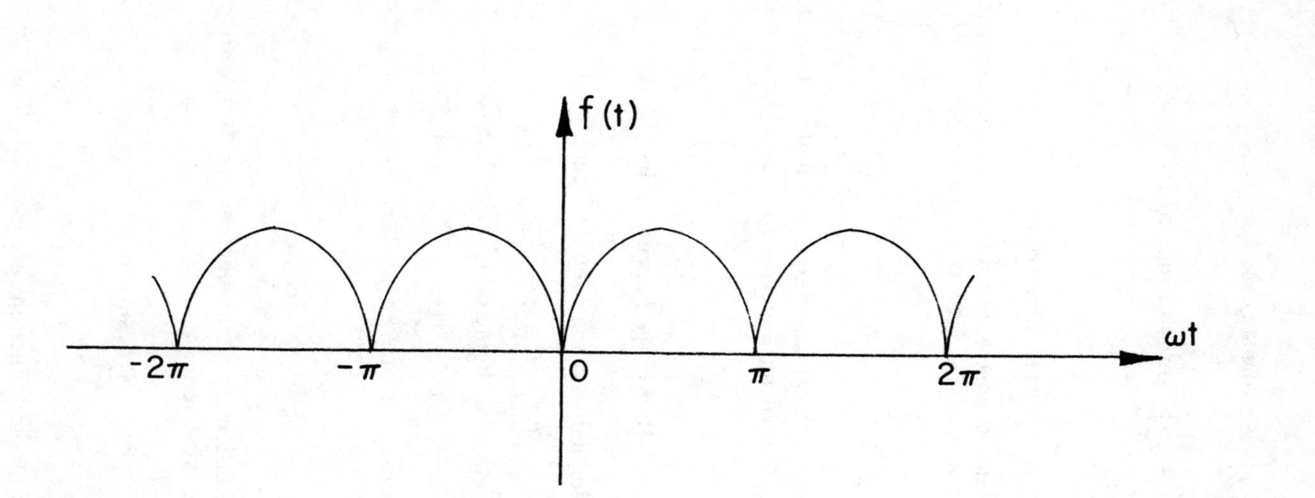

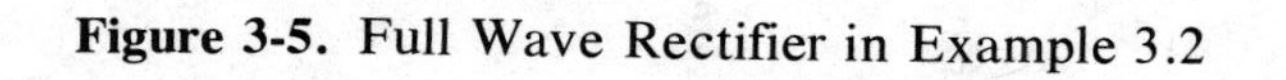

Figure 3-5. Full Wave Rectifier in Example 3.2

$$b_n = \frac{1}{\pi}\int_{-\pi}^{0} -\sin\omega t \sin m\omega t \, d(\omega t)$$

$$+ \frac{1}{\pi}\int_{0}^{\pi} \sin\omega t \sin m\omega t \, d(\omega t)$$

or

$$b_n = 0 \quad \text{for all values of } n \tag{3.26}$$

Therefore, in this problem because b_n are zeros, the series representation of $f(x)$ involves only cosine terms. The series becomes

$$f(t) = \frac{2}{\pi} - \frac{4}{\pi}\sum_{m=2}^{\infty} \frac{\cos m\omega t}{m^2 - 1}, \qquad m = 2, 4, 6, \ldots \tag{3.27}$$

Note: The original frequency ω has been eliminated, and the lowest frequency is now 2ω. Higher frequencies decay as $1/n^2$ as can be seen from Equation (3.27).

3.5 Even and Odd Functions—Half-range Fourier Series

The function $y = f(x)$ is defined as being even if (Figure 3-6a)

$$f(x) = f(-x) \tag{3.28}$$

The function $y = f(x)$ is defined as being odd if (Figure 3-6b)

$$f(-x) = -f(x) \tag{3.29}$$

The Fourier series expansion of an even function, $y = f(x)$, with period 2π is the following Fourier cosine series

$$f(x) = a_0 + \sum_{n=1}^{\infty} a_n \cos nx \tag{3.30}$$

The Fourier series expansion of an odd function, $y = f(x)$, with period 2π is the following Fourier sine series

$$f(x) = \sum_{n=1}^{\infty} b_n \sin mx \tag{3.31}$$

Note:

The sum and products of even functions are even functions.

The sum of odd functions is odd.

The product of two odd functions is even.

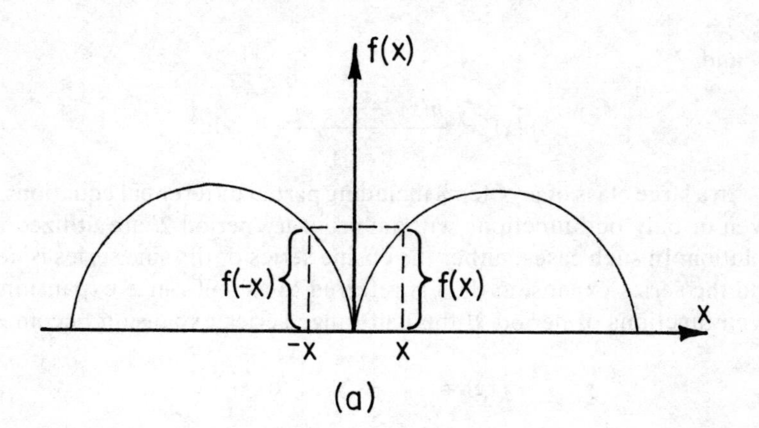

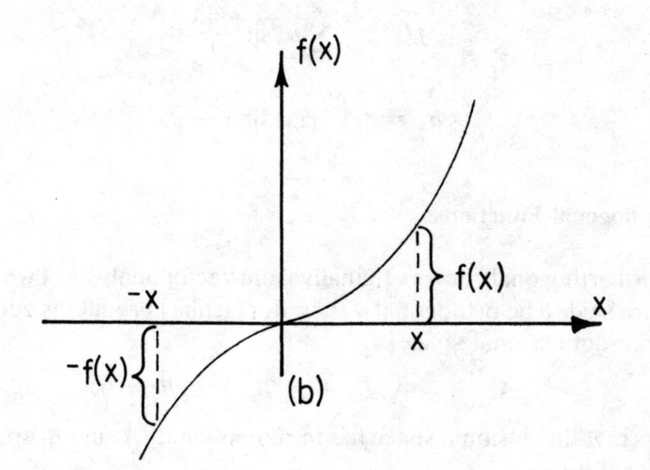

Figure 3-6. Even and Odd Functions

If $y = f(x)$ is odd, then $y = |f(x)|$ is even.
If $y = h(x)$ is any function defined for all x, then

$$g(x) = \frac{h(x) + h(-x)}{2} \quad \text{is even} \tag{3.32}$$

and

$$p(x) = \frac{h(x) - h(-x)}{2} \quad \text{is odd.} \tag{3.33}$$

In a large class of problems including partial differential equations, only even or only odd functions with an arbitrary period $2l$ are utilized in the solution. In such cases, either the cosine series or the sine series is needed and the series expansion then is referred to as half-range expansion. For even functions of period $2l$ the half-range series expansion becomes

$$f(x) = a_0 + \sum_{n=1}^{\infty} a_n \cos\frac{n\pi x}{l} \tag{3.34}$$

with

$$a_0 = \frac{1}{l}\int_0^l f(x)\,dx; \qquad a_n = \frac{2}{l}\int_0^l f(x)\cos\frac{n\pi x}{l}\,dx \tag{3.35a,b}$$

while for odd functions there results

$$f(x) = \sum_{n=1}^{\infty} b_n \sin\frac{n\pi x}{l} \tag{3.36}$$

$$b_n = \frac{2}{l}\int_0^l f(x)\sin\frac{n\pi x}{l}\,dx \tag{3.37}$$

3.6 Orthogonal Functions

The word orthogonal comes originally from vector analysis. Two vectors **A** and **B** are said to be orthogonal if their dot (scalar) product is zero, that is, (for three-dimensional space)

$$\mathbf{A} \cdot \mathbf{B} = A_x B_x + A_y B_y + A_z B_z = 0 \tag{3.38}$$

Vectors in n-dimensional space having components A_i and B_i are said to be orthogonal if

$$\sum_{i=1}^{i=n} A_i B_i = 0 \tag{3.39}$$

When the number of dimensions, n, becomes infinite, the components A_i and B_i become continuously distributed, and i is no longer a discrete number but rather a continuous variable x. The scalar product, then, as given by Equation (3.39) becomes an integral given as

$$\int_0^l A(x) B(x) \, dx = 0 \tag{3.40}$$

The property of orthogonality is indefinite unless the range of integration is specified, such as 0 to l. The analysis can be extended to include a set of functions $\{\phi_k(x), k = 1, 2, 3, \ldots\}$ having the properties

$$\int_a^b \phi_m(x) \phi_n(x) \, dx = 0 \qquad m \neq n \tag{3.41}$$

In a more general way, the functions $\phi_m(x)$ and $\phi_n(x)$ are said to be orthogonal with respect to a weight or density function $\omega(x)$ if

$$\int_a^b \omega(x) \phi_m(x) \phi_n(x) \, dx = 0 \qquad m \neq n \tag{3.42}$$

The norm N_m of the functions $\phi_m(x)$ on the interval $a \leq x \leq b$ is defined as

$$N_m = \left[\int_a^b \phi_m^2(x) \, dx \right]^{1/2} \tag{3.43}$$

An eigenfunction $g_n(x)$ is said to be normalized if

$$\int_a^b g_n^2(x) \, dx = 1 \tag{3.44}$$

Hence, any set of eigenfunctions can be normalized by dividing by the norm of the function, that is,

$$g_n(x) = \frac{\phi_m(x)}{N_m}.$$

A set of orthogonal eigenfunctions on an interval $a \leq x \leq b$, which is also normalized is referred to as an orthonormal set of functions on the interval $a \leq x \leq b$ having the property

$$\int_a^b g_m(x) g_n(x) \, dx = \begin{cases} 0 & m \neq n \\ 1 & m = n \end{cases} \tag{3.45}$$

3.7 Sturm-Liouville Systems

A large class of boundary value problems are represented by the linear homogeneous second-order differential equation

$$\frac{d}{dx}\left[p(x)\frac{dy}{dx}\right] + [q(x) + \lambda\omega(x)]y = 0 \tag{3.46}$$

on some interval $a \leq x \leq b$ and satisfying homogeneous boundary conditions of the form

$$h_1 y + K_1 \frac{dy}{dx} = 0 \quad \text{at} \quad x = a \tag{3.47}$$

$$h_2 y + K_2 \frac{dy}{dx} = 0 \quad \text{at} \quad x = b \tag{3.48}$$

In Equations (3.46) - (3.48) we have:

λ is a real parameter,
$\omega(x)$ is a weight function,
h_1, K_1, h_2, and K_2 are known real constants,
both h_1 and K_1 cannot be zero at the same time,
both h_2 and K_2 cannot be zero at the same time.

We like now to analyze the system stated by Equations (3.46) to (3.48) and examine the conditions under which this system yields orthogonal sets of functions. The system is referred to as the *Sturm-Liouville problem*.

Let $y = \phi_m(x)$ and $y = \phi_n(x)$ be eigenfunctions of the Sturm-Liouville problem (Equations (3.46) to (3.48)). Let the corresponding eigenvalues be λ_m and λ_n respectively. We invoke here that $\phi_m(x)$ and $\phi_n(x)$ are orthogonal with respect to $\omega(x)$ and that $d\phi_m(x)/dx$ and $d\phi_n(x)/dx$ are continuous in the interval $a \leq x \leq b$. Therefore, because $\phi_m(x)$ and $\phi_n(x)$ are solutions to Equation (3.46) we can write

$$\frac{d}{dx}\left[p\frac{d\phi_m}{dx}\right] + [q + \lambda_m\omega]\phi_m = 0 \tag{3.49}$$

$$\frac{d}{dx}\left[p\frac{d\phi_n}{dx}\right] + [q + \lambda_n\omega]\phi_n = 0 \tag{3.50}$$

Let us multiply the first of these equations by $\phi_n(x)$ and the second by $-\phi_m(x)$ and then add them; the result becomes

$$\begin{aligned}(\lambda_m - \lambda_n)\,\omega\,\phi_m\,\phi_n &= \phi_m\frac{d}{dx}\left[p\frac{d\phi_n}{dx}\right] - \phi_n\frac{d}{dx}\left[p\frac{d\phi_m}{dx}\right] \\ &= \frac{d}{dx}\left[\left(p\frac{d\phi_n}{dx}\right)\phi_m - \left(p\frac{d\phi_m}{dx}\right)\phi_n\right]\end{aligned} \tag{3.51}$$

Integrating Equation (3.51) from a to b yields

$$(\lambda_m - \lambda_n)\int_a^b \omega\,\phi_m\,\phi_n\,dx = \left[\left(p\frac{d\phi_n}{dx}\right)\phi_m - \left(p\frac{d\phi_m}{dx}\right)\phi_n\right]_a^b$$

$$= - \left[p\left\{ \left(\phi_n + \beta\frac{d\phi_n}{dx}\right)\frac{d\phi_m}{dx} - \left(\phi_m + \beta\frac{d\phi_m}{dx}\right)\frac{d\phi_n}{dx} \right\} \right]_a^b \tag{3.52}$$

where β has any arbitrary value, and $\beta\phi_n'\phi_m'$ has been added and subtracted on the right-hand side of Equation (3.52). Let us examine now Equation (3.52). It was indicated earlier that ϕ_m and ϕ_n are orthogonal with respect to ω. Therefore, from the properties of orthogonal functions, the left-hand side of Equation (3.52) should be zero. We note that $\lambda_m \neq \lambda_n$ because a distinct and different eigenvalue is associated with each eigenfunction. As our objective is to examine the conditions under which ϕ_m and ϕ_n are orthogonal, the right-hand side of Equation (3.52) determines these conditions. These conditions are the ones that make the right-hand side equal to zero.

The right-hand side is zero when

1. $y = 0$; hence $\phi_m = 0$ and $\phi_n = 0$ when $x = a$ or $x = b$.
2. $\dfrac{dy}{dx} = 0$; hence $\dfrac{d\phi_n}{dx} = 0$ and $\dfrac{d\phi_m}{dx} = 0$

   when $x = a$ or $x = b$.
3. $y + \beta\dfrac{dy}{dx} = 0$; hence $\left(\phi_n + \beta\dfrac{d\phi_n}{dx}\right) = 0$ and $\left(\phi_m + \beta\dfrac{d\phi_m}{dx}\right) = 0$

   when $x = a$ or $x = b$. Note: $\beta = 0$ yields condition 1 while $\beta = \infty$ yields condition 2.
4. $p = 0$ for $x = a$ or $x = b$.

   In this case, however, y and dy/dx are required to be finite at either $x = a$ or $x = b$; or $p\, dy/dx$ becomes zero at either of these points.
5. $y(a) = y(b)$; $\dfrac{dy}{dx}(a) = \dfrac{dy}{dx}(b)$; $p(a) = p(b)$.

   In this case, $\phi_m(x)$ and $\phi_n(x)$ are periodics with period $(b - a)$.

Therefore, if any of these conditions is satisfied, we will have then

$$\int_a^b \omega(x)\,\phi_m(x)\,\phi_n(x)\,dx = 0 \qquad m \neq n \tag{3.53}$$

Physical Interpretation of Orthogonality Conditions

Case 1. $y = 0$ when $x = a$ or $x = b$.

A differential equation that represents the spacial variation of the temperature distribution in a solid can be written as

$$\frac{d^2T}{dx^2} + \lambda T = 0 \tag{3.54}$$

$$T(0) = 0; \quad T(l) = 0 \tag{3.55}$$

Hence, in this example the temperature has to vanish at $x = a = 0$, and $x = b = l$ (Figure 3-7). A similar equation results from the solution of a vibrating string

$$\frac{d^2y}{dx^2} + \lambda y = 0 \tag{3.56}$$

$$y(0) = 0; \quad y(l) = 0 \tag{3.57}$$

The physical interpretation, here, is that the deflection of the string is zero at $x = a = 0$ and at $x = b = l$ (Figure 3-8). In other words, the string is hinged at $x = 0$ and at $x = l$.

In both problems, the solution of the differential equation is

$$\left.\begin{matrix} T(x) \\ y(x) \end{matrix}\right\} = \sin\sqrt{\lambda_n}x, \quad \lambda_n = \frac{n^2\pi^2}{l^2}; \quad n = 1, 2, 3, \ldots \tag{3.58}$$

When we compare the differential equation for these two boundary value problems with the general equation of the Sturm-Liouville system, Equations (3.46) - (3.48), we find that

$$p(x) = 1; \quad q(x) = 0; \quad \omega(x) = 1; \quad \frac{K_1}{h_1} = 0; \quad \frac{K_2}{h_2} = 0$$

Hence, the orthogonality condition, Equation (3.42), becomes

$$\int_0^l \sin\frac{n\pi x}{l} \sin\frac{m\pi x}{l}\, dx = 0 \quad m \neq n \tag{3.59}$$

Bessel Equation

Let us take now Bessel equation of the form

$$x^2\frac{d^2y}{dx^2} + x\frac{dy}{dx} + (\lambda^2x^2 - \nu^2)y = 0 \tag{3.60}$$

or its equivalent

$$\frac{d}{dx}\left(x\frac{dy}{dx}\right) + \left(-\frac{\nu^2}{x} + \lambda^2x\right)y = 0 \tag{3.61}$$

where ν is a non-negative integer. For each value of ν in Equation (3.61), there results a Sturm-Liouville equation with a parameter λ^2 instead of λ. When Equation (3.61) is compared with Equation (3.46) we find that

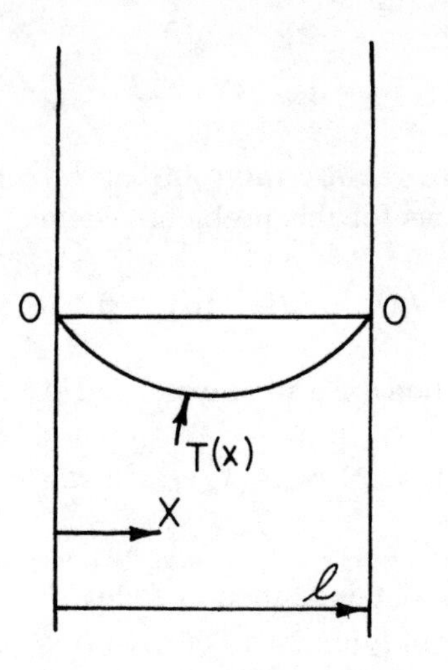

Figure 3-7. Temperature Distribution in a Slab

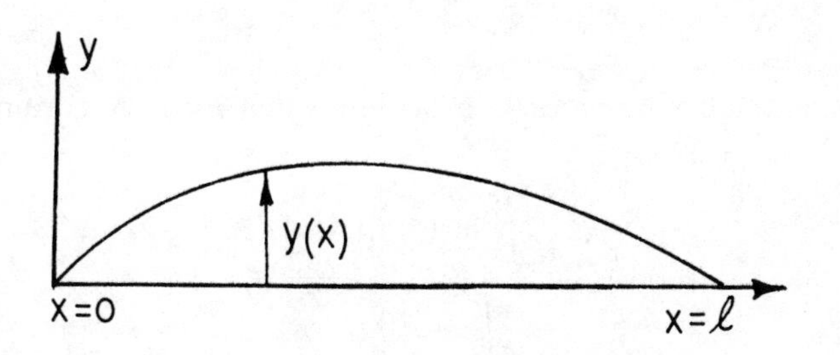

Figure 3-8. Vibrating String Fixed at Both Ends

$$p(x) = x; \quad q(x) = -\frac{\nu^2}{x}; \quad \omega(x) = x$$

Since $p(0) = 0$, in fact condition 4 is already satisfied at $x = a = 0$ provided y is finite at $x = 0$ and $x\, dy/dx \to 0$. Hence, if we have a homogeneous condition at $x = R$ (here the limit b is set equal to R) we obtain the problem's eigenfunctions orthogonal in the interval $0 \leq x \leq R$ with respect to the weight function $\omega(x) = x$ (Figure 3-9). It was found in chapter 2 that the solution to Equation (3.61), which is finite at $x = 0$, is

$$y(x) = J_\nu(\lambda x) \tag{3.62}$$

Applying the conditions of Case 1 at $x = R$ yields

$$0 = J_\nu(\lambda R) \tag{3.63}$$

Hence,

$$\lambda R = \alpha_{n\nu} \quad \text{or} \quad \lambda_n = \frac{\alpha_n}{R} \tag{3.64}$$

where $\alpha_n = \alpha_1, \alpha_2, \alpha_3, \ldots$ are the roots of $J_\nu(\alpha_{n\nu}) = 0$. Therefore, the orthogonality conditions for this problem become

$$\int_0^R x\, J_\nu(\lambda_n x)\, J_\nu(\lambda_m x)\, dx = 0 \qquad m \neq n \tag{3.65}$$

If an arbitrary function $f(x)$ is to be expanded in terms of $J_\nu(\lambda_n x)$ such as

$$f(x) = \sum_{m=1}^{\infty} A_n J_\nu(\lambda_n x) \qquad 0 < x < R \tag{3.66}$$

the coefficients A_n are obtained then as follows:

Multiply both sides of Equation (3.66) by $\omega(x)\, J_\nu(\lambda_m x)$ where, here, $\omega(x) = x$, and integrate the result between 0 and R to give

$$\int_0^R x f(x) J_\nu(\lambda_m x)\, dx = \sum_{n=1}^{\infty} A_n \int_0^R x\, J_\nu(\lambda_m x) J_\nu(\lambda_n x)\, dx \tag{3.67}$$

The right-hand side of Equation (3.67) is zero when $m \neq n$. Accordingly, we get

$$A_n = \frac{\int_0^R x f(x)\, J_\nu(\lambda_m x)\, dx}{\int_0^R x\, J_\nu^2(\lambda_m x)\, dx} \tag{3.68}$$

where

$$\int_0^R x\, J_\nu^2(\lambda_m x)\, dx = \frac{R^2}{2} J_{\nu+1}^2(\lambda_m R) \tag{3.69}$$

Case 2.

$$\frac{dy}{dx} = 0 \quad \text{when } x = a \text{ or } x = b.$$

For the one-dimensional temperature distribution in a slab which is *insulated* (Figure 3-10) at $x = 0$ and at $x = l$, the boundary value problem is written as

$$\frac{d^2T}{dx^2} + \lambda T = 0 \tag{3.70}$$

$$\frac{dT}{dx}(0) = 0; \qquad \frac{dT}{dx}(l) = 0 \tag{3.71}$$

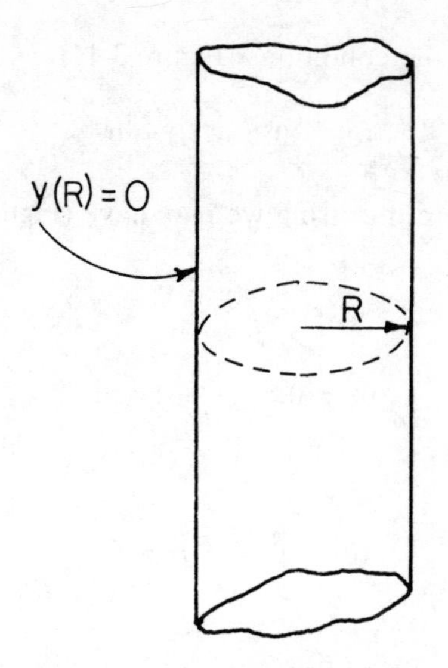

Figure 3-9. Cylinder with a Homogeneous Boundary Condition

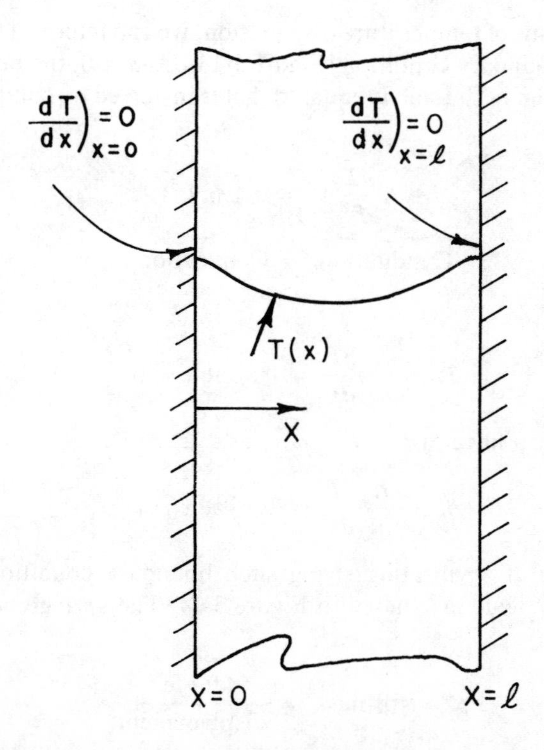

Figure 3-10. Temperature Distribution within an Insulated Slab

One may have as well the conditions (Figure 3-11)

$$\frac{dT}{dx}(0) = 0; \quad T(l) = 0 \tag{3.72}$$

In the case of a vibrating string we may have (Figure 3-12)

$$\frac{d^2y}{dx^2} + \lambda y = 0 \tag{3.73}$$

$$\frac{dy}{dx}(0) = 0; \quad \frac{dy}{dx}(l) = 0 \tag{3.74}$$

Another possibility is

$$\frac{dy}{dx}(0) = 0; \quad y(l) = 0 \tag{3.75}$$

Case 3.

$$y + \beta\frac{dy}{dx} = 0 \text{ when } x = a \text{ or } x = b.$$

Again in terms of temperature distribution, we can interpret this case as a convection boundary condition (Figure 3-13). At $x = 0$, the heat transfer by convection out of the slab is equal to that transferred by conduction at $x = 0$. Therefore,

$$\underbrace{+K\frac{dT}{dx}}_{\text{Conduction}} = \underbrace{h_1 T}_{\text{Convection}}$$

or

$$T - \frac{K}{h_1}\frac{dT}{dx} = 0 \quad \text{at } x = 0 \tag{3.76}$$

Similarly, we can have at $x = l$

$$T + \frac{K}{h_2}\frac{dT}{dx} = 0 \quad \text{at } x = l \tag{3.77}$$

In the case of a vibrating string such boundary conditions may be interpreted physically as shown in Figure 3-14. The spring constant K is equal to:

$$K = \text{stiffness} = \frac{\text{force}}{\text{displacement}}$$

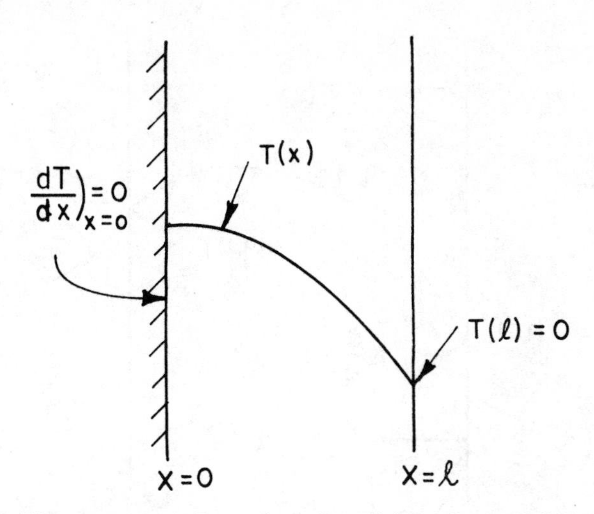

Figure 3-11. Homogeneous Boundary Conditions for a Slab Insulated on One Side

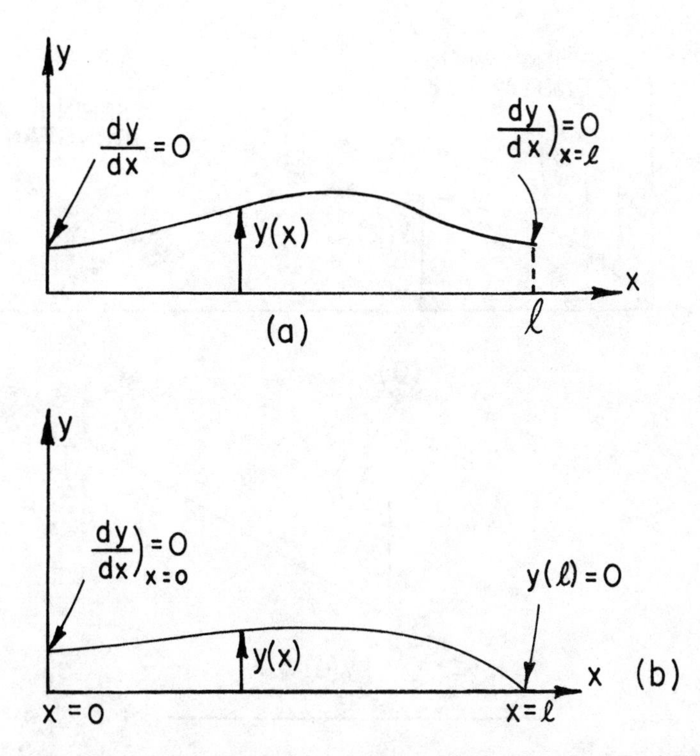

Figure 3-12. Vibrating String with: (a) Free Ends, (b) One End Free

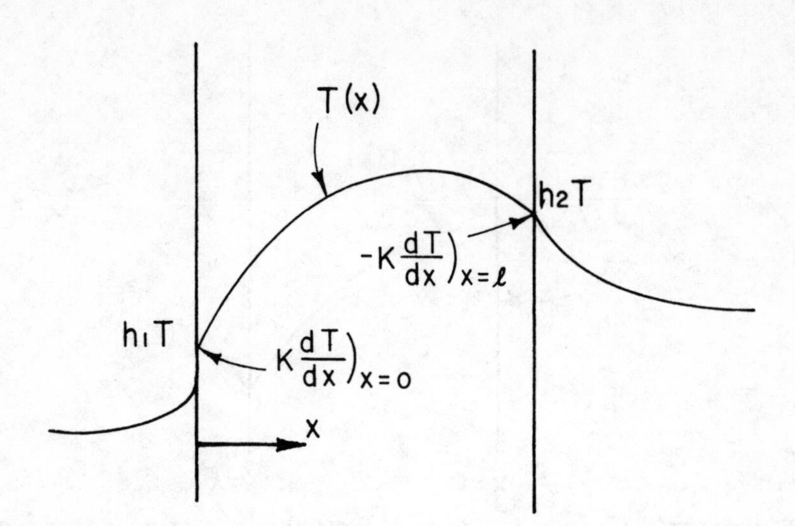

Figure 3-13. Temperature Distribution within a Slab with Convection Boundary Conditions

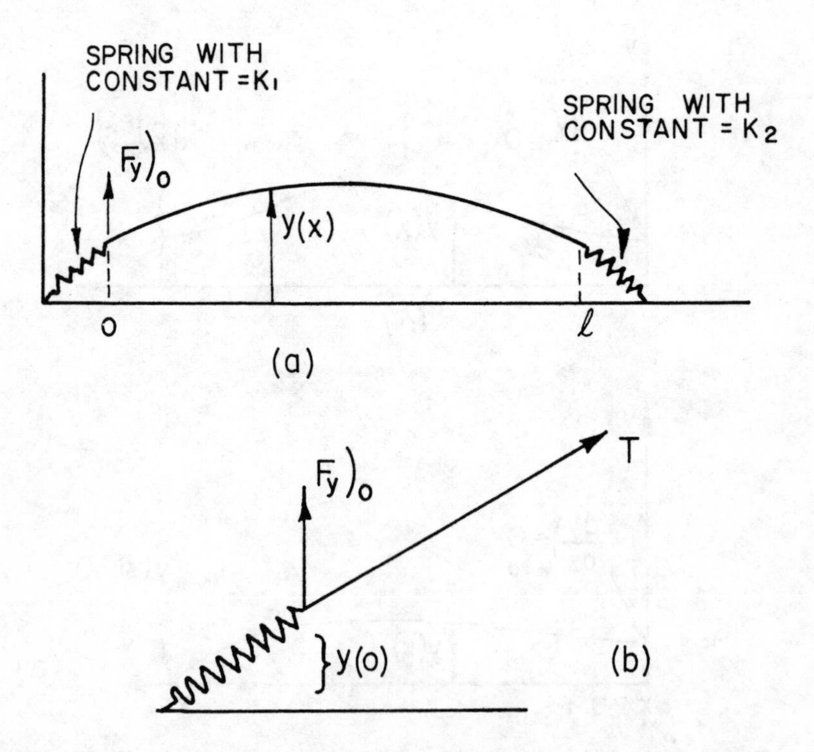

Figure 3-14. Vibrating String with Springs Attached at Its Ends

The displacement of the support at $x = 0$ is

$$y)_{x=0} = \frac{F_y)_0}{K_1} \tag{3.78}$$

but also

$$F_y)_0 = T\frac{dy}{dx}\bigg)_{x=0} \tag{3.79}$$

Equating Equations (3.78) and (3.79) yields

$$y - \frac{T}{K_1}\frac{dy}{dx}\bigg)_{x=0} = 0 \tag{3.80}$$

Similarly, we may have at $x = l$

$$y + \frac{T}{K_2}\frac{dy}{dx}\bigg)_{x=l} = 0 \tag{3.81}$$

Cases 4, 5.

$$p = 0 \text{ when } x = a \text{ or } x = b.$$

This case was explained for Bessel equation under Case 1. Let us take now Legendre's equation, which is written in the form

$$\frac{d}{dx}\left[(1 - x^2)\frac{dy}{dx}\right] + \lambda y = 0, \quad \lambda = n(n + 1) \tag{3.82}$$

Comparing this equation with Equation (3.46) we find

$$p(x) = 1 - x^2; \quad q(x) = 0; \quad \omega(x) = 1$$

Because $p = 0$ when $x = \pm 1$, no boundary conditions are required to form a Sturm-Liouville problem on the interval $-1 \leq x \leq 1$. It was found in chapter 2 that Legendre polynomials $P_n(x)$ are solutions to Equation (3.82), and therefore they constitute the eigenfunctions for the problem. In this case, then, the orthogonality relations become

$$\int_{-1}^{1} P_m(x)P_n(x)\,dx = 0 \quad m \neq n \tag{3.83}$$

When an arbitrary function $f(x)$ is to be expanded in terms of Legendre's polynomials, such as

$$f(x) = \sum_{n=0}^{\infty} A_n P_n(x) \tag{3.84}$$

the coefficients A_n are obtained as follows:

Multiply both sides of Equation (3.84) by $P_m(x)$ (here $\omega(x) = 1$) and integrate with respect to x between $x = -1$ and $x = +1$, that is,

$$\int_{-1}^{1} f(x)\, P_m(x)\, dx = \sum_{n=0}^{\infty} A_n \int_{-1}^{1} P_m(x)\, P_n(x)\, dx \tag{3.85}$$

The right-hand side in this equation is zero if $m \neq n$; hence,

$$A_n = \frac{\int_{-1}^{1} f(x)\, P_m(x)\, dx}{\int_{-1}^{1} P_m^2(x)\, dx} \tag{3.86}$$

where

$$\int_{-1}^{1} P_m^2(x)\, dx = \frac{2}{2m+1}; \qquad m = 0, 1, 2, \ldots \tag{3.87}$$

4 Partial Differential Equations in Engineering Sciences

4.1 Introduction

We concentrate in the following few chapters on the analysis and solutions of some important partial differential equations that arise in engineering applications. The equations are derived from basic principles of physics as applied to the various systems representing specified engineering problems.

In ordinary differential equations we found that the equation involving the derivatives of a function u of one variable x is an ordinary differential equation, such as

$$\frac{d^2u}{dx^2} + a_1\frac{du}{dx} + a_2u = f_1(x) \tag{4.1}$$

If the function u is dependent on two or more variables like x, y, and t, the derivatives, then, of u with respect to x, y, and t are partial derivatives. The equation involving such derivatives will then be a partial differential equation, that is,

$$\frac{\partial^2 u}{\partial x^2} + \frac{\partial^2 u}{\partial y^2} = \frac{\partial u}{\partial t} \tag{4.2}$$

The order of the partial differential equation is equal to the order of the highest derivative. A form of a second order equation with two independent variables x, y can be written as

$$a_{11}\frac{\partial^2 u}{\partial x^2} + 2a_{12}\frac{\partial^2 u}{\partial x\,\partial y} + a_{22}\frac{\partial^2 u}{\partial y^2} + F_1\left(x,y,u,\frac{\partial u}{\partial x},\frac{\partial u}{\partial y}\right) = 0 \tag{4.3}$$

A partial differential equation is designated as linear if the unknown function u or its derivatives appear to the first degree. In this case the coefficients a_{11}, a_{12}, and a_{22} in Equation (4.3) are functions of x and y only, or they are constants.

When each term in the equation contains the function u or its derivative, the equation is then homogeneous; if this is not the case, the equation is then nonhomogeneous, that is,

$$x\frac{\partial^2 u}{\partial x^2} + \frac{\partial^2 u}{\partial y^2} + \frac{\partial u}{\partial x} + u = 0 \quad \text{homogeneous} \tag{4.4}$$

$$\frac{\partial^2 u}{\partial x^2} + xy\frac{\partial^2 u}{\partial y^2} + \frac{\partial u}{\partial y} + u + f_1(x,y) = 0 \quad \text{nonhomogeneous} \tag{4.5}$$

$f_1(x,y)$ is the nonhomogeneous term and can be a constant.

A partial differential equation can be reduced to a simpler form through a certain transformation of variables. Such simpler forms are referred to as the canonical forms of the equation. We introduce such a transformation later on in this chapter.

4.2 Equations Solved by Direct Integration

When dealing with ordinary differential equations, we found that a simple second order equation of the form

$$\frac{d^2u}{dy^2} = 0$$

can be solved by direct integration as follows:

$$\frac{du}{dy} = C_1$$

$$u = C_1 y + C_2$$

where C_1 and C_2 are constants to be determined by specifying two conditions on u.

Let us solve now similar partial differential equations by direct integration.

Example 4.1 Solve the following equation:

$$\frac{\partial^2 u}{\partial y^2} = 0 \tag{4.6}$$

Solution: Implicit in this equation is the fact that u is a function of two variables (such as y and x); this is the reason it is written in terms of partial derivatives. The solution to this equation can be obtained as well by direct integration in the following way:

Integrating once gives

$$\frac{\partial u}{\partial y} = f_1(x) \tag{4.7}$$

"note the right-hand side was a constant for the case of ordinary differential equations"

Integrating again yields

$$u = \int f_1(x)\, dy + f_2(x)$$

or

$$u = y f_1(x) + f_2(x) \tag{4.8}$$

$f_1(x)$ and $f_2(x)$ are to be determined from two conditions to be imposed on u, for example,

1. when $y = 0 \quad u = \sin x \quad$ for all x

   Substituting in the general expression (4.8) yields

$$f_2(x) = \sin x \tag{4.9}$$

2. when $x = 0 \quad u = y + 2y^2$

   Substituting also in (4.8) gives

$$\begin{aligned} y + 2y^2 &= y f_1(0) + f_2(0) \\ &= y f_1(0) + 0 \end{aligned} \tag{4.10}$$

   Solving for $f_1(0)$ we obtain

$$\frac{2y^2 + y}{y} = f_1(0)$$

   or

$$f_1(0) = 2y + 1 \tag{4.11}$$

Therefore, $f_1(x)$ is any function of x that has a value of $2y + 1$ when $x = 0$.

Example 4.2 Let us take the following equation:

$$\frac{\partial^2 u}{\partial x \, \partial t} = 10 \tag{4.12}$$

Solution: Integrating with respect to x gives

$$\frac{\partial u}{\partial t} = 10x + f_1(t) \tag{4.13}$$

Integrating now with respect to t yields

$$u = \int 10x \, dt + \int f_1(t) \, dt + F_2(x)$$

or

$$u = 10xt + F_1(t) + F_2(x) \tag{4.14}$$

where we set $\int f_1(t) \, dt = F_1(t)$ to be another arbitrary function of t. To evaluate $F_1(t)$ and $F_2(x)$ we need to impose two conditions, such as

1. when $x = 0 \quad u = 0$

$$\therefore 0 = F_1(t) + F_2(0) \tag{4.15}$$

2. when $t = 0 \quad u = \cos x$

$$\therefore \cos x = F_1(0) + F_2(x) \tag{4.16}$$

From Equation (4.16) we have

$$F_2(x) = \cos x - F_1(0) \tag{4.17}$$

From Equation (4.15) we have, for any t including $t = 0$

$$F_1(t) = -F_2(0) = F_1(0) \tag{4.18}$$

Substituting Equations (4.17) and (4.18) in the general solution for u, Equation (4.14) gives

$$u = 10xt + \cos x$$

Example 4.3 Solve the following differential equation

$$\frac{\partial^2 u}{\partial x\, \partial t} = \frac{\partial u}{\partial x} \tag{4.19}$$

Solution: Let $p = \partial u/\partial x$. Hence, Equation (4.19) becomes

$$\frac{\partial p}{\partial t} = p \tag{4.20}$$

Separating the variables and integrating give

$$\ln p = t + G_1(x)$$

or

$$p = e^t e^{G_1(x)} = G_2(x) e^t \tag{4.21}$$

where we set $e^{G_1(x)} = G_2(x)$. Therefore,

$$p = \frac{\partial u}{\partial x} = G_2(x)\, e^t \tag{4.22}$$

Integrating Equation (4.22) results in the following solution for u

$$u = \int e^t G_2(x)\, dx + G_3(t)$$

or

$$u = G_4(x)\, e^t + G_3(t) \tag{4.23}$$

In Equation (4.23) we set $\int G_2(x)\, dx = G_4(x)$.

4.3 Classification of Partial Differential Equations—Canonical Form

Let us consider Equation (4.3) again and write it in the following form

$$a_{11} u_{xx} + 2a_{12} u_{xy} + a_{22} u_{yy} = H(x, y, u, u_x, u_y) \tag{4.24}$$

Here, we introduced the following usual notations:

$$\frac{\partial u}{\partial x} = u_x, \qquad \frac{\partial^2 u}{\partial x^2} = u_{xx}, \qquad \frac{\partial^2 u}{\partial x \, \partial y} = u_{xy}, \qquad \text{etc.}$$

In Equation (4.24), a_{11}, a_{12}, and a_{22} are functions of x and y; H is a continuous function of its arguments. Our objective is to reduce Equation (4.24) by some transformation into one of the following forms:

$$u_{\xi\xi} + u_{\eta\eta} = G, \qquad u_{\xi\xi} = G, \qquad u_{\xi\eta} = G \tag{4.25}$$

ξ and η are new variables of x and y, and G is a function of ξ, η, u, u_ξ, and u_η. The forms in Equation (4.25) are referred to as the canonical forms of the partial differential Equation (4.24). It is very much simpler to deal with equations in their canonical forms rather than the more general form appearing in Equation (4.24).

To transform a differential equation into a canonical form we need certain transformation variables. Let these variables be represented in the following form

$$\xi = \phi(x,y), \qquad \eta = \psi(x,y) \tag{4.26}$$

The functional relationship is not known yet. Our objective is to obtain an equation in terms of ξ and η equivalent to the initial equation that will have the simplest form.

When the derivatives in Equation (4.24) are transformed into the new variables, we obtain the following expressions:

$$u_x = u_\xi \, \xi_x + u_\eta \, \eta_x \tag{4.27}$$

$$u_y = u_\xi \, \xi_y + u_\eta \, \eta_y \tag{4.28}$$

$$u_{xx} = u_{\xi\xi} \xi_x^2 + 2u_{\xi\eta} \xi_x \eta_x + u_{\eta\eta} \eta_x^2 + u_\xi \xi_{xx} + u_\eta \eta_{xx} \tag{4.29}$$

$$u_{xy} = u_{\xi\xi} \xi_x \xi_y + u_{\xi\eta}(\xi_x \eta_y + \xi_y \eta_x) + u_{\eta\eta} \eta_x \eta_y + u_\xi \xi_{xy} + u_\eta \eta_{xy} \tag{4.30}$$

$$u_{yy} = u_{\xi\xi} \xi_y^2 + 2u_{\xi\eta} \xi_y \eta_y + u_{\eta\eta} \eta_y^2 + u_\xi \xi_{yy} + u_\eta \eta_{yy} \tag{4.31}$$

Substituting Equations (4.27) to (4.31) into Equation (4.24) yields

$$A_{11} u_{\xi\xi} + 2A_{12} u_{\xi\eta} + A_{22} u_{\eta\eta} + F = 0 \tag{4.32}$$

where

$$A_{11} = a_{11}\xi_x^2 + 2a_{12}\xi_x\xi_y + a_{22}\xi_y^2 \tag{4.33}$$

$$A_{12} = a_{11}\xi_x\eta_x + a_{12}(\xi_x\eta_y + \eta_x\xi_y) + a_{22}\xi_y\eta_y \tag{4.34}$$

$$A_{22} = a_{11}\eta_x^2 + 2a_{12}\eta_x\eta_y + a_{22}\eta_y^2 \tag{4.35}$$

and

$$F = F(\xi, \eta, u, u_\xi, u_\eta) \tag{4.36}$$

Looking at Equation (4.32) we can see that if some of the coefficients A_{11}, A_{12}, or A_{22} are zero the equation will be greatly simplified. Hence, ξ and η can be chosen such that A_{11} is zero. In this case, we have then from Equation (4.33)

$$a_{11}\xi_x^2 + 2a_{12}\xi_x\xi_y + a_{22}\xi_y^2 = 0 \tag{4.37}$$

Therefore, the problem of selecting the new independent variables ξ (and η) depends upon the solution of Equation (4.37). This is so because if $\xi = \phi(x,y)$ is a solution to Equation (4.37), then the coefficient A_{11} will be zero. Accordingly, $\phi(x,y) = C$ (constant) is a general integral of the equation

$$a_{11}\,dy^2 - 2a_{12}\,dx\,dy + a_{22}\,dx^2 = 0 \tag{4.38}$$

Equation (4.38) is obtained by substituting $\phi(x,y)$ for ξ in Equation (4.37), having in mind that $\phi(x,y) = C$.

Equation (4.38) is referred to as the characteristic equation for Equation (4.24).

Similarly, if $\psi(x,y) = C$ (another constant) is another integral of Equation (4.38), the coefficient of $u_{\eta\eta}$ will then be zero.

Solving Equation (4.38) for dy/dx results in

$$\frac{dy}{dx} = \frac{a_{12} \pm \sqrt{a_{12}^2 - a_{11}a_{22}}}{a_{11}} \tag{4.39}$$

The sign of the quantity under the radical sign determines the type of partial differential equation as to whether it is hyperbolic, elliptic, or parabolic type. When

$$a_{12}^2 - a_{11}a_{22} > 0 \quad \text{the equation is hyperbolic} \tag{4.40a}$$

$$a_{12}^2 - a_{11}a_{22} < 0 \quad \text{the equation is elliptic} \tag{4.40b}$$

$$a_{12}^2 - a_{11}a_{22} = 0 \quad \text{the equation is parabolic} \tag{4.40c}$$

Canonical Form for the Hyperbolic, Parabolic, and Elliptic Equations

Hyperbolic: In this case, $a_{12}^2 - a_{11}a_{22} > 0$, and the right-hand side of

Equation (4.39) is real with two different values which result in the general integrals

$$\phi(x,y) = C \quad \text{and} \quad \psi(x,y) = C$$

Hence, by choosing

$$\xi = \phi(x,y) \quad \text{and} \quad \eta = \psi(x,y)$$

Equation (4.32) is reduced to the form

$$u_{\xi\eta} = \theta_1(\xi, \eta, u, u_\xi, u_\eta) \tag{4.41}$$

where

$$\theta_1 = -\frac{F}{2A_{12}}$$

Equation (4.41) represents one canonical form for equations of the hyperbolic type.

There is another canonical form for the equation of the hyperbolic type, which can be obtained as follows: Let

$$\alpha = \frac{\xi + \eta}{2}, \qquad \beta = \frac{\xi - \eta}{2}$$

then

$$u_{\xi\eta} = \tfrac{1}{4}(u_{\alpha\alpha} - u_{\beta\beta})$$

and Equation (4.41) becomes

$$u_{\alpha\alpha} - u_{\beta\beta} = 4\theta_1 \tag{4.42}$$

which is the second canonical form.

Parabolic: In this case, $a_{12}^2 - a_{11}a_{22} = 0$, and Equation (4.39) yields one real general integral $\phi(x,y) =$ constant. Therefore, in this case we let

$$\xi = \phi(x,y)$$

as obtained from Equation (4.39), and

$$\eta = \eta(x,y)$$

η is *any* function of x, y which is independent of $\phi(x,y)$.

Performing the differentiation and the changing of variables yields the canonical form for the parabolic type as

$$u_{\eta\eta} = \theta_2(\xi, \eta, u, u_\xi, u_\eta) \tag{4.43}$$

where

$$\theta_2 = -\frac{F}{A_{22}}$$

Elliptic: Here, $a_{12}^2 - a_{11}a_{22} < 0$, and Equation (4.39) yields complex integrals as

$$\phi(x,y) = C \quad \text{and} \quad \overline{\phi}(x,y) = C$$

where $\bar{\phi}$ is the complex conjugate of ϕ. If we set $\xi = \phi(x,y)$ and $\eta = \bar{\phi}(x,y)$ we can reduce Equation (4.32) to the canonical form. However, in order to avoid complex variables, we can set

$$\alpha = \frac{\phi + \overline{\phi}}{2} \quad \text{and} \quad \beta = \frac{\phi - \phi}{2i}$$

and obtain the canonical form as

$$u_{\alpha\alpha} + u_{\beta\beta} = \theta_3(\alpha, \beta, u, u_\alpha, u_\beta) \tag{4.44}$$

where

$$\theta_3 = -\frac{F}{A_{22}}$$

In conclusion, we state that the canonical forms for the second order partial differential equations represented by Equation (4.24) are:

$$u_{xy} = \Phi \quad \text{Hyperbolic}$$

or

$$u_{xx} - u_{yy} = \Phi \quad \text{Hyperbolic} \tag{4.45}$$

$$u_{xx} = \Phi \quad \text{Parabolic} \tag{4.46}$$

$$u_{xx} + u_{yy} = \Phi \quad \text{Elliptic} \tag{4.47}$$

Example 4.4 Classify the following partial differential equation and reduce it to the canonical form:

$$u_{xx} + 2u_{xy} + u_{yy} = 0 \tag{4.48}$$

Solution: Comparing this equation with Equation (4.24) yields

$$a_{11} = 1 \quad a_{12} = 1 \quad a_{22} = 1 \quad H = 0$$

From Equation (4.39) we have

$$\frac{dy}{dx} = \frac{1 \pm \sqrt{1-1}}{1} = 1 \tag{4.49}$$

Therefore, the equation is parabolic. Integrating Equation (4.49) yields

$$y = x + C \quad \text{or} \quad y - x = C \tag{4.50}$$

Hence, we choose

$$\xi = \phi(x,y) = y - x \tag{4.51}$$

We need another function of x,y which is independent of ξ. Let us choose

$$\eta = \psi(x,y) = x \quad \text{(independent of } \xi) \tag{4.52}$$

Accordingly, the new variables are

$$\xi = y - x, \qquad \eta = x$$

The derivatives in terms of the new variables become:

$$\xi_x = -1, \quad \xi_y = 1, \quad \eta_x = 1, \quad \eta_y = 0 \tag{4.53}$$
$$\text{(second derivatives} = 0)$$

$$u_{xx} = u_{\xi\xi} - 2u_{\xi\eta} + u_{\eta\eta}, \quad u_{xy} = -u_{\xi\xi} + u_{\xi\eta}, \quad u_{yy} = u_{\xi\xi} \tag{4.54}$$

Substituting in Equation (4.48) yields

$$u_{\eta\eta} = 0 \tag{4.55}$$

This equation can be solved by direct integration. Integrating once gives

$$u_\eta = F_1(\xi) \tag{4.56}$$

Integrating again

$$u(\xi, \eta) = \int F_1(\xi)\, d\eta + F_2(\xi) \tag{4.57}$$

or

$$u(\xi, \eta) = \eta F_1(\xi) + F_2(\xi) \tag{4.58}$$

Introducing the definitions of ξ and η in Equation (4.58), we get

$$u(x,y) = xF_1(y - x) + F_2(y - x) \tag{4.59}$$

Example 4.5 Classify the following equation and reduce it to canonical form:

$$u_{xx} + y\, u_{yy} = 0 \tag{4.60}$$

Solution: Comparing this equation with Equation (4.24) gives

$$a_{11} = 0, \quad a_{22} = y, \quad a_{12} = 0 \tag{4.61}$$

Substituting in Equation (4.39), we obtain

$$\frac{dy}{dx} = \pm\sqrt{-y} \tag{4.62}$$

Integrating now Equation (4.62) gives

$$-2(-y)^{1/2} = \pm x + C \tag{4.63}$$

Therefore, the transformation variables are

$$\xi = 2(-y)^{1/2} + x, \qquad \eta = 2(-y)^{1/2} - x \tag{4.64}$$

According to Equation (4.62), the classification of Equation (4.60) is dependent upon the value of y. Using the transformation variables given by Equation (4.64) yields

$$\xi_x = 1, \quad \xi_y = (-y)^{-1/2}, \quad \eta_x = -1, \quad \eta_y = (-y)^{-1/2} \tag{4.65}$$

$$\xi_{xx} = 0, \quad \xi_{yy} = \tfrac{1}{2}(-y)^{-3/2}, \quad \eta_{xx} = 0, \quad \eta_{yy} = \tfrac{1}{2}(-y)^{-3/2} \tag{4.66}$$

Correspondingly, when these derivatives are used in Equations (4.29) and (4.31) to obtain u_{xx} and u_{yy}, Equation (4.60) becomes

$$u_{\xi\eta} = -\frac{1}{2}\frac{1}{\xi+\eta}[u_\xi + u_\eta]$$

Equations of the Hyperbolic Type

5.1 Wave Equation

The theory of wave motion characterizes a large number of common physical phenomena in nature. Among such phenomena are the propagation of a pressure disturbance in a fluid, the transmission of electromagnetic waves in a medium, the longitudinal vibration of a beam, the vibration of a tightly stretched string and of membranes, the motion of tidal waves in a canal, and many others. In all these phenomena the disturbance propagates at speeds that depend upon the medium properties. We derive the partial differential equations governing these physical entities and solve them citing the physical interpretation of the solutions.

In the following sections we consider a few very fundamental problems in the theory of wave motion, and we show that the differential equations governing the wave motion in all these problems are very much similar.

5.2 Vibrating String

Let us consider a perfectly elastic string stretched to a length l and fixed at points $x = 0$ and $x = l$ (Figure 5-1). We assume that the string has a uniform mass per unit length, m, and offers no resistance to bending; the tension, T, on the string is quite large such that we can relatively ignore the effect of gravity, and the deflection $u(x,t)$ at any location and time is much less than the string length l. The problem is to find the deflection $u(x,t)$ at any location and at any time once the string is deflected from its undisturbed position along the x-axis and released. Summing up vertical forces on MN we obtain

$$F_u = T \sin\theta_2 - T \sin\theta_1 \tag{5.1}$$

For small oscillations, we can write

$$\sin\theta_2 = \tan\theta_2 = \left(\frac{\partial u}{\partial x}\right)_{x+\Delta x} \tag{5.2}$$

$$\sin\theta_1 = \tan\theta_1 = \left(\frac{\partial u}{\partial x}\right)_{x} \tag{5.3}$$

Therefore we obtain

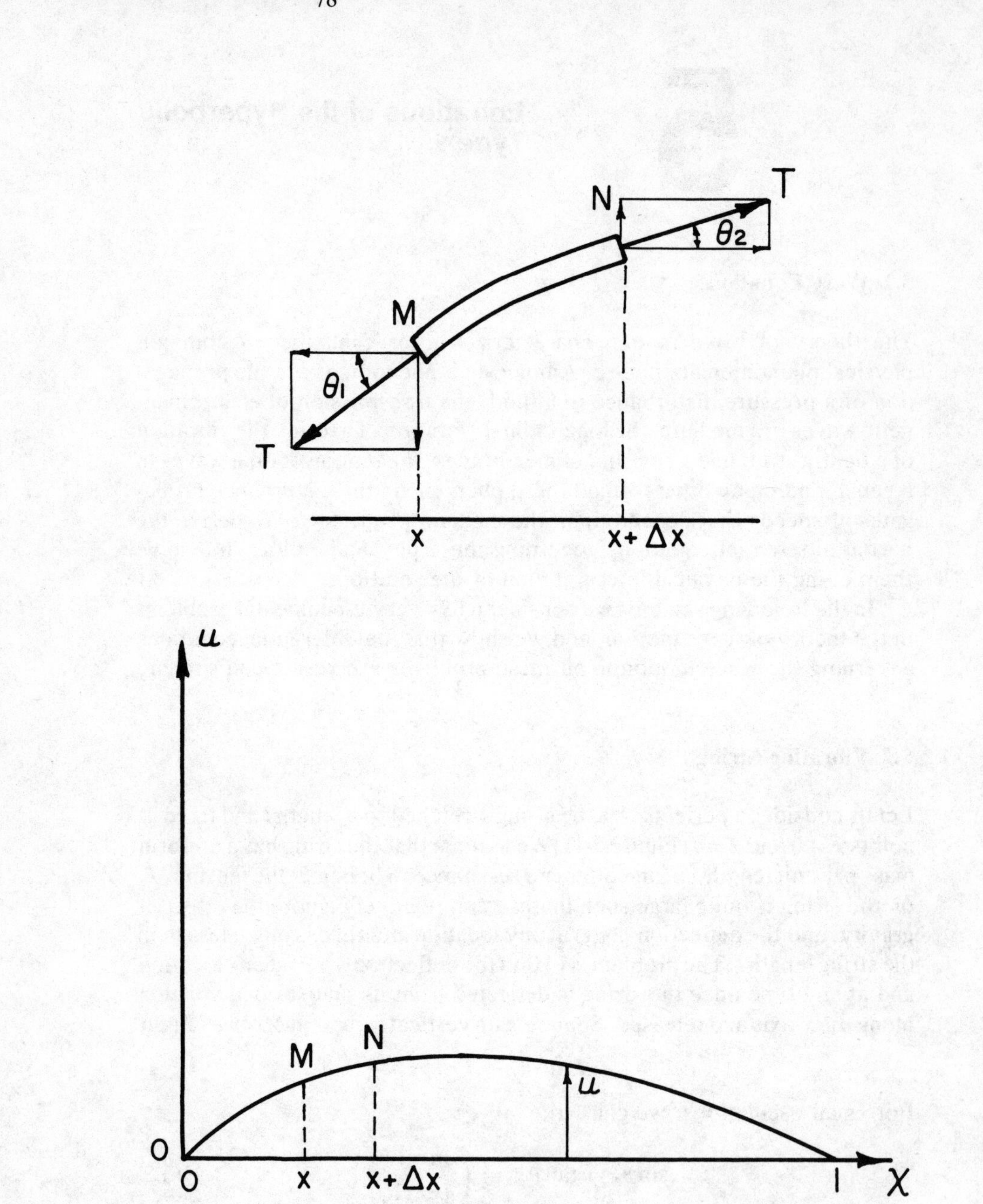

Figure 5-1. Vibrating String

$$F_u = T\left[\frac{\partial u}{\partial x}\bigg|_{x+\Delta x} - \frac{\partial u}{\partial x}\bigg|_{x}\right] \tag{5.4}$$

Introducing Newton's Second Law of Motion

$$F_u = \text{mass} \times \text{acceleration} \tag{5.5}$$

we can write then

$$T\left[\frac{\partial u}{\partial x}\bigg|_{x+\Delta x} - \frac{\partial u}{\partial x}\bigg|_{x}\right] = m\Delta x \frac{\partial^2 u}{\partial t^2} \tag{5.6}$$

where $m\Delta x$ is the mass of section MN and $\partial^2 u/\partial t^2$ is the acceleration of that section. Let us rewrite Equation (5.6) as follows

$$\left[\frac{\partial u}{\partial x}\bigg|_{x+\Delta x} - \frac{\partial u}{\partial x}\bigg|_{x}\right]\Big/\Delta x = \frac{m}{T}\frac{\partial^2 u}{\partial t^2} \tag{5.7}$$

When we let Δx approach zero the left-hand side of Equation (5.7) becomes $\partial^2 u/\partial x^2$ and Equation (5.7) becomes

$$\frac{\partial^2 u}{\partial x^2} = \frac{1}{c^2}\frac{\partial^2 u}{\partial t^2} \tag{5.8}$$

where we set

$$c^2 = T/m \tag{5.9}$$

c^2 was chosen instead of c to emphasize that c is a positive quantity. The constant c has the dimensions of length/time and therefore the dimensions of velocity. It will be shown in Article 5.8 that c is the velocity of the wave as it propagates in the medium.

Nonhomogeneous Equation of the Vibrating String

When an external force f per unit length is applied on the string parallel to the transverse deflection u (Figure 5-2), Equation (5.6) becomes

$$-f\Delta x + T\left[\frac{\partial u}{\partial x}\bigg|_{x+\Delta x} - \frac{\partial u}{\partial x}\bigg|_{x}\right] = m\Delta x \frac{\partial^2 u}{\partial t^2} \tag{5.10}$$

Dividing by $T\Delta x$ and letting Δx approach zero we obtain

$$\frac{\partial^2 u}{\partial x^2} - \frac{f}{T} = \frac{1}{c^2}\frac{\partial^2 u}{\partial t^2} \tag{5.11}$$

When the external force f varies along the string and varies with time as well, Equation (5.11) becomes

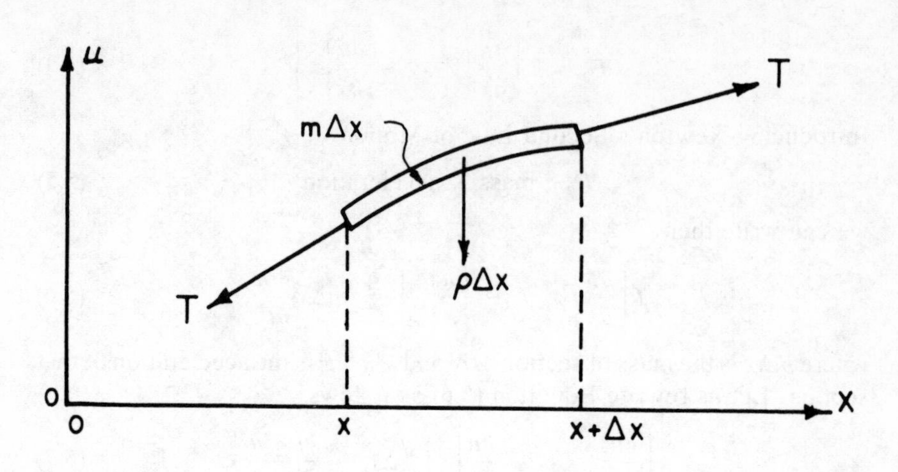

Figure 5-2. Vibrating String Subject to an External Force

$$\frac{\partial^2 u}{\partial x^2} + F(x,t) = \frac{1}{c^2}\frac{\partial^2 u}{\partial t^2} \tag{5.12}$$

where we set

$$F(x,t) = \frac{-f(x,t)}{T}$$

Equation (5.12) is referred to as the nonhomogeneous one-dimensional wave equation.

Vibration with Damping

When an element of machinery vibrates, it loses energy due to the internal friction and due to the frictional forces between the vibrating element and the surrounding medium. Hence, if a string is vibrating in air there will be loss of energy of vibration due to these two effects. Let us consider that these frictional or damping forces, F_d, are proportional to the velocity of the element. Therefore, we can write the damping force per unit length in the form

$$F_d = -\beta\frac{\partial u}{\partial t} \tag{5.13}$$

where β is a positive damping constant. F_d should be added to Equation (5.6) to give

$$T\left[\frac{\partial u}{\partial x}\bigg|_{x+\Delta x} - \frac{\partial u}{\partial x}\bigg|_{x}\right] - \beta\,\Delta x\frac{\partial u}{\partial t} = m\,\Delta x\frac{\partial^2 u}{\partial t^2} \tag{5.14}$$

Dividing by $T\Delta x$ and letting Δx approach zero we get

$$\frac{\partial^2 u}{\partial x^2} = \frac{1}{c^2}\frac{\partial^2 u}{\partial t^2} + \gamma\frac{\partial u}{\partial t} \tag{5.15}$$

where we set $\gamma = \beta/T$ as another damping constant. We like to note that γ in Equation (5.15) is not constant but varies to some extent with frequency. We have also ignored the inertial effect of the vibrating air around the string. These two effects are usually insignificant.

5.3 Longitudinal Vibrations of a Beam

We derive in this section the differential equations governing the displacement $u(x,t)$ of a beam vibrating along its axis (x-axis); see Figure 5-3. An initial displacement of the beam along the x-axis causes the sections of the beam to move in the x direction. Let us set at time t the displacement at section x as $u(x,t)$ and that at $x + \Delta x$ as $u(x + \Delta x,t)$. Therefore a length Δx is stretched by a quantity given by

$$\Delta x\epsilon = u(x + \Delta x,t) - u(x,t) \tag{5.16}$$

where ϵ is the strain of the element Δx. We consider that the beam behaves according to Hooke's Law. Therefore the force exerted at section x of the beam is

$$(F)_x = -AE\epsilon = -AE\frac{[u(x + \Delta x,t) - u(x,t)]}{\Delta x} \tag{5.17}$$

where A is the beam cross-sectional area and E is the modulus of elasticity. When we let Δx approach zero, Equation (5.17) becomes

$$(F)_x = -AE\frac{\partial u}{\partial x}(x,t) \tag{5.18}$$

Similarly the force at $x + \Delta x$ becomes

$$(F)_{x+\Delta x} = +AE\frac{\partial u}{\partial x}(x + \Delta x,t) \tag{5.19}$$

Applying Newton's Second Law to the element Δx, we obtain

$$\rho A\,\Delta x\frac{\partial^2 u}{\partial t^2} = AE\frac{\partial u}{\partial x}(x + \Delta x,t) - AE\frac{\partial u}{\partial x}(x,t) \tag{5.20}$$

or

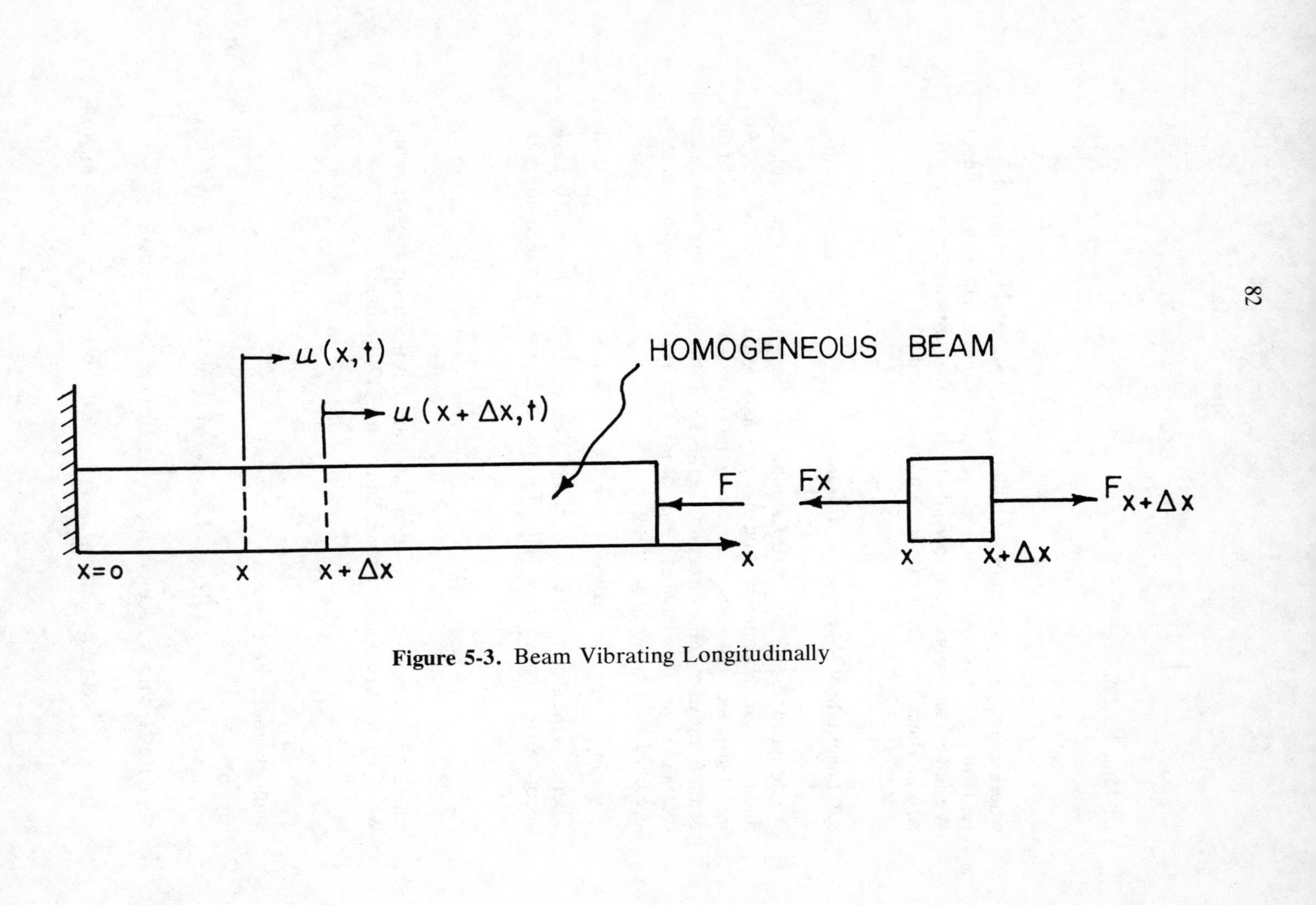

Figure 5-3. Beam Vibrating Longitudinally

$$\frac{\rho}{E}\frac{\partial^2 u}{\partial t^2} = \frac{\dfrac{\partial u}{\partial x}(x + \Delta x,t) - \dfrac{\partial u}{\partial x}(x,t)}{\Delta x} \tag{5.21}$$

When Δx approaches zero in Equation (5.21), the result becomes

$$\frac{\partial^2 u}{\partial x^2} = \frac{1}{c^2}\frac{\partial^2 u}{\partial t^2} \tag{5.22}$$

where

$$c^2 = \frac{E}{\rho} = \frac{\text{modulus of elasticity}}{\text{density}} \tag{5.23}$$

Equation (5.22) is the wave equation representing in this case the propagation of a sudden disturbance (or a wave) in displacement with a velocity c along the axis of the beam.

5.4 Transverse Vibration of Beams

In this section we derive the differential equation governing the free transverse vibration of a uniform cantilever beam (Figure 5-4). The resulting equation occurs in many engineering problems such as the calculation of the stability of rotating shafts and the study of the vibration of ships.

We consider that the deflection of the beam is very small and that the length of the neutral axis does not change on bending, that is, $\Delta x = \Delta L$. The faces at x and $x + \Delta x$ (initially parallel) of a segment of length Δx after bending will form an angle $d\beta$ (Figure 5-4). The value of $d\beta$ is related to the deflections at x and at $x + \Delta x$ by the following relation

$$d\beta = \left.\frac{\partial y}{\partial x}\right|_{x} - \left.\frac{\partial y}{\partial x}\right|_{x+\Delta x} \tag{5.24}$$

From Figure 5-4 we see that a section at a distance δ from the neutral axis undergoes a deformation equal to $\delta\, d\beta$. Applying Hooke's Law, we can write the tensile force acting on that section to be

$$dN = Eb\, d\delta \frac{\delta\, d\beta}{\Delta x} \tag{5.25}$$

Using Equation (5.24) in Equation (5.25) yields

$$dN = -Eb\frac{\partial^2 y}{\partial x^2}\delta\, d\delta \tag{5.26}$$

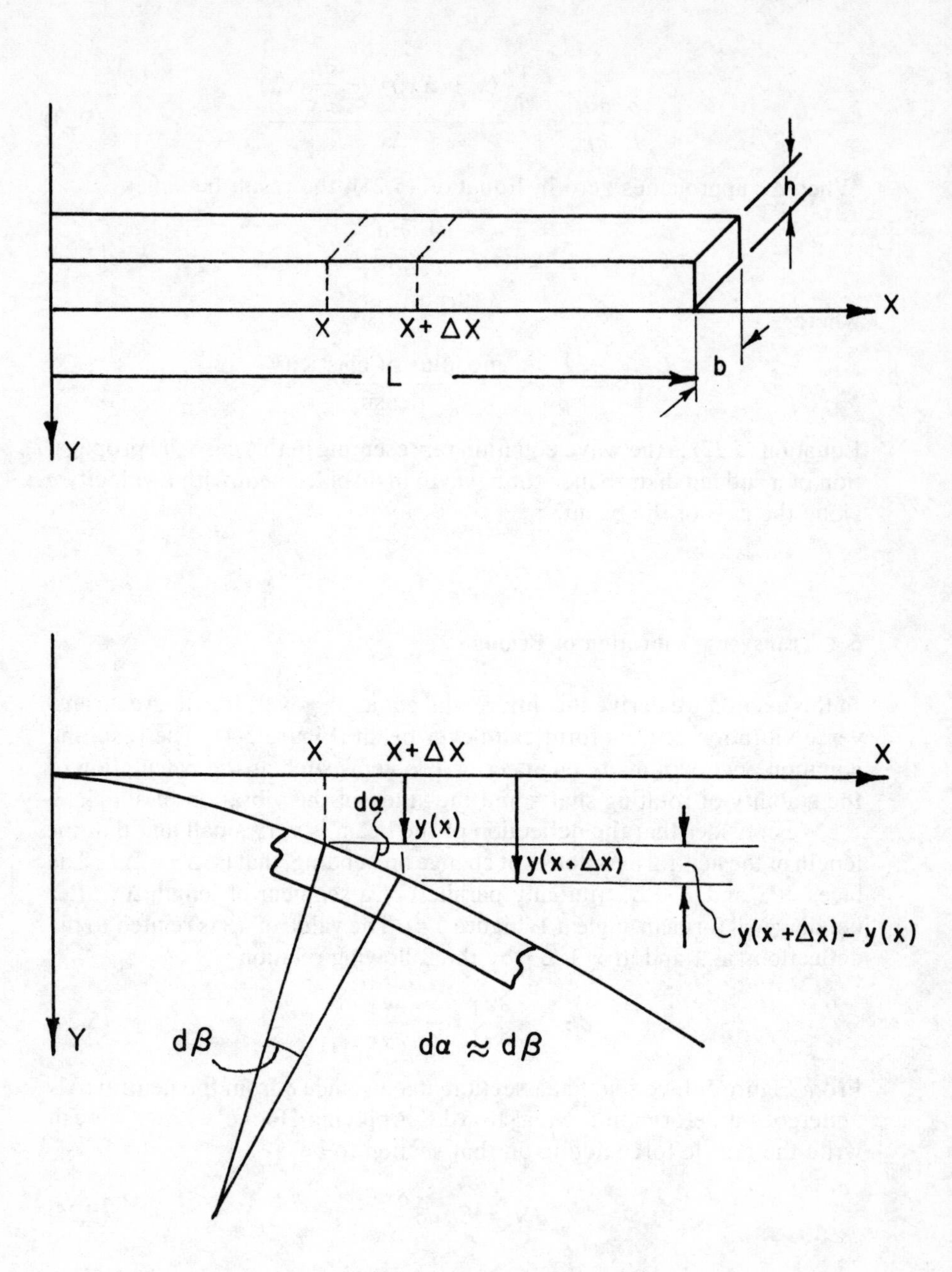

Figure 5-4. Beam Vibrating Transversely

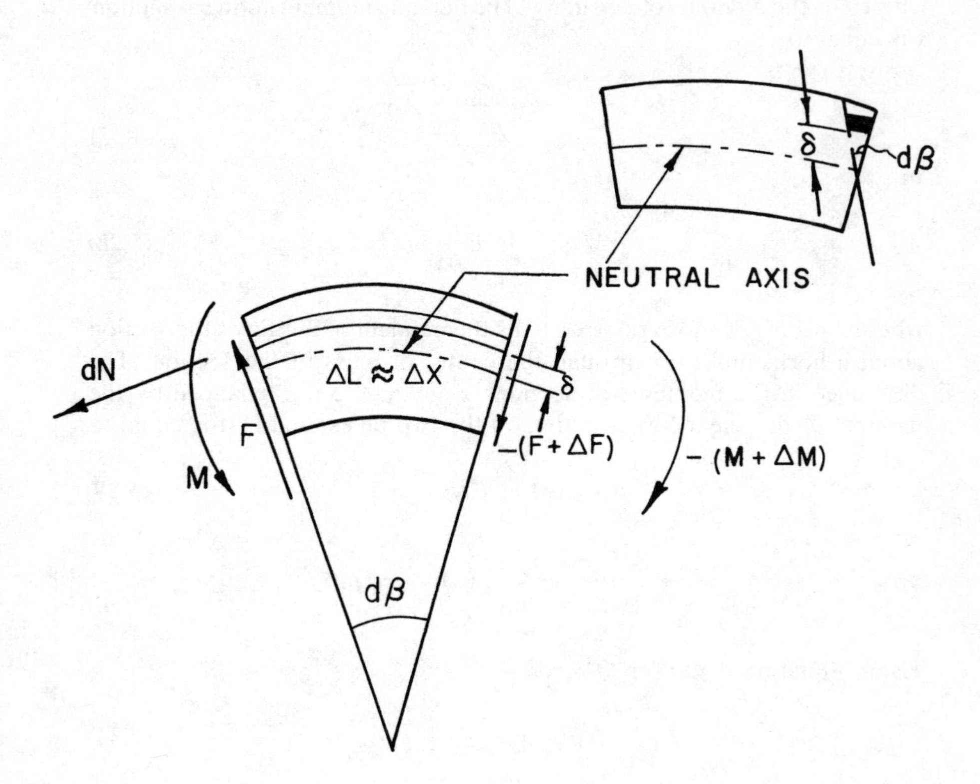

N = Tensile force

F = Shear force

M = Bending moment

where E is the modulus of elasticity. The bending moment acting at section x is equal to

$$M|_x = -Eb \frac{\partial^2 y}{\partial x^2} b \int_{-h/2}^{h/2} \delta^2 \, d\delta \tag{5.27}$$

or

$$M|_x = -E \frac{\partial^2 y}{\partial x^2} J \tag{5.28}$$

where $J = bh^3/12$ and is referred to as the moment of inertia of the section about a horizontal axis through the center of mass of the section. The difference in the moments at sections x and $x + \Delta x$ is balanced by the moment of the shear forces acting on the two faces, and is it is equal to

$$\Delta M = F \Delta x \tag{5.29}$$

or

$$F(x,t) = \lim_{\Delta x \to 0} \frac{\Delta M}{\Delta x} = \frac{\partial M}{\partial x} \tag{5.30}$$

Using Equation (5.28) for M gives

$$F(x,t) = -EJ \frac{\partial^3 y}{\partial x^3} \tag{5.31}$$

Hence the resultant shear force acting on the element is ΔF and is given by

$$\Delta F = \lim_{\Delta \to 0} \frac{\Delta F}{\Delta x} \Delta x = \frac{\partial F}{\partial x} \Delta x = -EJ \frac{\partial^4 y}{\partial x^4} \Delta x \tag{5.32}$$

Using Newton's Second Law, ΔF as given by Equation (5.32) is equated to the product of the mass and the acceleration of the element. The result becomes

$$\rho A \frac{\partial^2 y}{\partial t^2} \Delta x = -EJ \frac{\partial^4 y}{\partial x^4} \Delta x \tag{5.33}$$

or

$$\frac{\partial^2 y}{\partial t^2} + a^2 \frac{\partial^4 y}{\partial x^4} = 0 \tag{5.34}$$

where $a^2 = EJ/\rho A$, ρ is the density and A is the cross-sectional area of the beam.

5.5 Transmission Line Equation

Among the applications of the wave equation is the determination of the voltage and the current distributions along transmission lines or along telephone lines. The voltage and the current are functions of the distance x and time t. Let us denote the parameters of the line by:

R = resistance per unit length

L = inductance per unit length

G = conductance per unit length

C = capacitance per unit length

We consider the case of a single-wire transmission line as shown in Figure 5-5. If $V(x,t)$ and $I(x,t)$ are the voltage and the current, respectively, at location x, they are related as follows:

$$V(x,t) - V(x+\Delta x,t) = IR\Delta x + L\Delta x\frac{\partial I}{\partial t} \tag{5.35a}$$

$$I(x,t) - I(x+\Delta x,t) = GV\Delta x + C\Delta x\frac{\partial V}{\partial t} \tag{5.36a}$$

Dividing Equations (5.35a) and (5.36a) by Δx yields

$$\frac{V(x,t) - V(x+\Delta x,t)}{\Delta x} = IR + L\frac{\partial I}{\partial t} \tag{5.35b}$$

$$\frac{I(x,t) - I(x+\Delta x,t)}{\Delta x} = GV + C\frac{\partial V}{\partial t} \tag{5.36b}$$

Letting Δx approach zero in Equations (5.35b) and (5.36b), we obtain

$$-\frac{\partial V}{\partial x} = IR + L\frac{\partial I}{\partial t} \tag{5.37}$$

$$-\frac{\partial I}{\partial x} = GV + C\frac{\partial V}{\partial t} \tag{5.38}$$

Equations (5.37) and (5.38) are simultaneous partial differential equations for the voltage $V(x,t)$ and the current $I(x,t)$ along the transmission line. We can eliminate the current I by differentiating Equation (5.37) with respect to x and Equation (5.38) with respect to t to obtain

$$-\frac{\partial^2 V}{\partial x^2} = R\frac{\partial I}{\partial x} + L\frac{\partial}{\partial x}\left(\frac{\partial I}{\partial t}\right) \tag{5.39}$$

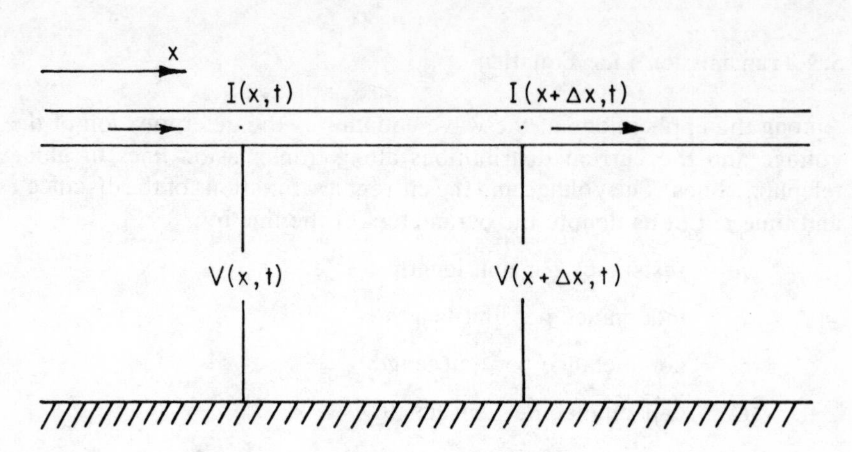

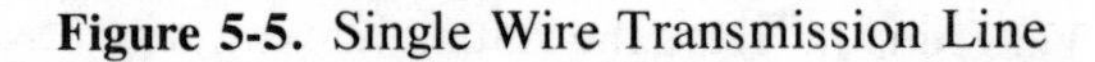

Figure 5-5. Single Wire Transmission Line

$$-\frac{\partial}{\partial t}\left(\frac{\partial I}{\partial x}\right) = G\frac{\partial V}{\partial t} + C\frac{\partial^2 V}{\partial t^2} \tag{5.40}$$

Eliminating the cross derivatives in Equations (5.39) and (5.40) and using Equation (5.38) to eliminate $\partial I/\partial x$ we obtain

$$\frac{\partial^2 V}{\partial x^2} = LC\frac{\partial^2 V}{\partial t^2} + (LG+RC)\frac{\partial V}{\partial t} + RGV \tag{5.41}$$

Similarly, on eliminating the voltage (by differentiating Equation (5.37) with respect to t and Equation (5.38) with respect to x and eliminating the cross derivatives and $\partial V/\partial x$) the result becomes

$$\frac{\partial^2 I}{\partial x^2} = LC\frac{\partial^2 I}{\partial t^2} + (LG+RC)\frac{\partial I}{\partial t} + RGI \tag{5.42}$$

When the transmission lines are well constructed, the shunt conductance is usually negligible; therefore, in this case the foregoing equations are simplified to the following

$$\frac{\partial^2 V}{\partial x^2} = LC\frac{\partial^2 V}{\partial t^2} + RC\frac{\partial V}{\partial t} \tag{5.43}$$

$$\frac{\partial^2 I}{\partial x^2} = LC\frac{\partial^2 I}{\partial t^2} + RC\frac{\partial I}{\partial t} \tag{5.44}$$

Equations (5.43) and (5.44) are of the same form as Equation (5.15) which is the form of the damped wave equation.

When the losses are negligible such that

$$G = R = 0$$

the equations are reduced in form to the following one-dimensional undamped wave equation obtained earlier

$$\frac{\partial^2 V}{\partial x^2} = \frac{1}{c^2}\frac{\partial^2 V}{\partial t^2} \tag{5.45}$$

$$\frac{\partial^2 I}{\partial x^2} = \frac{1}{c^2}\frac{\partial^2 I}{\partial t^2} \tag{5.46}$$

where c, the velocity of propagation of the wave, is given in this case by

$$c = \frac{1}{\sqrt{LC}} \tag{5.47}$$

5.6 Acoustic Equation—Sound Waves in a Gas

The speed of sound in a medium is essentially the speed of a small disturbance generated in the medium by any source that can create a change in pressure and hence changes in other properties in the medium such as rapid fluctuation in density and temperature. The speed of sound therefore enters the gasdynamic equations as the velocity of small disturbances. When the changes in the density ρ_0 and pressure p_0 of the gaseous medium are small relative to the average or to the undisturbed values, and when the fluid velocity V is small relative to the speed of sound, the gasdynamic equations are reduced to equations describing the propagation of sound waves.

Initially we consider that the medium has undisturbed state of rest governed by uniform properties ρ_0, p_0 and $V = 0$. The pressure disturbance or acoustic motions produce small perturbations or deviations from this uniform state. Hence, the local quantities can be stated as

$$p = p_0 + \hat{p} \qquad \rho = \rho_0 + \hat{\rho} \qquad \mathbf{V} = 0 + \mathbf{V} \tag{5.48}$$

Here we emphasize that $\hat{p} \ll p_0$ and $\hat{\rho} \ll \rho_0$ and $V \ll$ speed of sound. When one studies the motion of small disturbances with the restrictions stated earlier, one will be, in essence, linearizing the equations of motion. Because the uniform fluid is initially at rest, and as the amplitude of the disturbance is very small, the convective derivatives are negligible in comparison with the unsteady derivatives. With the further restriction of inviscid flow we can write the momentum equation and the equation of continuity of mass as follows:

$$\frac{\partial \mathbf{V}}{\partial t} + \mathbf{V}\cdot\nabla\mathbf{V} + \frac{1}{\rho}\nabla p = 0 \tag{5.49}$$

$$\frac{\partial \rho}{\partial t} + \mathbf{V}\cdot\nabla\rho + \rho\nabla\cdot\mathbf{V} = 0 \tag{5.50}$$

When the coefficients of the derivatives are replaced by their undisturbed values (this implies ignoring second-order terms) and dropping the convective terms relative to the time derivative we obtain:

$$\rho_0\frac{\partial \mathbf{V}}{\partial t} + \nabla p = 0 \tag{5.51}$$

$$\frac{\partial \rho}{\partial t} + \rho_0\nabla\cdot\mathbf{V} = 0 \tag{5.52}$$

If the flow is assumed inviscid and heat conduction is not present, then there is no entropy generation and the entropy of the medium remains constant $s = s_0$. In this case the pressure p is a function of the density ρ such that the pressure gradient is proportional to the density gradient in the form

$$\nabla p = \left(\frac{dp}{d\rho}\right)_s \nabla\rho = c^2\nabla\rho \tag{5.53}$$

where

$$c^2 = \left(\frac{dp}{d\rho}\right)_s \tag{5.54}$$

Substituting Equation (5.53) in Equation (5.51) yields

$$\rho_0\frac{\partial \mathbf{V}}{\partial t} + c^2\nabla\rho = 0 \tag{5.55}$$

Equations (5.52) and (5.55) represent a system of linear equations describing the sound wave.

Differentiating Equation (5.52) with respect to time gives

$$\frac{\partial^2 \rho}{\partial t^2} + \rho_0\nabla\cdot\frac{\partial \mathbf{V}}{\partial t} = 0 \tag{5.56}$$

and for the one-dimensional case Equation (5.56) becomes

$$\frac{\partial^2 \rho}{\partial t^2} + \rho_0\frac{\partial^2 \mathbf{V}}{\partial x\,\partial t} = 0 \tag{5.57}$$

Invoking also the one-dimensional case on Equation (5.55) and differentiating it with respect to the space variable x we obtain

$$\rho_0\frac{\partial^2 \mathbf{V}}{\partial x\,\partial t} + c^2\frac{\partial^2 \rho}{\partial x^2} = 0 \tag{5.58}$$

Eliminating the cross derivative between equations (5.57) and (5.58) results in the following wave equation which represents the propagation with a velocity c of a small disturbance in the density of a compressible medium

$$\frac{\partial^2 \rho}{\partial x^2} = \frac{1}{c^2} \frac{\partial^2 \rho}{\partial t^2} \tag{5.59}$$

When we assume that the gaseous medium can be treated as an ideal gas and further consider the changes in the pressure p and density ρ to be related by the isentropic relation

$$p/\rho^\gamma = \text{constant} \tag{5.60}$$

where γ is the ratio of the specific heat at constant pressure to specific heat at constant value, we can then relate the speed c of propagation of the wave to the gas properties in the following way:

The logarithmic form of Equation (5.60) is

$$\ln p - \gamma \ln \rho = \text{constant} \tag{5.61}$$

Differentiating Equation (5.61) yields

$$\frac{dp}{p} = \gamma \frac{d\rho}{\rho} \tag{5.62}$$

or

$$\left(\frac{\partial p}{\partial \rho}\right)_s = \frac{\gamma p}{\rho} \tag{5.63}$$

For an ideal gas, the equation of state is given by

$$p = \rho RT \tag{5.64}$$

or

$$p/\rho = RT \tag{5.65}$$

where R is the gas constant and T is temperature. Substituting Equation (5.65) in Equation (5.63) gives

$$\left(\frac{\partial p}{\partial \rho}\right)_s = c^2 = \gamma RT \tag{5.66}$$

For most gases, γ is practically constant and hence c, for a certain gas, is strictly a function of the temperature of that gas.

Because the change in pressure is proportional to the change in density, we can also write the wave equation as given by Equation (5.59) in terms of the pressure p as

$$\frac{\partial^2 p}{\partial x^2} = \frac{1}{c^2}\frac{\partial^2 p}{\partial t^2} \tag{5.67}$$

where Equation (5.67) represents now the propagation of a disturbance in pressure of a compressible medium.

5.7 Gravitational Oscillations—Tidal Waves in Canals

We shall consider in this section the small oscillations, under gravity, of a liquid having a free surface. The word tidal has been used to describe the waves in which the motion of the fluid is mainly horizontal. We take as an example the propagation of such waves along a straight canal with a horizontal bed; see Figure 5-6, and we invoke the assumption that the motion is independent of the width "z direction" of the canal. Hence, we can take the analysis per unit width.

When the vertical acceleration of the fluid particles is neglected, the pressure, then, at any point in the xy plane can be represented by the following expression:

$$p(x,y) = p_a + \rho g(h_0 + \eta - y) \tag{5.68}$$

where p_a = external pressure on liquid surface, ρ = liquid density and g = gravitational acceleration.

From Equation (5.68) we can write

$$\frac{\partial p}{\partial x} = \rho g \frac{\partial \eta}{\partial x} \tag{5.69}$$

Applying the continuity equation to the strips shown in Figure 5-7 yields the following:

$$\text{volume of initial strip at } x = h_0\,dx$$

$$\text{volume of strip at } \xi = (h_0 + \eta)\left[dx + \frac{\partial \xi}{\partial x}dx\right]$$

Because the fluid is considered incompressible, the two volumes are equal. Hence, equating the two volumes gives

$$h_0\,dx = h_0\,dx + h_0\frac{\partial \xi}{\partial x}dx + \eta\,dx + \eta\frac{\partial \xi}{\partial x}dx$$

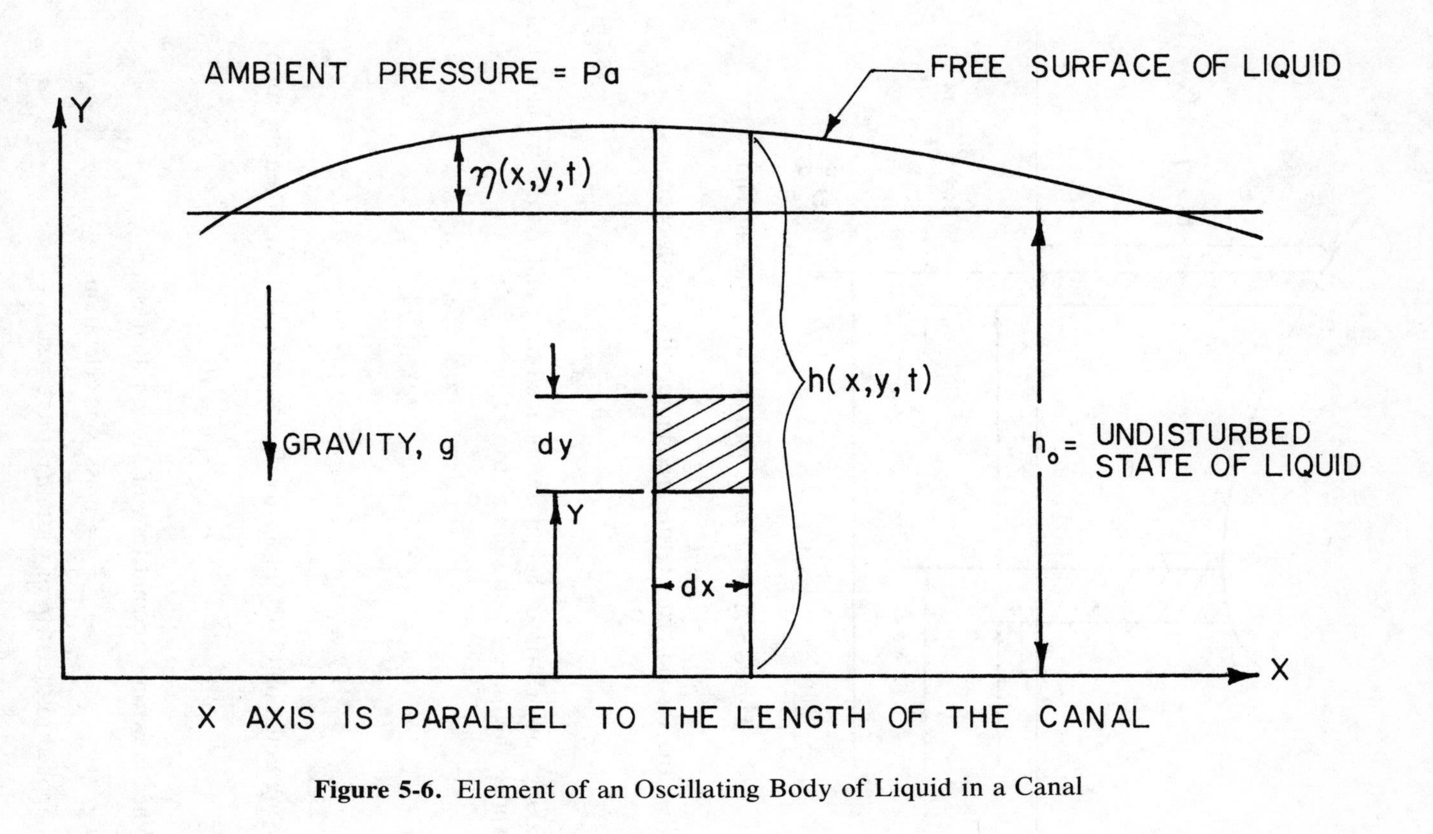

Figure 5-6. Element of an Oscillating Body of Liquid in a Canal

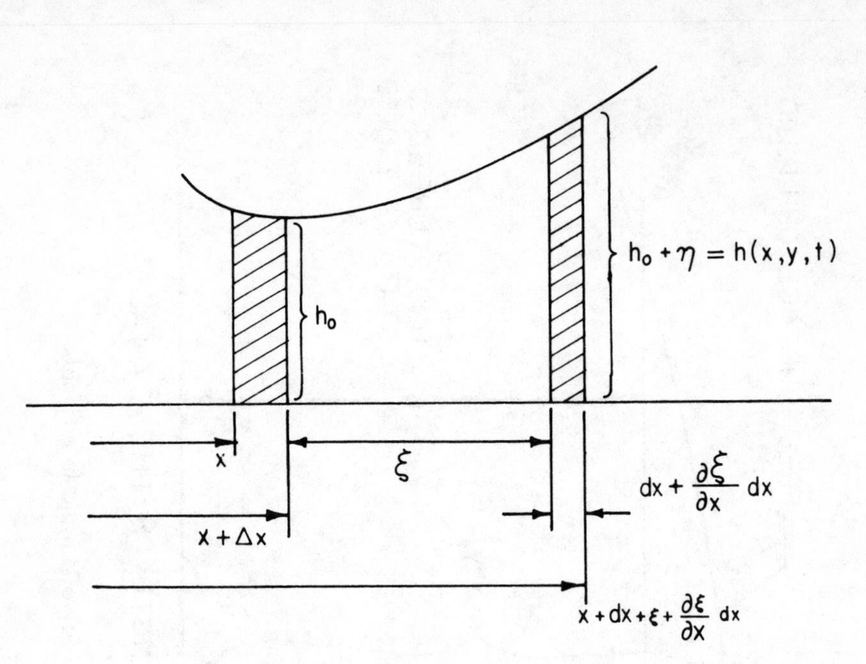

Figure 5-7. Strip of Liquid at Two Different Times

When the second order terms are ignored, the result becomes

$$\eta = -h_0 \frac{\partial \xi}{\partial x} \tag{5.70}$$

Taking now a strip of height dy bounded by x and $x + dx$, we can then express the force F_x on such a strip as

$$F_x = p\, dy\,|_x - p\, dy\,|_{x+\Delta x}$$

or

$$F_x = p\, dy - \left(p + \frac{\partial p}{\partial x} dx\right) dy = -\frac{\partial p}{\partial x} dx\, dy \tag{5.71}$$

Using Equation (5.69) for $\partial p / \partial x$, Equation (5.71) takes then the form

$$F_x = -\rho g \frac{\partial \eta}{\partial x} dx\, dy \tag{5.72}$$

When Newton's Second Law of Motion, that is,

$$F_x = \text{mass} \times \text{acceleration}$$

is applied to the strip of width dx, there results

$$-\rho g \frac{\partial \eta}{\partial x} dx\, dy = \rho\, dx\, dy \frac{\partial^2 \xi}{\partial t^2} \tag{5.73}$$

Equation (5.70) can be used now in Equation (5.73) to replace $\partial\eta/\partial x$. The result becomes

$$g h_0 \frac{\partial^2 \xi}{\partial x^2} = \frac{\partial^2 \xi}{\partial t^2}$$

or

$$\frac{\partial^2 \xi}{\partial x^2} = \frac{1}{c^2} \frac{\partial^2 \xi}{\partial t^2} \tag{5.74}$$

where for this problem c is given by

$$c = \sqrt{g h_0}$$

Equation (5.74) is again the well-known wave equation found earlier for other physical problems.

We turn next to the methods and analysis of solution to the wave equation.

5.8 Solutions of Equations of the Hyperbolic Type

Solutions of the wave equation of the form

$$\frac{\partial^2 \phi}{\partial x^2} = \frac{1}{c^2} \frac{\partial^2 \phi}{\partial t^2}$$

are commonly made in either of two ways:

1. In terms of eigenfunctions which usually represent standing waves. This method is one of the most widely used in the solution of partial differential equations.
2. In terms of traveling waves of arbitrary shape.

The first method, usually, results from the method of separation of variables, which we will discuss first.

The Method of Separation of Variables

Let us consider the problem of vibration of an elastic string of length l. The governing differential equation for the deflection is Equation (5.8), and it is

$$\frac{\partial^2 u}{\partial x^2} = \frac{1}{c^2}\,\frac{\partial^2 u}{\partial t^2} \tag{5.75}$$

where, as explained earlier,

$$c^2 = T/m$$

To solve Equation (5.75) for a specific physical problem we should have:

1. Two boundary conditions, "the highest order in space derivative is two."
2. Two initial conditions, "the highest order in time derivative is two."

If we consider first a string that is hinged at $x = 0$ and $x = l$ (Figure 5-8), the two boundary conditions become

$$u(0,t) = 0, \qquad u(l,t) = 0 \qquad \text{for all } t \tag{5.76a,b}$$

Because the form of the motion of the string depends upon the shape of the initial deflection as well as upon the initial velocity imparted to the string, we have to specify the two initial conditions. Let

$$u(x,0) = f(x) \qquad \text{initial displacement} \tag{5.77a}$$

$$\frac{\partial u}{\partial t}(x,0) = g(x) \qquad \text{initial velocity} \tag{5.77b}$$

Now we have a complete problem posed as follows:

$$\frac{\partial^2 u}{\partial x^2} = \frac{1}{c^2}\frac{\partial^2 u}{\partial t^2} \tag{5.75}$$

$$\left.\begin{aligned} u(0,t) &= 0 \qquad (5.76a) \\ u(l,t) &= 0 \qquad (5.76b) \end{aligned}\right\} \text{Homogeneous boundary conditions}$$

$$u(x,0) = f(x) \tag{5.77a}$$

$$\frac{\partial u}{\partial t}(x,0) = g(x) \tag{5.77b}$$

Let the solution of Equation (5.75) be represented as a product of two functions, $X(x)$ and $G(t)$ where

$X(x)$ is a function of x only

$G(t)$ is a function of t only

Hence,

$$u(x,t) = X(x)\,G(t) \tag{5.78}$$

Let us find

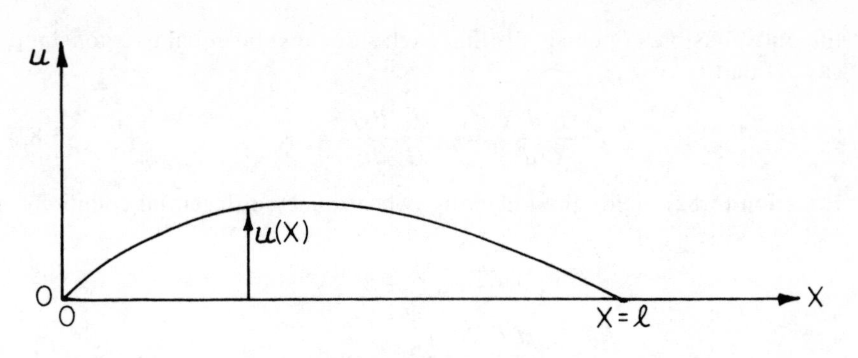

Figure 5-8. Vibrating String Hinged at Both Ends

$$\frac{\partial^2 u}{\partial x^2} \quad \text{and} \quad \frac{\partial^2 u}{\partial t^2}$$

for Equation (5.78)

$$\frac{\partial^2 u}{\partial x^2} = G(t)\frac{d^2X(x)}{dx^2} \tag{5.79a}$$

$$\frac{\partial^2 u}{\partial t^2} = X(x)\frac{d^2G(t)}{dt^2} \tag{5.79b}$$

Substituting Equations (5.79a) and (5.79b) in Equation (5.75) yields

$$G(t)\frac{d^2X(x)}{dx^2} = \frac{1}{c^2}X(x)\frac{d^2G(t)}{dt^2} \tag{5.80}$$

Dividing Equation (5.80) by XG we obtain

$$\frac{1}{X}\frac{d^2X}{dx^2} = \frac{1}{c^2G}\frac{d^2G}{dt^2} \tag{5.81}$$

In Equation (5.81) we have:

1. The left-hand side involves functions that depend on x only.
2. The right-hand side involves functions that depend on t only.
3. Equation (5.81) must be satisfied identically for $0 < x < l$, $t > 0$.
4. In fixing a certain value of x and varying t (or vice versa), we find that each side of Equation (5.81) should remain constant.
5. Any change in x will not change the right-hand member of Equation (5.81) (which involves functions dependent on t only).
6. Any change in t will not change the left-hand member of Equation (5.81) (which involves functions dependent on x only).

Therefore, each side of Equation (5.81) must be independent of x and t, and the only possible conclusion is that each side must be equal to a constant, say K, that is,

$$\frac{1}{X}\frac{d^2X}{dx^2} = \frac{1}{c^2G}\frac{d^2G}{dt^2} = K \tag{5.82}$$

Equation (5.82) yields the following two ordinary differential equations

$$\frac{d^2X}{dx^2} - KX = 0 \tag{5.83}$$

$$\frac{d^2G}{dt^2} - Kc^2G = 0 \tag{5.84}$$

Because K is still arbitrary, and in order to properly judge the selection of the sign of this constant, we have to consider separately the cases when K is positive, zero, or negative.

First let us verify that the boundary conditions of the problem, as given by Equations (5.76a) and (5.76b), should be satisfied by $X(x)$. The solution to our problem is now of the form

$$u(x,t) = X(x)\,G(t) \tag{5.85}$$

Applying the boundary conditions yields

$$u(0,t) = 0 = X(0)\,G(t) \qquad \text{for all } t \tag{5.86}$$

$$u(l,t) = 0 = X(l)\,G(t) \qquad \text{for all } t \tag{5.87}$$

Therefore, the function $X(x)$ must satisfy the boundary conditions

$$X(0) = 0, \qquad X(l) = 0 \tag{5.88}$$

We note here that if the boundary conditions were not homogeneous, this step will fail.

K is Positive

If K is positive, say $K = \lambda^2$, the general solution to Equation (5.83) becomes

$$X(x) = Ae^{\lambda x} + Be^{-\lambda x}$$

Applying the boundary conditions given by Equation (5.88) gives

$$0 = A + B \quad \text{or} \quad A = -B$$

$$0 = Ae^{\lambda l} + Be^{-\lambda l}$$

or

$$0 = A[e^{\lambda l} - e^{-\lambda l}]$$

Because λ is a positive and real quantity

$$e^{\lambda l} - e^{-\lambda l} \neq 0$$

Therefore $A = B = 0$, and consequently

$$X(x) \equiv 0 \quad \text{and} \quad u(x,t) \equiv 0$$

which is a trivial solution and of no interest.

K is Zero

In this case the general solution to Equation (5.83) is

$$X(x) = Ax + B$$

with

$$X(0) = 0, \qquad B = 0$$

with

$$X(l) = 0 = Al, \qquad A = 0$$

Therefore, here also $A = B = 0$ and

$$X(x) \equiv 0 \qquad \text{and} \qquad u(x,t) \equiv 0$$

This solution is another trivial solution which is of no interest. We are left then with negative values for K.

K is Negative

For negative K, let $K = -\lambda^2$. Equations (5.83) and (5.84) become

$$\frac{d^2X}{dx^2} + \lambda^2 X = 0 \tag{5.89}$$

$$\frac{d^2G}{dt^2} + \lambda^2 c^2 G = 0 \tag{5.90}$$

The general solution of Equation (5.89) is

$$X(x) = A \cos \lambda x + B \sin \lambda x \tag{5.91}$$

Applying the boundary conditions: $X(0) = 0$, $X(l) = 0$ we obtain

$$X(0) = A = 0 \tag{5.92}$$

and

$$X(l) = B \sin \lambda l = 0 \tag{5.93}$$

If $B = 0$, we end up with another trivial solution. Therefore, $B \neq 0$ and

$$\sin \lambda l = 0 \tag{5.94}$$

Hence,

$$\lambda l = n\pi \quad \text{or} \quad \lambda_n = \frac{n\pi}{l}, \quad n = 1, 2, 3, \ldots \tag{5.95}$$

In this problem, the eigenvalues λ_n start with $n = 1$ rather than $n = 0$. For $n = 0$, $\sin n\pi x/l$ is identically zero for any value of x, and this results in the trivial solution $u(x,t) \equiv 0$. λ_n's are the eigenvalues for which a solution exists. The corresponding eigenfunctions are the solution $X(x)$ given as

$$X_n(x) = \sin \frac{n\pi x}{l} \tag{5.96}$$

where we set the constant $B = 1$ without any loss in generality because $X(x)$ will eventually be multiplied by $G(t)$ which will involve two arbitrary constants.

Thc solution for $G(t)$ from Equation (5.90) becomes

$$G_n(t) = C_n \cos \frac{n\pi c}{l} t + D_n \sin \frac{n\pi c}{l} t \tag{5.97}$$

where C_n and D_n are yet to be determined.

Returning now to the original problem stated by Equations (5.75) to (5.77) with the solution represented as a product of two functions $X(x)\,G(t)$ we can write

$$u_n(x,t) = X_n(x)\,G_n(t) = \left[C_n \cos \frac{n\pi ct}{l} + D_n \sin \frac{n\pi ct}{l} \right] \sin \frac{n\pi x}{l}$$

$$n = 1, 2, 3, \ldots \tag{5.98}$$

These functions represent particular solutions of our original problem, and *they satisfy the differential equation and the boundary conditions* stated. We can see, however, that a *single* solution from Equation (5.98) *does not satisfy the initial conditions of the problem*. Because of the linearity and homogeneity of the differential equation of this problem, the sum of the particular solutions $u_n(x)$ is also a solution. Therefore, to satisfy the imposed initial conditions, we consider this sum, which is an infinite series given as

$$u(x,t) = \sum_{n=1}^{\infty} u_n(x,t)$$

$$u(x,t) = \sum_{n=1}^{\infty}\left(C_n \cos\frac{n\pi c}{l}t + D_n \sin\frac{n\pi c}{l}t\right)\sin\frac{n\pi x}{l} \tag{5.99}$$

Equation (5.99) satisfies the differential equation and the two boundary conditions as well. To verify that, let us take the following derivatives to Equation (5.99)

$$\frac{\partial^2 u}{\partial x^2} = \sum_{n=1}^{\infty}\left(C_n \cos\frac{n\pi c}{l}t + D_n \sin\frac{n\pi c}{l}t\right)\left(-\frac{n^2\pi^2}{l^2}\right)\sin\frac{n\pi x}{l}$$

$$\frac{\partial^2 u}{\partial t^2} = \sum_{n=1}^{\infty} -\left(\frac{n^2\pi^2 c^2}{l^2}\right)\left\{C_n \cos\frac{n\pi c}{l}t + D_n \sin\frac{n\pi c}{l}t\right\}\sin\frac{n\pi x}{l}$$

When these two expressions are substituted in Equation (5.75), we readily see that the differential equation is satisfied. The boundary conditions, that is, $u(0,t) = 0 = u(l,t)$ are satisfied because for any integer n ($n = 1, 2, 3, \ldots$),

$$\sin\frac{n\pi x}{l} = 0 \quad \text{for} \quad x = 0 \quad \text{and} \quad x = l.$$

The initial conditions, Equations (5.77a and 5.77b), are now applied to the general solution as given by Equation (5.99).

$$u(x,0) = f(x) = \sum_{n=1}^{\infty} u_n(x,0)$$

or

$$f(x) = \sum_{n=1}^{\infty} C_n \sin\frac{n\pi x}{l} \tag{5.100}$$

and

$$\frac{\partial u}{\partial t}(x,0) = g(x) = \sum_{n=1}^{\infty}\frac{\partial u_n}{\partial t}(x,0)$$

or

$$g(x) = \sum_{n=1}^{\infty}\left(\frac{n\pi c}{l}\right) D_n \sin\frac{n\pi}{l}x \tag{5.101}$$

Equations (5.100) and (5.101) represent expansions of arbitrary functions $f(x)$ and $g(x)$ in terms of Fourier sine series. The coefficients C_n and D_n are therefore given by (see chapter 3, Article 3.4)

$$C_n = \frac{2}{l}\int_0^l f(x)\sin\frac{n\pi x}{l}\,dx, \quad n = 1, 2, 3, \ldots \tag{5.102}$$

and

$$\left(\frac{n\pi c}{l}\right)D_n = \frac{2}{l}\int_0^l g(x) \sin\frac{n\pi x}{l}\,dx$$

or

$$D_n = \frac{2}{n\pi c}\int_0^l g(x) \sin\frac{n\pi x}{l}\,dx, \quad n = 1, 2, 3, \ldots \tag{5.103}$$

Therefore, the expression for $u(x,t)$, as given by Equation (5.99), with the coefficients given by Equations (5.102) and (5.203), constitutes the solution of the problem provided:

1. The series representing $u(x,t)$ converges.
2. The series representing

$$\frac{\partial^2 u}{\partial x^2} \text{ and } \frac{\partial^2 u}{\partial t^2},$$

as obtained by termwise differentiation of Equation (5.99), converge and are continuous.

It was shown in chapter 3 that these conditions are fulfilled for Fourier series. Furthermore, the solution does satisfy the initial and boundary conditions of the problem.

In the event the initial velocity $(\partial u(x,0)/\partial t)$ is zero, $g(x)$, then, is zero and hence D_n. The solution in this case takes the following form:

$$u(x,t) = \sum_{n=1}^{\infty} C_n \cos\frac{n\pi c}{l}t \sin\frac{n\pi x}{l}, \quad n = 1, 2, \ldots \tag{5.104}$$

with C_n given by Equation (5.102).

Physical Interpretation of the Solution

Let us take the function $u_n(x,t)$ given by Equation (5.98) and consider that the initial velocity is zero, which means D_n is zero. We have then

$$u_n(x,t) = C_n \cos\frac{n\pi c}{l}t \sin\frac{n\pi x}{l} \tag{5.105}$$

From Equation (5.105) we can say that any point on the string identified by $x = x_0$ executes harmonic vibration given by

$$u_n(x_0,t) = \left(C_n \sin\frac{n\pi x_0}{l}\right) \cos\frac{n\pi c}{l}t \tag{5.106}$$

where

$$C_n \sin \frac{n\pi x_0}{l}$$

is the amplitude of vibration. The frequency is given by

$$f_n = \frac{1}{2\pi}\left(\frac{n\pi c}{l}\right) = \frac{nc}{2l} = \frac{n}{2l}\sqrt{\frac{T}{m}} \quad \text{cycles/unit time} \qquad (5.107)$$

This type of motion, as was indicated earlier, is called a standing wave. Each term of Equation (5.105) or Equation (5.106) is referred to as a normal mode of the string. When $n = 1$, we have the first normal mode or the fundamental mode (or first harmonic). For $n > 1$, we have the overtones, 1st, 2nd, . . . , overtones, or 2nd, 3rd, . . . , harmonic.

From Equation (5.105) we see that

$$\sin \frac{n\pi x}{l} = 0 \quad \text{when } x = \frac{l}{n}, \frac{2l}{n}, \ldots, \left(\frac{n-1}{n}\right) l$$

These points, which remain stationary during the entire process, are called nodes of the standing wave $u_n(x,t)$. The first three normal modes are shown in Figure 5-9.

As time progresses, the shape of these modes changes from the solid curve to the dotted curve in Figure 5-9 and then back again to the solid curve. The time for the complete cycle is the period, and the reciprocal of the period is the frequency given earlier by Equation (5.107) as

$$f_n = \frac{nc}{2l}$$

We like to note that because all the frequencies are integer multiples of the lowest frequency $f_1 (f_1 = c/2l)$, the vibration of the string results in a musical tone. The sound of the string is a superposition of simple tones ($u_n(x,t)$) corresponding to the standing waves into which the vibration can be resolved. Such simple tones can be resolved experimentally by means of resonators.

The sound of the string depends upon the method of excitation that is essentially the imposed initial conditions given for this problem by Equations (5.100) and (5.101) as

$$u(x,0) = f(x) = \sum_{n=1}^{\infty} C_n \sin \frac{n\pi x}{l}$$

and

$$u_t(x,0) = g(x) = \sum_{n=1}^{\infty} \left(\frac{n\pi c}{l}\right) D_n \sin \frac{n\pi x}{l}$$

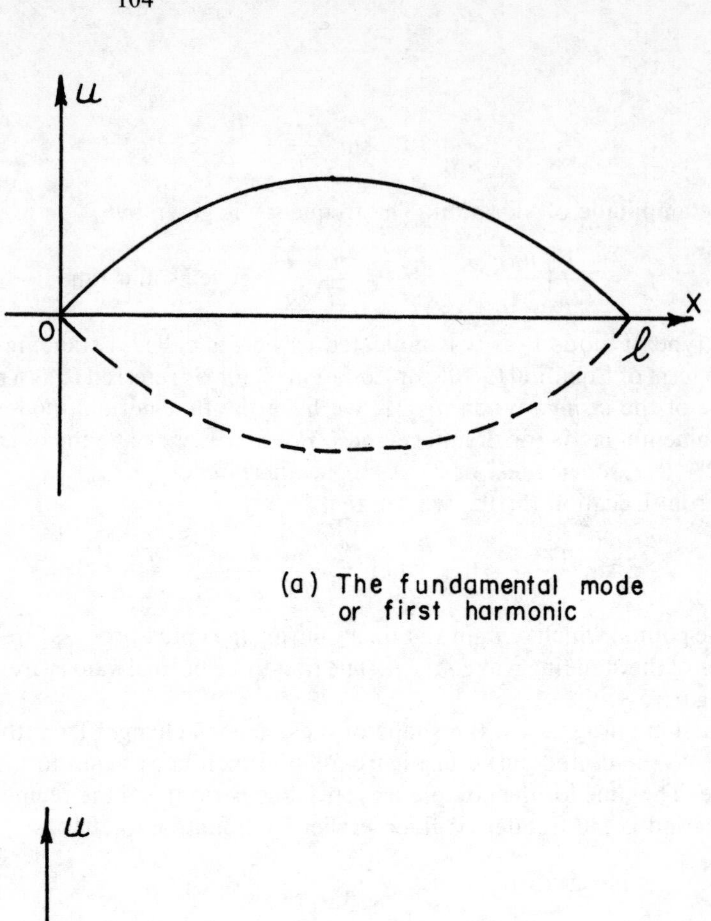

(a) The fundamental mode
or first harmonic

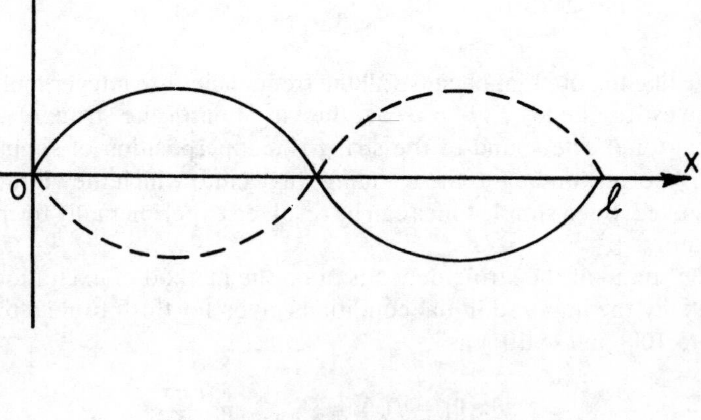

(b) The second normal mode
or second harmonic
or first overtone

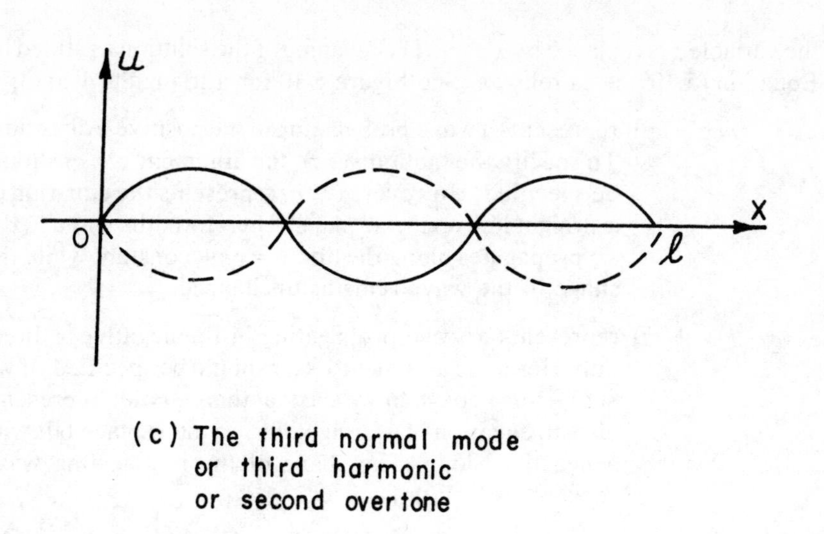

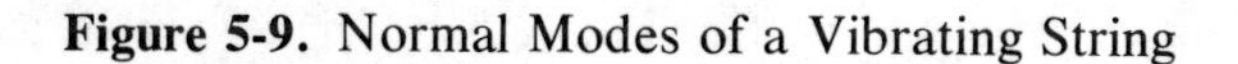
Figure 5-9. Normal Modes of a Vibrating String

and, therefore, the sound generated depends upon the coefficients C_n and D_n.

The solution given by Equation (5.104) as

$$u(x,t) = \sum_{n=1}^{\infty} C_n \cos \frac{n\pi c}{l} t \sin \frac{n\pi x}{l} \tag{5.104}$$

can be written in a closed form as follows: Let us use the following identity

$$\cos \frac{n\pi c}{l} t \sin \frac{n\pi x}{l} = \frac{1}{2}\left[\sin \left\{\frac{n\pi}{l}(x - ct)\right\} + \sin \left\{\frac{n\pi}{l}(x + ct)\right\}\right]$$

in Equation (5.104). The solution, then, can be written as

$$u(x,t) = \frac{1}{2}\left[\sum_{n=1}^{\infty} C_n \sin \left\{\frac{n\pi}{l}(x - ct)\right\} + \sum_{n=1}^{\infty} C_n \sin \left\{\frac{n\pi}{l}(x + ct)\right\}\right] \tag{5.108}$$

or

$$u(x,t) = \frac{1}{2}[F(x - ct) + F(x + ct)] \tag{5.109}$$

where $F(x \pm ct)$ is the Fourier sine series given in Equation (5.108) in which the variable x is replaced by $x \pm ct$. The meaning of the solution as stated by Equation (5.109) is as follows (See Figure 5-10 for a triangular wave):

$F(x - ct)$ represents a wave propagating in the positive x direction. To specify the motion $u(x,t)$, the argument $x - ct$ should be specified. However, $x - ct$ represents the equation of a straight line on the xt plane. Therefore, the wave $F(x - ct)$ propagates along the line $x = ct +$ constant while the shape of the wave remains unchanged.

$F(x + ct)$ represents a wave propagating in the negative x direction. Here, the argument $x + ct$ should be specified. If we set $x + ct =$ constant we see that this equation represents also an equation of a straight line on the xt plane but with a negative slope. The shape of the propagating wave remains as well unchanged.

$u(x,t)$ the solution to the problem, is equal to the average of the two waves. See Figure 5-10 for a wave of a triangular shape.

Example 5.1 A beam of uniform cross-sectional area A, modulus of elasticity E, density ρ, and of length l is subjected to a force F axially at its free end as shown in Figure 5-11. Find the expression representing the axial vibration of the beam once the load F is removed.
Solution: The differential equation and the initial and boundary conditions are

$$\frac{\partial^2 u}{\partial t^2} = c^2 \frac{\partial^2 u}{\partial x^2}, \quad c = \sqrt{\frac{E}{\rho}} \tag{5.110}$$

$$u(x,0) = f(x) = Fx/EA \quad 0 \leq x \leq l \tag{5.111}$$

$$\frac{\partial u}{\partial t}(x,0) = 0 \quad \text{``no initial velocity''} \tag{5.112}$$

Boundary Conditions

$$u(0,t) = 0 \quad \text{``side at } x = 0 \text{ is anchored''} \tag{5.113}$$

$$\frac{\partial u}{\partial x}(l,t) = 0 \quad \text{``side at } x = l \text{ is free''} \tag{5.114}$$

We employ the method of separation of variables to solve this problem. Let

$$u(x,t) = X(x)\, G(t) \tag{5.115}$$

Substituting Equation (5.115) in Equation (5.110) yields two ordinary differential equations—one for X and one for G as follows:

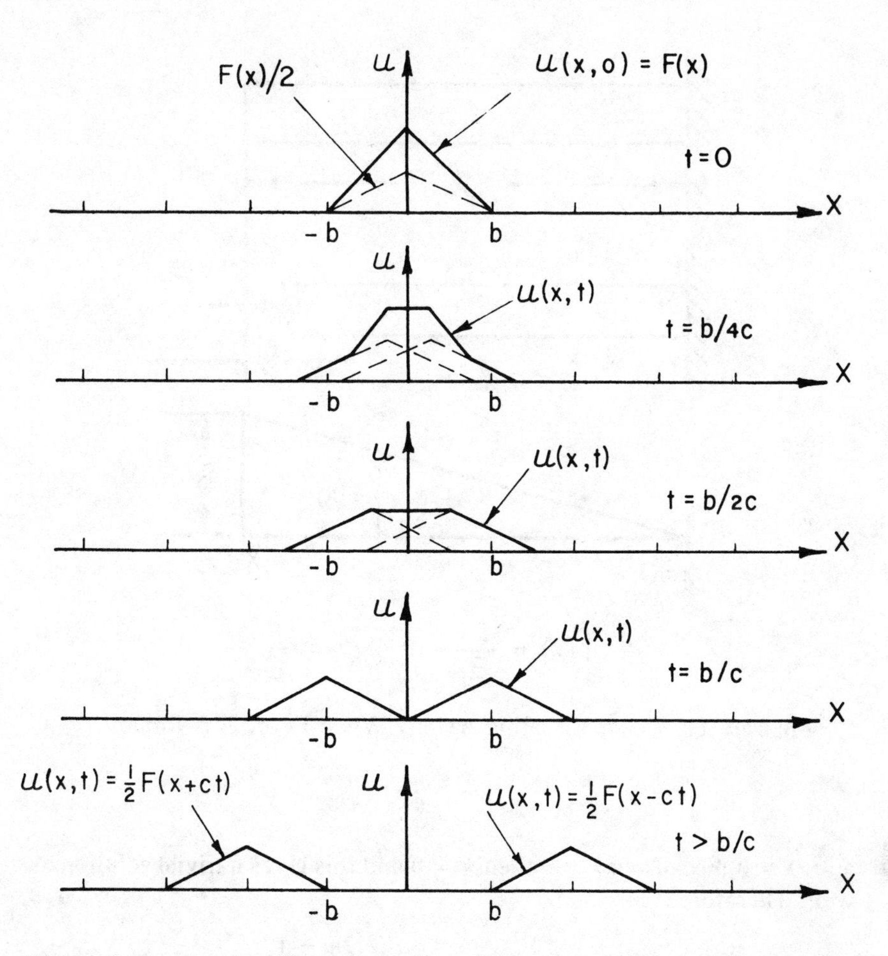

Figure 5-10. Graphical Interpretation of Equation (5.109) for a Triangular Wave

$$\frac{d^2X}{dx^2} + \lambda^2 X = 0 \tag{5.116}$$

$$X(0) = 0, \qquad \frac{dX}{dx}(l) = 0 \tag{5.117}$$

The solution for $X(x)$ is

$$X = A \cos \lambda x + B \sin \lambda x \tag{5.118}$$

Applying the boundary conditions (5.117) to Equation (5.118) yields

$$A = 0, \qquad B\lambda \cos \lambda l = 0$$

B cannot be zero or, as analyzed before, we have a trivial solution $u = 0$ for

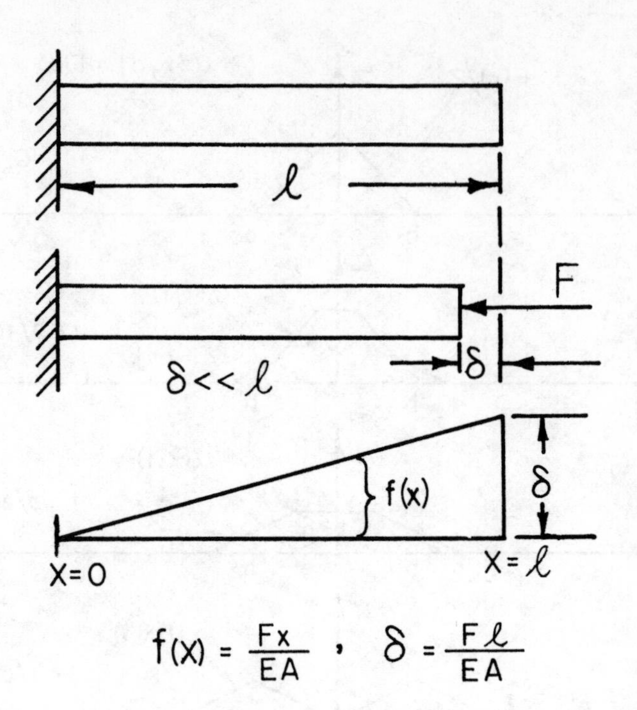

Figure 5-11. Beam Vibrating Axially When Force F is Removed

all t. $\lambda \neq 0$, because if $\lambda = 0$ then $X = 0$ and this gives a trivial solution as well. Therefore,

$$\cos \lambda l = 0, \quad \lambda_n l = \left(\frac{2n+1}{2}\right)\pi \tag{5.119}$$

and the eigenvalues are

$$\lambda_n = \left(\frac{2n+1}{2l}\right)\pi \quad n = 0, 1, 2, \ldots \tag{5.120}$$

The eigenvalues here start with $n = 0$ because for $n = 0$, λ_0 does not yield a trivial solution, namely $u(x,t) = 0$. (See Equation (5.95) and the note that follows that equation.)

Hence,

$$X_n(x) = \sin\left(\frac{2n+1}{2l}\right)\pi x \tag{5.121}$$

The differential equation for $G(t)$ is

$$\frac{d^2G}{dt^2} + \lambda_n^2 c^2 G = 0 \tag{5.122}$$

The general solution for $G(t)$ becomes: "for $\lambda_n = [(2n + 1)/2l]\pi$"

$$G_n(t) = C_n \cos\left(\frac{2n + 1}{2l}\right)\pi ct + D_n \sin\left(\frac{2n + 1}{2l}\right)\pi ct \tag{5.123}$$

Applying now initial condition (5.112) results in

$$\frac{\partial u}{\partial t}(x,0) = 0 = X\frac{\partial G_n}{\partial t}(0)$$

or

$$D_n = 0$$

Hence,

$$G_n(t) = C_n \cos\left(\frac{2n + 1}{2l}\right)\pi ct$$

The form of the solution $u(x,t)$ for this problem becomes

$$u_n(x,t) = XG$$

or

$$u_n(x,t) = C_n \cos\left(\frac{2n + 1}{2l}\right)\pi ct \sin\left(\frac{2n + 1}{2l}\right)\pi x \tag{5.124}$$

To satisfy the given initial condition (5.111), we should sum Equation (5.124) over all λ's; this yields

$$u(x,t) = \sum_{n=0}^{\infty} C_n \cos\left(\frac{2n + 1}{2l}\right)\pi ct \sin\left(\frac{2n + 1}{2l}\right)\pi x \tag{5.125}$$

The next step is to evaluate Fourier coefficients C_n. Applying initial condition (5.111) there results

$$\frac{Fx}{EA} = \sum_{n=0}^{\infty} C_n \sin\left(\frac{2n + 1}{2l}\right)\pi x \tag{5.126}$$

or

$$C_n = \frac{2}{l}\int_0^l \frac{Fx}{EA} \sin\left(\frac{2n + 1}{2l}\right)\pi x \, dx$$

$$= \frac{8Fl}{EA\pi^2} \frac{(-1)^n}{(2n + 1)^2} \tag{5.127}$$

Finally the solution becomes

$$u(x,t) = \frac{8Fl}{EA\pi^2}\sum_{n=0}^{\infty}\frac{(-1)^n}{(2n+1)^2}\cos\left(\frac{2n+1}{2l}\right)\pi ct$$

$$\times \sin\left(\frac{2n+1}{2l}\right)\pi x \tag{5.128}$$

A point at $x = x_0$ on the beam vibrates with an amplitude of

$$C_n \sin\left(\frac{2n+1}{2l}\right)\pi x_0$$

and a frequency of

$$\frac{1}{2\pi}\left(\frac{2n+1}{2l}\right)\pi c \quad \text{cycles/sec.}$$

Example 5.2 Consider a string of length l fixed at both ends. If the string is plucked at its midpoint as shown in Figure 5-12 and then released, determine the subsequent motion of the string.

Solution: We have to determine in this problem the displacement of all the points on the string as a function of time. Therefore, we have to determine $u(x,t)$.

In Article 5.8 we arrived at the general solution to this problem as given by Equation (5.99). However, because there is no initial velocity, the solution is reduced to that given by Equation (5.104) as

$$u(x,t) = \sum_{n=1}^{\infty} C_n \cos\frac{n\pi c}{l}t \sin\frac{n\pi x}{l} \qquad n = 1, 2, \ldots \tag{5.129}$$

with C_n given by Equation (5.102) as

$$C_n = \frac{2}{l}\int_0^l f(x)\sin\frac{n\pi x}{l}dx \qquad n = 1, 2, \ldots \tag{5.130}$$

In this problem, the initial deflection $f(x)$ is given by

$$f(x) = \begin{cases} \dfrac{2hx}{l} & 0 \leqslant x \leqslant \dfrac{l}{2} \quad (5.131a) \\ \dfrac{2h}{l}(l-x) & \dfrac{l}{2} \leqslant x \leqslant l \quad (5.131b) \end{cases}$$

Evaluating now C_n yields

$$C_n = \frac{2}{l}\left[\int_0^{l/2}\left(\frac{2hx}{l}\right)\sin\frac{n\pi x}{l}dx\right.$$

$$\left. + \int_{l/2}^{l}\left(\frac{2h}{l}\right)(l-x)\sin\frac{n\pi x}{l}dx\right] \tag{5.132}$$

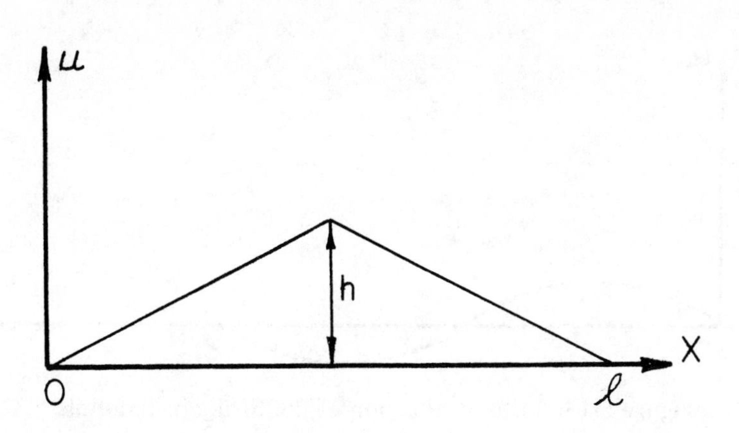

Figure 5-12. Initial Deflection of the String in Example 5.2

Performing the integration results in

$$C_n = \frac{8h}{\pi^2 n^2} \sin \frac{n\pi}{2} \quad \text{for } n \text{ odd} \tag{5.133a}$$

$$C_n = 0 \quad \text{for } n \text{ even} \tag{5.133b}$$

With

$$\sin \frac{n\pi}{2} = (-1)^{(n-1)/2} \tag{5.134}$$

The solution $u(x,t)$ can be written as

$$u(x,t) = \frac{8h}{\pi^2} \sum_{n=1}^{\infty} \frac{(-1)^{(n-1)/2}}{n^2} \sin \frac{n\pi x}{l} \cos \frac{n\pi ct}{l} \qquad n = 1, 3, 5, \ldots$$

$$= \frac{8h}{\pi^2}\left[\sin \frac{\pi x}{l} \cos \frac{\pi ct}{l} - \frac{1}{9} \sin \frac{3\pi x}{l} \cos \frac{3\pi ct}{l} + \ldots\right] \tag{5.135}$$

In this problem, no even harmonics are excited. Furthermore, the amplitude of the second harmonic is 1/9th the amplitude of the fundamental. Because the energy emitted by an oscillating string as sound is proportional to the square of the amplitude, it is evident then that the fundamental tone will be heard much louder than the harmonics.

Example 5.3 Repeat Example 5.2, however, let the initial displacement of the string be represented by (Figure 5-13)

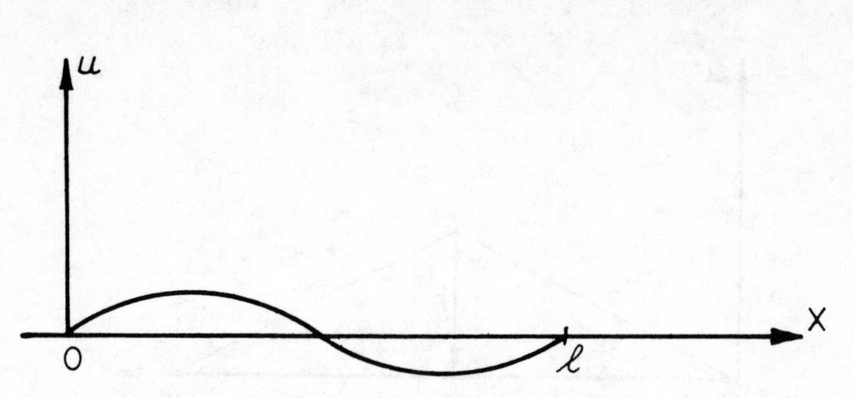

Figure 5-13. Initial Deflection of the String in Example 5.3

$$u(x,0) = f(x) = \sin \frac{2\pi x}{l}$$

Solution: Here again the solution as obtained by the method of separation of variables is represented by Equation (5.104) and it is

$$u(x,t) = \sum_{n=1}^{\infty} C_n \cos \frac{n\pi c}{l} t \sin \frac{n\pi x}{l} \tag{5.136}$$

We need now to evaluate C_n where C_n is given by

$$C_n = \frac{2}{l}\int_0^l f(x) \sin \frac{n\pi x}{l} dx \tag{5.137}$$

With $f(x) = \sin 2\pi x/l$ we obtain

$$C_n = \frac{2}{l}\int_0^l \sin \frac{2\pi x}{l} \sin \frac{n\pi x}{l} dx \tag{5.138}$$

Because $\sin 2\pi x/l$ and $\sin n\pi x/l$ are orthogonal, the integral is zero for all values of n except $n = 2$. Hence,

$$C_n = C_2 = \frac{2}{l}\int_0^l \left(\sin \frac{2\pi x}{l} \right)^2 dx = \left(\frac{2}{l}\right)\left(\frac{l}{2}\right) = 1 \tag{5.139}$$

Therefore, the solution degenerates into one term; it is the term for $n = 2$, that is,

$$u(x,t) = \sin \frac{2\pi x}{l} \cos \frac{2\pi c}{l} t \tag{5.140}$$

The physical meaning of this solution is that the string vibrates only with the second normal mode or second harmonic. No other harmonics are present.

Example 5.4 Analyze the vibration in Example 5.3 for the case where damping is important. Examine the effect of damping on the amplitude and frequency of oscillation. Consider that the differential equation in this case (Equation (5.15)) is written as

$$c^2\frac{\partial^2 u}{\partial x^2} = \frac{\partial^2 u}{\partial t^2} + 2\beta\frac{\partial u}{\partial t} \qquad \beta = \text{constant} \tag{5.141}$$

and the vibration is underdamped.

Solution: From the method of separation of variables we let

$$u(x,t) = X(x)\,G(t) \tag{5.142}$$

Substituting Equation (5.142) in Equation (5.141) yields the following two ordinary differential equations

$$\frac{d^2X}{dx^2} + \lambda^2 X = 0 \tag{5.143}$$

with

$$X(0) = X(l) = 0 \tag{5.144}$$

and

$$\frac{d^2G}{dt^2} + 2\beta\frac{dG}{dt} + \lambda^2 c^2 G = 0 \tag{5.145}$$

Equation (5.143) with boundary conditions (5.144) yields

$$X(x) = \sin \lambda_n x \qquad \lambda_n = \frac{n\pi}{l}, n = 1, 2, 3, \ldots \tag{5.146}$$

Equation (5.145) has the following solution (see chapter 1)

$$G(t) = e^{-\beta t}[A \cos \omega_n t + B \sin \omega_n t] \tag{5.147}$$

where

$$\omega = \sqrt{\lambda^2 c^2 - \beta^2} = \sqrt{\frac{n^2\pi^2c^2}{l^2} - \beta^2} \tag{5.148}$$

Therefore, the expression for $u(x,t)$ takes the following form

$$u(x,t) = \sum_{n=1}^{\infty} e^{-\beta t}[A_n \cos \omega_n t + B_n \sin \omega_n t] \sin \frac{n\pi}{l}x \tag{5.149}$$

Invoking that the initial velocity is zero we obtain

$$\frac{\partial u}{\partial t}(x,0) = 0 = \sum_{n=1}^{\infty} \{e^{-\beta t}[-\omega_n A_n \sin \omega_n t + \omega_n B_n \cos \omega_n t]$$

$$-\beta e^{-\beta t}[A_n \cos \omega_n t + B_n \sin \omega_n t]\}_{t=0} \sin \frac{n\pi x}{l} \quad (5.150)$$

Equation (5.150) yields

$$(\omega_n B_n - \beta A_n) \sin \frac{n\pi x}{l} = 0 \quad (5.151)$$

or

$$\omega_n B_n - \beta A_n = 0 \quad \text{and} \quad B_n = \frac{\beta A_n}{\omega_n} \quad (5.152)$$

Therefore, Equation (5.149) becomes

$$u(x,t) = \sum_{n=1}^{\infty} A_n e^{-\beta t}\left[\cos \omega_n t + \frac{\beta}{\omega_n} \sin \omega_n t\right] \sin \frac{n\pi x}{l} \quad (5.153)$$

Applying the initial displacement as stated in Example 5.3 yields

$$u(x,0) = \sin \frac{2\pi x}{l} = \sum_{n=1}^{\infty} A_n \sin \frac{n\pi x}{l} \quad (5.154)$$

Therefore, as in Example 5.3, $n = 2$, $A_n = 1$, and the solution becomes

$$u(x,t) = e^{-\beta t}\left[\cos\left(\sqrt{\frac{4\pi^2c^2}{l^2} - \beta^2}\right)t\right.$$

$$\left. + \frac{\beta}{\sqrt{(4\pi^2c^2/l^2) - \beta^2}} \sin\left(\sqrt{\frac{4\pi^2c^2}{l^2} - \beta^2}\right)t\right] \sin \frac{2\pi x}{l} \quad (5.155)$$

Equation (5.155) shows that the presence of damping causes a decrease in frequency and a decrease in the amplitude of oscillation with time.

Example 5.5 In stringed musical instruments we identify usually the plucked string (like a harp or a guitar), the struck string (like a piano), and the bowed string (like a violin). In a plucked string instrument, the string is subjected to an initial displacement and no initial velocity. In a struck string instrument, the vibration is excited by a blow that imparts an initial velocity to the string but no initial displacement.

Let a string of length l, fixed at both ends, be excited by a blow (or an impulse) distributed uniformly over a segment of a string of width 2δ. Find the expression representing the vibration of the string (Figure 5-14).

Solution: This problem represents a string excited by imparting a constant

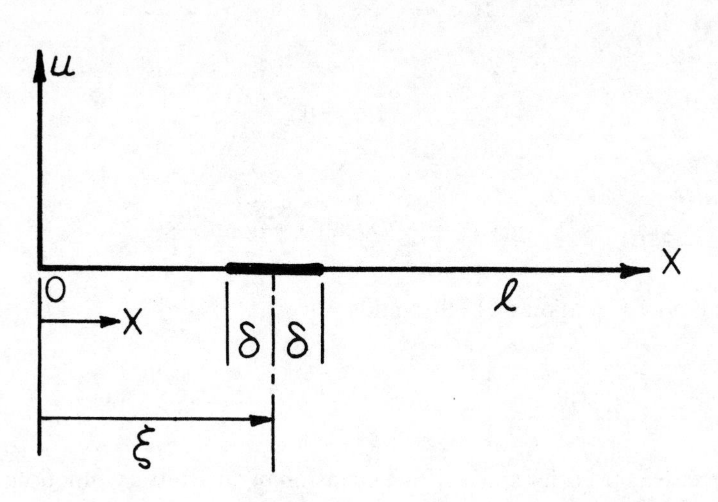

Figure 5-14. A String Struck by a Hammer of Width 2δ

initial velocity, say v_0, by a rigid hammer having a width 2δ striking at location ξ.

The differential equation and the problem conditions can be stated as follows:

$$\frac{\partial^2 u}{\partial t^2} = c^2 \frac{\partial^2 u}{\partial x^2} \tag{5.156}$$

$$u(x,0) = 0 \tag{5.157a}$$

$$\frac{\partial u}{\partial t}(x,0) = \begin{cases} v_0 & |x - \xi| < \delta \\ 0 & |x - \xi| > \delta \end{cases} \tag{5.157b}$$

$$\begin{aligned} u(0,t) &= 0 \\ u(l,t) &= 0 \end{aligned} \tag{5.158}$$

From separation of variables and the imposed boundary conditions the general form of solution is that given earlier by Equation (5.99) and it is

$$u(x,t) = \sum_{n=1}^{\infty} \left[C_n \cos \frac{n\pi c}{l} t + D_n \sin \frac{n\pi c}{l} t \right] \sin \frac{n\pi x}{l} \tag{5.159}$$

Applying the initial conditions given by Equation (5.157a) gives

$$u(x,0) = 0 = \sum C_n \sin \frac{n\pi x}{l}$$

or

$$C_n = 0$$

Hence,

$$u(x,t) = \sum_{n=1}^{\infty} D_n \sin \frac{n\pi c}{l} t \sin \frac{n\pi x}{l} \tag{5.160}$$

Using now Equation (5.157b) yields

$$\frac{\partial u}{\partial t}(x,0) = \begin{cases} v_0, \ |x - \xi| < \delta \\ \\ 0, \ |x - \xi| > \delta \end{cases} = \sum_{n=1}^{\infty} D_n \frac{n\pi c}{l} \sin \frac{n\pi x}{l} \tag{5.161}$$

Hence, from Fourier sine series expansion of an arbitrary function we can write

$$D_n = \frac{2}{l}\left(\frac{l}{n\pi c}\right)\int_{\xi-\delta}^{\xi+\delta} v_0 \sin \frac{n\pi x}{l} dx \tag{5.162}$$

Performing the integration yields

$$D_n = \frac{4v_0 l}{n^2\pi^2 c} \sin \frac{n\pi\xi}{l} \sin \frac{n\pi\delta}{l} \tag{5.163}$$

Therefore, the expression representing the vibration of the string becomes

$$u(x,t) = \frac{4v_0 l}{\pi^2 c} \sum_{n=1}^{\infty} \frac{1}{n^2} \sin \frac{n\pi\xi}{l} \sin \frac{n\pi\delta}{l} \times \left(\sin \frac{n\pi x}{l} \sin \frac{n\pi c}{l} t\right) \tag{5.164}$$

From this expression we see that the width of the interval 2δ has a considerable effect on the amplitude and hence on the sound (or energy) of a vibrating string. Because the amplitude has the term $\sin n\pi\xi/l$ in it, we can say that if the center of the blow of the hammer happens to occur at a node of the nth harmonic, the energy or the contribution from the corresponding harmonic is zero.

Example 5.6 Repeat Example 5.5 but consider that the string is excited by a convex hammer of width 2δ imparting initial velocity to the string according to

$$\frac{\partial u}{\partial t}(x,0) = \begin{cases} v_0 \cos \dfrac{(x - \xi)}{\delta} \cdot \dfrac{\pi}{2} & |x - \xi| < \delta \\ 0 & |x - \xi| > \delta \end{cases}$$

Solution: The solution to this problem is similar to that in Example 5.5. The difference lies in the evaluation of the coefficient D_n. Hence, from Example 5.5 we have

$$u(x,t) = \sum_{n=1}^{\infty} D_n \sin \frac{n\pi c}{l} t \sin \frac{n\pi x}{l} \tag{5.165}$$

$$D_n = \frac{2}{l} \cdot \left(\frac{l}{n\pi c}\right) \int_{\xi-\delta}^{\xi+\delta} v_0 \cos \frac{(x-\xi)}{\delta} \frac{\pi}{2} \sin \frac{n\pi x}{l} dx \tag{5.166}$$

Performing the integration yields

$$D_n = \frac{8v_0\delta}{\pi^2 c} \frac{1}{n} \frac{\cos \dfrac{n\pi\delta}{l} \sin \dfrac{n\pi}{l} c}{1 - \left(\dfrac{2\delta n}{l}\right)^2} \tag{5.167}$$

The expression for $u(x,t)$ becomes

$$u(x,t) = \frac{8v_0\delta}{\pi^2 c} \sum_{n=1}^{\infty} \frac{1}{n} \frac{\cos \dfrac{n\pi}{l}\delta \sin \dfrac{n\pi}{l}\xi}{1 - \left(\dfrac{2\delta n}{l}\right)^2} \times \sin \frac{n\pi}{l} x \sin \frac{n\pi}{l} ct \tag{5.168}$$

5.9 Nonhomogeneous Boundary Conditions

Let us consider now the case in which the boundary condition is not homogeneous. Physically this means, in the case of a vibrating string, that the ends of the string have finite deflections which may be constant or functions of time. See Figure 5-15.

In the absence of vibration, the string will have deflections that vary from $u = u_1$ at $x = 0$ to $u = u_2$ at $x = l$. We refer to this deflection as the steady state solution, and it is represented by the dashed line in Figure 5-15.

The differential equation and initial and boundary conditions of the problem can be stated as follows:

$$\frac{\partial^2 u}{\partial x^2} = \frac{1}{c^2} \frac{\partial^2 u}{\partial t^2} \tag{5.169}$$

$$u(x,0) = f(x) \tag{5.170a}$$

$$\frac{\partial u}{\partial t}(x,0) = g(x) \tag{5.170b}$$

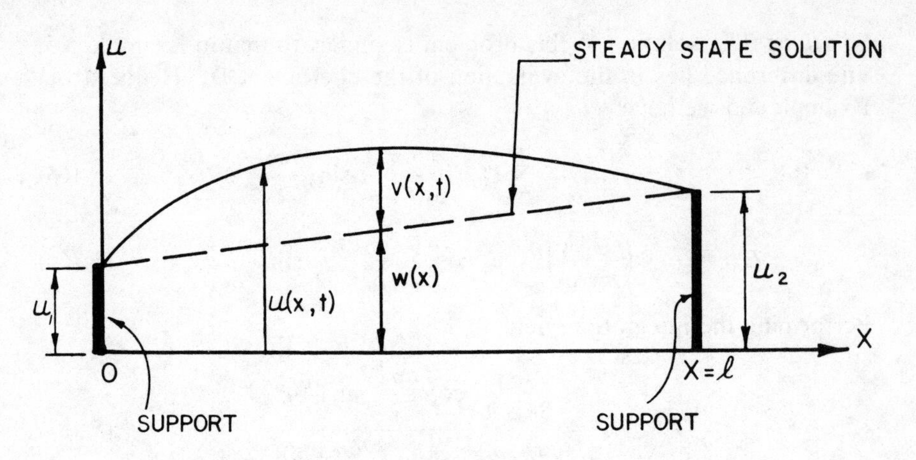

Figure 5-15. Nonhomogeneous Boundary Conditions and the Steady State Solution for a Vibrating String

$$u(0,t) = u_1 \tag{5.171a}$$

$$u(l,t) = u_2 \tag{5.171b}$$

Let us take now the case where u_1 and u_2 are constants. The method of separation of variables requires that the boundary conditions at $x = 0$ and $x = l$ be homogeneous. Hence, we cannot directly apply the method to Equation (5.169). Looking at Figure 5-15, and having in mind that the differential equation is linear, we can write

$$u(x,t) = v(x,t) + w(x) \tag{5.172}$$

with

$$\frac{\partial^2 u}{\partial x^2} = \frac{\partial^2 v}{\partial x^2} + \frac{d^2 w}{dx^2} \tag{5.173}$$

and

$$\frac{\partial^2 u}{\partial t^2} = \frac{\partial^2 v}{\partial t^2} + \cancelto{0}{\frac{\partial^2 w}{\partial t^2}} \tag{5.174}$$

Substituting Equations (5.173) and (5.174) in Equation (5.169) yields

$$\frac{\partial^2 v}{\partial x^2} + \frac{d^2 w}{dx^2} = \frac{1}{c^2}\frac{\partial^2 v}{\partial t^2} \tag{5.175}$$

For $v(x,t)$ to satisfy the wave equation, we should have

$$\frac{\partial^2 v}{\partial x^2} = \frac{1}{c^2}\frac{\partial^2 v}{\partial t^2} \tag{5.176}$$

Hence,

$$\frac{d^2w}{dx^2} = 0 \tag{5.177}$$

Therefore, we have split the original differential equation for $u(x,t)$ into two differential equations: one for $v(x,t)$ and one for $w(x)$. The equation for $w(x)$ is an ordinary differential equation and can be solved by direct integration. To solve the equation for $v(x,t)$ by the method of separation of variables, we should impose homogeneous boundary conditions on $v(x,t)$. This is quite possible now as the steady state solution $w(x)$ can take care of the stated boundary conditions. The initial and boundary conditions become

$$u(x,0) = v(x,0) + w(x) = f(x) \tag{5.178}$$

$$\frac{\partial u}{\partial t}(x,0) = \frac{\partial v}{\partial t}(x,0) + \cancelto{0}{\frac{\partial w}{\partial t}} = g(x) \tag{5.179}$$

and

$$u(0,t) = v(0,t) + w(0) = u_1 \tag{5.180}$$

$$u(l,t) = v(l,t) + w(l) = u_2 \tag{5.181}$$

Hence, from Equation (5.178) and (5.179) we obtain

$$v(x,0) = f(x) - w(x) \tag{5.182}$$

$$\frac{\partial v}{\partial t}(x,0) = g(x) \tag{5.183}$$

and from Equations (5.180) and (5.181) we obtain

$$w(0) = u_1 \tag{5.184}$$

$$w(l) = u_2 \tag{5.185}$$

$$v(0,t) = 0 \tag{5.186}$$

$$v(l,t) = 0 \tag{5.187}$$

Therefore, we have reduced our original problem into the following two problems:

$$\begin{array}{c|c}
\dfrac{\partial^2 v}{\partial x^2} = \dfrac{1}{c^2}\dfrac{\partial^2 v}{\partial t^2} & \dfrac{d^2w}{dx^2} = 0 \\
v(x,0) = f(x) - w(x) & w(0) = u_1 \\
\dfrac{\partial v}{\partial t}(x,0) = g(x) & w(l) = u_2
\end{array}$$

$$v(0,t) = 0 \qquad \text{which yields}$$

$$v(l,t) = 0 \qquad w(x) = u_1 + (u_2 - u_1)\frac{x}{l}$$

The boundary value problem posed by $v(x,t)$ can now be solved by the method of separation of variables because it is equivalent to the problem posed by Equations (5.75) to (5.77) in Article 5.8. The solution (as presented by Equation (5.99)) becomes

$$\begin{aligned} v(x,t) &= \sum_{n=1}^{\infty} v_n(x,t) \\ &= \sum_{n=1}^{\infty}\left(C_n \cos \frac{n\pi c}{l}t + D_n \sin \frac{n\pi c}{l}t\right) \sin \frac{n\pi x}{l} \end{aligned} \tag{5.188}$$

and

$$C_n = \frac{2}{l}\int_0^l [f(x) - w(x)] \sin \frac{n\pi x}{l} dx \quad n = 1, 2, 3, \ldots \tag{5.189}$$

$$D_n = \frac{2}{n\pi c}\int_0^l g(x) \sin \frac{n\pi x}{l} dx \quad n = 1, 2, 3, \ldots \tag{5.190}$$

The solution for $u(x,t)$ becomes

$$u(x,t) = v(x,t) + u_1 + (u_2 - u_1)\frac{x}{l} \tag{5.191}$$

We like to note that the two nonhomogeneous boundary conditions, namely $u = u_1$ at $x = 0$ and $u = u_2$ at $x = l$, could have been handled by two steady state equations. Although it is not necessary in this problem, we will see that in some problems we have to do that. $w(x)$, then, is split into two solutions to be obtained by solving the following differential equations: let

$$w(x) = w_1(x) + w_2(x) \tag{5.192}$$

where w_1 and w_2 satisfy

$$\frac{d^2w_1}{dx^2} = 0 \qquad \frac{d^2w_2}{dx^2} = 0$$

$$w_1(0) = u_1 \qquad w_2(0) = 0$$

$$w_1(l) = 0 \qquad w_2(l) = u_2$$

or $\qquad\qquad$ or

$$w_1 = u_1\left(1 - \frac{x}{l}\right) \qquad w_2 = u_2\frac{x}{l}$$

The sum of the two solutions for $w_1(x)$ and $w_2(x)$ as evident yields the expression for $w(x)$ as obtained earlier; see Figure 5-16.

5.10 Nonhomogeneous Differential Equations

We have shown in Article 5.2 that a nonhomogeneous wave equation arises when an external force is applied on a vibrating string parallel to the direction of its deflection. Let us treat now the case in which the external force is a function of x. We further make the problem a little more general by imposing nonhomogeneous boundary conditions. Taking $q(x) = Q(x)/m$, where $Q(x)$ is the force per unit length, the differential equation and the initial and boundary conditions can be written then as

$$c^2\frac{\partial^2 u}{\partial x^2} + q(x) = \frac{\partial^2 u}{\partial t^2} \quad \text{(see Figure 5-17)} \tag{5.193}$$

$$u(x,0) = f(x) \tag{5.194a}$$

$$\frac{\partial u}{\partial t}(x,0) = g(x) \tag{5.194b}$$

$$u(0,t) = u_1 \tag{5.195a}$$

$$u(l,t) = u_2 \tag{5.195b}$$

In this problem we have two sources of nonhomogeneities: one is in the differential equation and it is a function of x only, and the other is in the boundary conditions. Therefore, we have to reduce the problem into a set of problems which can then be solved by any of the methods we have covered so far. This problem, in the absence of vibration, exhibits also a steady state deflection, $w(x)$, which varies from u_1 at $x = 0$ to u_2 at $x = l$ in a manner that depends upon the external function $q(x)$. Therefore, let us write

$$u(x,t) = v(x,t) + w(x) \tag{5.196}$$

Substituting Equation (5.196) in Equation (5.193) we obtain

$$c^2\left[\frac{\partial^2 v}{\partial x^2} + \frac{d^2 w}{dx^2}\right] + q(x) = \frac{\partial^2 v}{\partial t^2} \tag{5.197}$$

Choosing $v(x,t)$ to satisfy the homogeneous wave equation, namely

$$c^2\frac{\partial^2 v}{\partial x^2} = \frac{\partial^2 v}{\partial t^2} \tag{5.198}$$

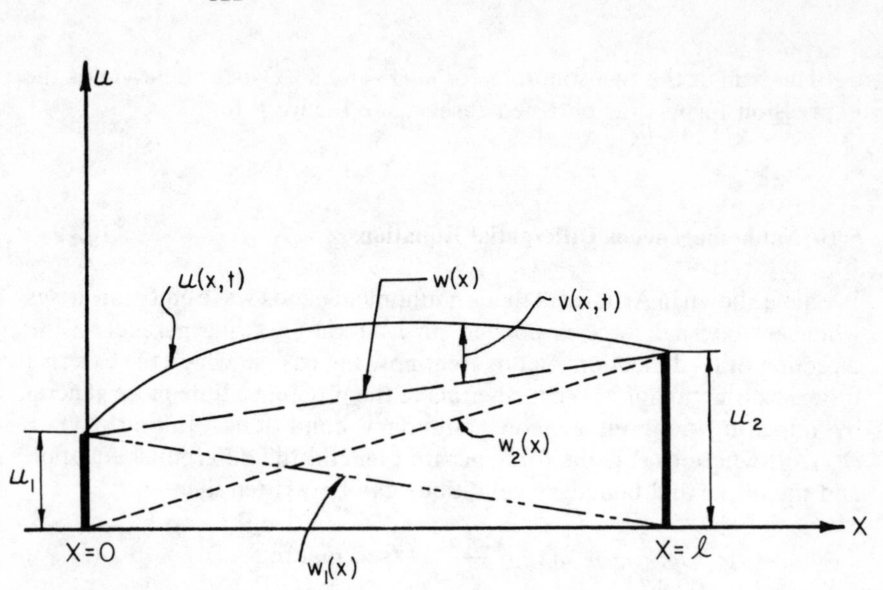

Figure 5-16. Nonhomogeneous Boundary Conditions and Alternate Steady State Solutions for a Vibrating String

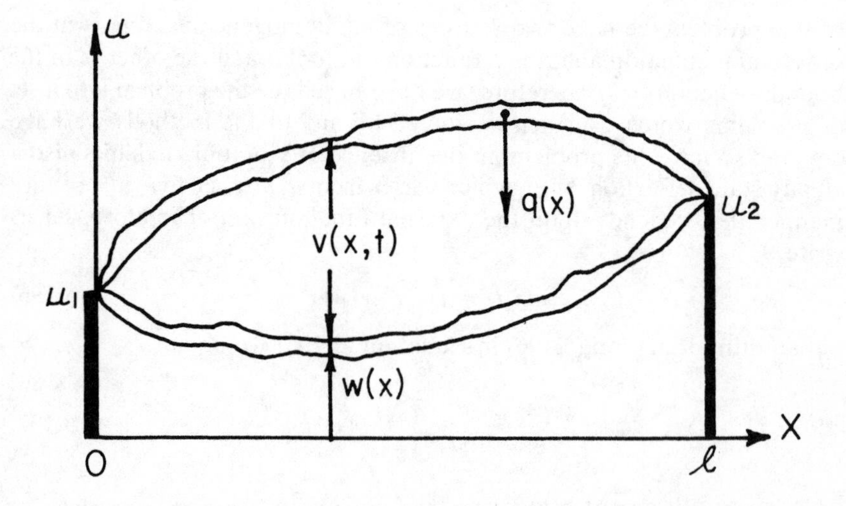

Figure 5-17. Nonhomogeneous Boundary Conditions for a Vibrating String Subject to an External Force $q(x)$

There results

$$c^2\frac{d^2w}{dx^2} + q(x) = 0 \tag{5.199}$$

Here again we have split the original nonhomogeneous differential equation into two equations: one for $v(x,t)$, and it is homogeneous, and the other for $w(x)$, which takes care of the nonhomogeneities in both the original differential equation and the boundary conditions.

The initial and boundary conditions become

$$\left.\begin{aligned} u(x,0) &= v(x,0) + w(x) = f(x) && (5.200)\\ \frac{\partial u}{\partial t}(x,0) &= \frac{\partial v}{\partial t}(x,0) + \cancelto{0}{\frac{\partial w}{\partial t}} = g(x) && (5.201) \end{aligned}\right\} 0 \leq x \leq l$$

and

$$\left.\begin{aligned} u(0,t) &= v(0,t) + w(0) = u_1 && (5.202)\\ u(l,t) &= v(l,t) + w(l) = u_2 && (5.203) \end{aligned}\right\} t > 0$$

From Equations (5.200) and (5.201) we get

$$v(x,0) = f(x) - w(x) \tag{5.204}$$

$$\frac{\partial v}{\partial t}(x,0) = g(x) \tag{5.205}$$

From Equations (5.202) and (5.203) we can write

$$w(0) = u_1 \tag{5.206}$$

$$w(l) = u_2 \tag{5.207}$$

$$v(0,t) = 0 \tag{5.208}$$

$$v(l,t) = 0 \tag{5.209}$$

Here again we reduced the original problem into the following two problems:

$$\frac{\partial^2 v}{\partial x^2} = \frac{1}{c^2}\frac{\partial^2 v}{\partial t^2}$$

$$\left.\begin{aligned} v(x,0) &= f(x) - w(x)\\ \frac{\partial v}{\partial t}(x,0) &= g(x) \end{aligned}\right\} 0 \leq x \leq l$$

$$\left.\begin{aligned} v(0,t) &= 0\\ v(l,t) &= 0 \end{aligned}\right\} t > 0$$

$$\frac{d^2w}{dx^2} + \frac{q(x)}{c^2} = 0$$

$$w(0) = u_1$$

$$w(l) = u_2$$

The solution for $w(x)$ can be obtained by direct integration. If $q(x)$ is a constant, say q_0, $w(x)$ becomes

$$w(x) = u_1 + (u_2 - u_1)\frac{x}{l} + \frac{q_0 x^2}{2c^2}\left[\frac{l}{x} - 1\right] \tag{5.210}$$

The solution for $v(x,t)$ is given by Equations (5.188) to (5.190). $u(x,t)$ is then equal to the sum of $v(x,t)$ and $w(x)$ as

$$u(x,t) = v(x,t) + w(x)$$

Example 5.7 Find the expression describing the transverse vibration of a rod of rectangular cross section clamped at one end as shown in Figure 5-18.

Solution: The differential equation for this problem has been derived in Article 5.4 and it is

$$\frac{\partial^2 u}{\partial t^2} + c^2\frac{\partial^4 u}{\partial x^4} = 0 \tag{5.211}$$

where

$$c^2 = \frac{EJ}{\rho A}$$

Equation (5.211) requires two initial and four boundary conditions to define the motion of the rod.

Initial conditions:

$$u(x,0) = f(x) \tag{5.212}$$

$$\frac{\partial u}{\partial t}(x,0) = g(x) \tag{5.213}$$

Boundary conditions:

$u(0,t) = 0$ zero deflection at the clamped end (5.214)

$\frac{\partial u}{\partial x}(0,t) = 0$ elastic curve has a zero slope at the clamped end (5.215)

$\frac{\partial^2 u}{\partial x^2}(l,t) = 0$ the moment is zero at the free end, $x = l$ (5.216)

$\frac{\partial^3 u}{\partial x^3}(l,t) = 0$ the shear force is zero at the free end, $x = l$ (5.217)

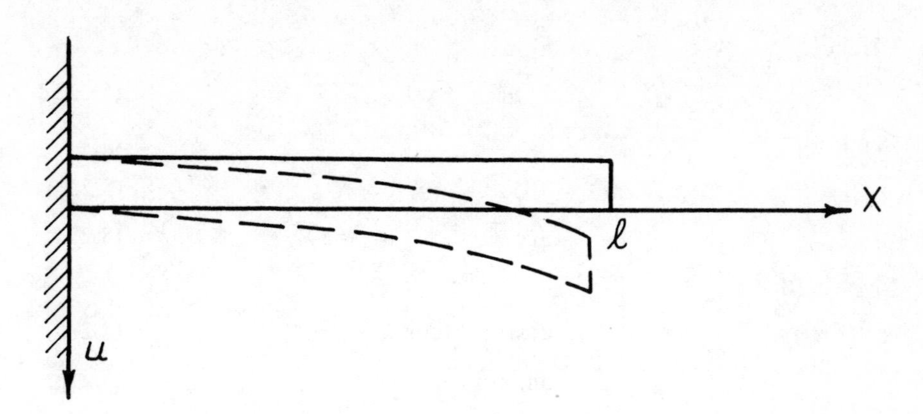

Figure 5-18. Beam Vibrating Transversely

Let us use the method of separation of variables, and therefore, we let

$$u(x,t) = X(x)\,G(t) \tag{5.218}$$

Substituting Equation (5.218) in Equation (5.211) yields

$$\frac{1}{c^2}\frac{d^2G}{dt^2} = -\frac{1}{X}\frac{d^4X}{dx^4} = -\lambda \tag{5.219}$$

Equation (5.219) then gives

$$\frac{d^4X}{dx^4} - \lambda X = 0 \tag{5.220}$$

with

$X(0) = 0$ from Equation (5.214)

$\dfrac{dX}{dx}(0) = 0$ from Equation (5.215)

$\dfrac{d^2X}{dx^2}(l) = 0$ from Equation (5.216)

$\dfrac{d^3X}{dx^3}(l) = 0$ from Equation (5.217)

Equation (5.220) has the following auxiliary equation for the four roots of the solution

$$m^4 - \lambda = 0 \tag{5.221}$$

Hence,

$$m^2 = \pm\sqrt{\lambda} \tag{5.222}$$

and

$$m_1 = +\sqrt[4]{\lambda}, \quad m_2 = -\sqrt[4]{\lambda} \tag{5.223}$$

$$m_3 = +i\sqrt[4]{\lambda}, \quad m_4 = -i\sqrt[4]{\lambda} \tag{5.224}$$

Noting that

$$\cosh z = \cos iz \tag{5.225}$$

$$\sinh z = -i \sin iz \tag{5.226}$$

the solution to Equation (5.220) can be written as

$$X(x) = A \cosh\sqrt[4]{\lambda}\,x + B \sinh\sqrt[4]{\lambda}\,x$$
$$+ C \cos\sqrt[4]{\lambda}\,x + D \sin\sqrt[4]{\lambda}\,x \tag{5.227}$$

Applying boundary conditions (5.214) and (5.215) we obtain

$$A = -C \quad \text{and} \quad B = -D$$

Therefore, Equation (5.227) can be written as

$$X(x) = A[\cosh\sqrt[4]{\lambda}\,x - \cos\sqrt[4]{\lambda}\,x]$$
$$+ B[\sinh\sqrt[4]{\lambda}\,x - \sin\sqrt[4]{\lambda}\,x] \tag{5.228}$$

We have two more boundary conditions (5.216) and (5.217) to generate the eigenvalues λ's. Applying boundary conditions (5.216) and (5.217) to Equation (5.228) results in

$$A[\cosh \sqrt[4]{\lambda}\,l + \cos \sqrt[4]{\lambda}\,l] + B[\sinh \sqrt[4]{\lambda}\,l + \sin \sqrt[4]{\lambda}\,l] = 0 \tag{5.229}$$

and

$$A[\sinh \sqrt[4]{\lambda}\,l - \sin \sqrt[4]{\lambda}\,l] + B[\cosh \sqrt[4]{\lambda}\,l + \cos \sqrt[4]{\lambda}\,l] = 0 \tag{5.230}$$

For a nontrivial solution for A and B, the determinant of the coefficients of A and B in Equations (5.229) and (5.230) should be zero. Setting the determinant equal to zero gives

$$\sinh^2 \sqrt[4]{\lambda}\,l - \sin^2 \sqrt[4]{\lambda}\,l = \cosh^2 \sqrt[4]{\lambda}\,l + \cos^2 \sqrt[4]{\lambda}\,l$$
$$+ 2(\cosh \sqrt[4]{\lambda}\,l)(\cos \sqrt[4]{\lambda}\,l) \tag{5.231}$$

Using the identities

$$\cosh^2 z - \sinh^2 z = 1 \quad \text{and} \quad \sin^2 z + \cos^2 z = 1$$

Equation (5.231) takes the form

$$(\cosh \sqrt[4]{\lambda}\, l)(\cos \sqrt[4]{\lambda}\, l) = -1 \tag{5.232}$$

or

$$\cosh\beta \cos\beta = -1 \quad \text{where} \quad \beta = \sqrt[4]{\lambda}\, l \tag{5.233}$$

Equation (5.233) yields the eigenvalues to the problem. The equation can be solved numerically for β. Some of the roots to Equation (5.233) are

$$\beta_1 = 1.875$$

$$\beta_2 = 4.694$$

$$\beta_3 = 7.854$$

For $n > 3$ the eigenvalues can be obtained from the following approximate equation

$$\beta_n \approx (\pi/2)(2n - 1) \quad n > 3 \tag{5.234}$$

5.11 Vibration of a String with Concentrated Masses Attached to It

In this section we treat the vibration of a string to which concentrated masses M_i are attached at locations $x = x_i\,(i = 1, 2, \ldots)$; see Figure 5-19. For problems of this nature, the wave equation is written separately for each portion of the string stretched between two adjacent masses, that is, M_i and $M_{i\pm1}$. The general form of the solution for each region of the string "the segment between two masses" is then obtained by the method of separation of variables. However, the condiiions at $x = x_i$ are to be determined. We select for convenience and simplicity one mass attached to a string at a location $x = b$.

Since the string is continuous, the displacements u_1 and u_2 at $x = b$ should be equal. Hence, one condition is

$$u(b - 0,t) = u(b + 0,t)$$

or

$$u_1 = u_2 \quad \text{at } x = b \tag{5.235}$$

Another condition is obtained by applying Newton's Second Law for the Mass as follows; see Figure 5-20.

$$T \sin \theta_2 - T \sin \theta_1 = M \frac{\partial^2 u_M}{\partial t^2} \tag{5.236}$$

where u_M is equal to u at the location of the mass M. Because we are

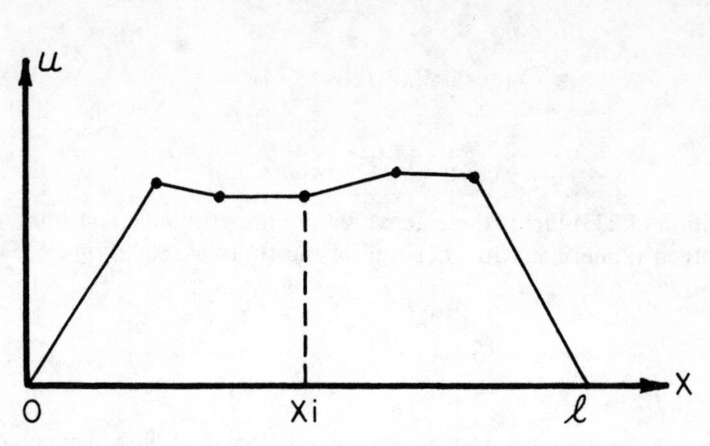

Figure 5-19. String with Concentrated Masses Attached to It

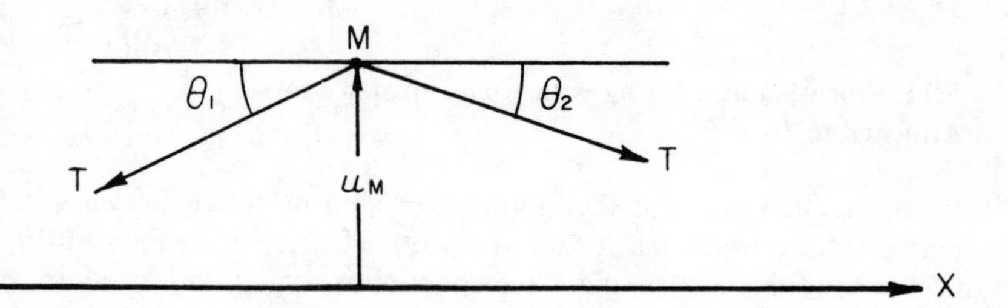

Figure 5-20. Forces on a String with Mass *M* Attached

invoking small displacements, we use the approximations stated in Equations (5.2) and (5.3), namely

$$\sin\theta_1 \approx \tan\theta_1 = \frac{\partial u_1}{\partial x} \tag{5.237}$$

$$\sin\theta_2 \approx \tan\theta_2 = \frac{\partial u_2}{\partial x} \tag{5.238}$$

Using Equations (5.237) and (5.238) in Equation (5.236) we obtain the second condition as

$$M\frac{\partial^2 u}{\partial x^2}(b,t) = T\left[\frac{\partial u_2}{\partial x} - \frac{\partial u_1}{\partial x}\right] \tag{5.239}$$

where $u(b,t)$ represents the displacement of the mass M at $x = b$.

Therefore, at the point b, where the concentrated load is applied, the displacement $u(x,t)$ remains continuous while the first derivative of the displacement $\partial u/\partial x$ becomes discontinuous. We can summarize the formulation of the problem as follows:

$$\frac{\partial^2 u_1}{\partial x^2} = \frac{1}{c^2}\frac{\partial^2 u_1}{\partial t^2}, \qquad 0 \le x \le b$$

$$u_1(0,t) = 0$$

$$u_1(b,t) = u_2(b,t)$$

$$M\frac{\partial^2 u_1}{\partial x^2}(b,t) = T\left[\frac{\partial u_2}{\partial x} - \frac{\partial u_1}{\partial x}\right]$$

$$u_1(x,0) = u(x,0) = f(x), \qquad 0 \le x \le b$$

$$\frac{\partial u_1}{\partial t}(x,0) = \frac{\partial u}{\partial t}(x,0) = g(x), \qquad 0 \le x \le b$$

$$\frac{\partial^2 u_2}{\partial x^2} = \frac{1}{c^2}\frac{\partial^2 u_2}{\partial t^2}, \qquad b \le x \le l$$

$$u_2(l,t) = 0$$

$$u_2(b,t) = u_1(b,t)$$

$$M\frac{\partial^2 u_2}{\partial x^2}(b,t) = T\left[\frac{\partial u_2}{\partial x} - \frac{\partial u_1}{\partial x}\right]$$

$$u_2(x,0) = u(x,0) = f(x), \qquad b \le x \le l$$

$$\frac{\partial u_2}{\partial t}(x,0) = \frac{\partial u}{\partial t}(x,0) = g(x), \qquad b \le x \le l$$

where $f(x)$ is the initial displacement and $g(x)$ is the initial velocity.

Example 5.8 Analyze the vibration of a string of length l to which a mass M is attached at the point $x = l/2$ (Figure 5-21).
Solution: The displacements u_1 and u_2 satisfy the following differential equations:

$$\frac{\partial^2 u_1}{\partial x^2} = \frac{1}{c^2}\frac{\partial^2 u_1}{\partial t^2}, \qquad \frac{\partial^2 u_2}{\partial x^2} = \frac{1}{c^2}\frac{\partial^2 u_2}{\partial t^2} \tag{5.240}$$

Using the method of separation of variables, the general forms of solution for u_1 and u_2 become (see Article 5.8)

$$u_{1,n}(x,t) = (A_1 \cos \lambda_n x + B_1 \sin \lambda_n x)(C_1 \cos \lambda_n ct + D_1 \sin \lambda_n ct) \tag{5.241}$$

and

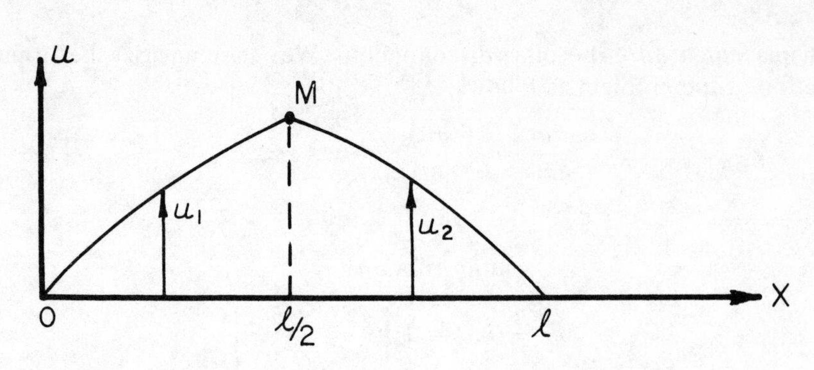

Figure 5-21. String with Mass M Attached at Location $l/2$

$$u_{2,n}(x,t) = (A_2 \cos \lambda_n x + B_2 \sin \lambda_n x)(C_1 \cos \lambda_n ct + D_1 \sin \lambda_n ct) \tag{5.242}$$

where λ_n is the same for both u_1 and u_2. This has to be true in order for the whole string to vibrate as one unit and not break apart. Applying the boundary condition at $x = 0$, namely

$$u(0,t) = 0 \quad \text{yields} \quad A_1 = 0$$

Therefore, we obtain for $u_{1,n}$

$$u_{1,n} = (C \cos \lambda_n ct + D \sin \lambda_n ct) \sin \lambda_n x \tag{5.243}$$

which can be written as

$$u_{1,n} = C_0 \cos (\lambda_n ct - \delta) \sin \lambda_n x \tag{5.244}$$

where

$$\delta = \tan^{-1} \frac{D}{C}. \tag{5.245}$$

Similarly, $u_{2,n}$, which satisfies $u(l,t) = 0$, can be written as

$$u_{2,n} = D_0 \cos (\lambda_n ct - \delta) \sin \lambda_n (l - x) \tag{5.246}$$

Using now

$$u_{1,n}[(l/2),t] = u_{2,n}[(1/2),t] \tag{5.247}$$

yields

$$C_0 \sin \frac{\lambda_n l}{2} = D_0 \sin \frac{\lambda_n l}{2} \tag{5.248}$$

or

$$C_0 = D_0 \tag{5.249}$$

Using also

$$T\left[\frac{\partial u_{2,n}}{\partial x} - \frac{\partial u_{1,n}}{\partial x}\right]_{l/2} = M\frac{\partial^2 u_{1,n}}{\partial t^2} = M\frac{\partial^2 u_{2,n}}{\partial t^2} \tag{5.250}$$

we get

$$TD_0\left[-\lambda_n \cos\frac{\lambda_n l}{2} - \lambda_n \cos\frac{\lambda_n l}{2}\right]\cos(\lambda_n ct - \delta)$$

$$= -D_0 M\lambda_n^2 c^2 \sin\frac{\lambda_n l}{2}\cos(\lambda_n ct - \delta) \tag{5.251}$$

or

$$2T\lambda_n \cos\frac{\lambda_n l}{2} = M\lambda_n^2 c^2 \sin\frac{\lambda_n l}{2}$$

or, with $T = c^2 m$ we have

$$\frac{2m}{M} = \lambda_n \tan\frac{\lambda_n l}{2} \tag{5.252}$$

The values of λ_n are obtained from this equation either numerically or graphically. Figure 5-22 shows a graphical determination of the roots of Equation (5.252). These roots are the points of intersection of the curves

$$y = \tan\frac{\lambda_n l}{2} \tag{5.253}$$

and

$$y = \frac{2m}{M\lambda_n} = \left(\frac{ml}{M}\right)\frac{1}{\lambda_n l/2} \tag{5.254}$$

in which $\lambda_n l/2$ is taken as the independent variable.

Therefore, the frequency of vibration is given by

$$f_n = \frac{\lambda_n c}{2\pi} \tag{5.255}$$

where λ_n are obtained from Equation (5.252).

To complete the solution of the problem, we have to specify the initial conditions. Let

$$u(x,0) = f(x) \tag{5.256}$$

$$\frac{\partial u}{\partial t}(x,0) = 0 \tag{5.257}$$

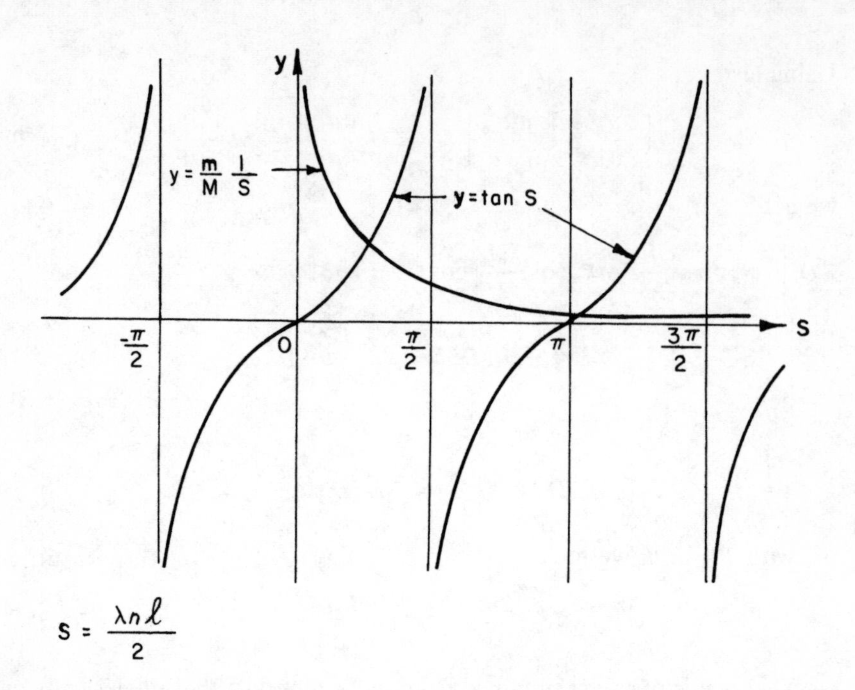

Figure 5-22. Graphical Determination of the Roots of Equation (5.252) in Example 5.8

To satisfy the initial conditions, the following series for Equations (5.244) and (5.246) are then considered

$$u_1(x,t) = \sum_{n=1}^{\infty} C_n \cos(\lambda_n ct - \delta) \sin \lambda_n x \tag{5.258}$$

and

$$u_2(x,t) = \sum_{n=1}^{\infty} D_n \cos(\lambda_n ct - \delta) \sin \lambda_n (l - x) \tag{5.259}$$

Applying now Equations (5.256) and (5.257) yields

$$f(x) = \sum_{n=1}^{\infty} C_n \sin \lambda_n x \qquad 0 < x < \frac{l}{2} \tag{5.260}$$

$$f(x) = \sum_{n=1}^{\infty} D_n \sin \lambda_n (l - x) \qquad \frac{l}{2} < x < l \tag{5.261}$$

The orthogonality conditions for the case of a loaded string[a] are given as follows for eigenfunctions $X_m(x)$

[a] A.N. Tikhonov and A.A. Samarskii, *Equations of Mathematical Physics* (New York: Pergamon Press, 1963), chap. 2, pp. 157-60.

$$\int_0^l X_n(x)\, X_m(x)\, w(x)\, dx \; + \; \sum_{i=1}^{k} M_i\, X_n(x_i)\, X_m(x_i) \; = \; 0 \qquad m \neq n \tag{5.262}$$

In this case, when a function $f(x)$ is expanded into a series of the form

$$f(x) \; = \; \sum_{n=1}^{\infty} C_n\, X_n(x) \tag{5.263}$$

the expansion coefficients are defined by

$$C_n \; = \; \frac{\displaystyle\int_0^l f(x)\, X_n(x)\, w(x)\, dx \; + \; \sum_{i=1}^{k} M_i f(x_i)\, X_n(x_i)}{\displaystyle\int_0^l X_n^2(x)\, w(x)\, dx \; + \; \sum_{i=1}^{k} M_i\, X_n^2(x_i)} \tag{5.264}$$

Therefore, for the present problem the coefficients C_n and D_n in Equations (5.260) and (5.261) can be represented as

$$C_n \; = \; \frac{\displaystyle\int_0^{l/2} f(x) \sin \lambda_n x\, dx \; + \; Mf\left(\frac{l}{2}\right) \sin \frac{\lambda_n l}{2}}{\displaystyle\int_0^{l/2} \sin^2 \lambda_n x\, dx \; + \; M \sin^2 \frac{\lambda_n l}{2}} \tag{5.265}$$

and

$$D_n \; = \; \frac{\displaystyle\int_{l/2}^{l} f(x) \sin \lambda_n (l - x)\, dx \; + \; Mf\left(\frac{l}{2}\right) \sin \frac{\lambda_n l}{2}}{\displaystyle\int_{l/2}^{l} \sin^2 \lambda_n (l - x)\, dx \; + \; M \sin^2 \frac{\lambda_n l}{2}} \tag{5.266}$$

Condition (5.257) is used now to evaluate δ in Equation (5.258) or in Equation (5.259). This results in $\delta = 0$. This can easily be deduced from Equation (5.243) when we set $\partial u_{1,n}/\partial t = 0$. This results in $D = 0$, and hence, from Equation (5.245), $\delta = 0$.

5.12 Solution in Terms of Traveling Waves—D'Alembert's Formula

In this section we present another important and well-known method of solution to the wave equation. The method lends itself to being better suited for infinite and semi-infinite systems. Let us take first the case of an infinite string subjected to certain initial disturbances. Our objective is to study and analyze the physical behavior of the string following the imposed initial conditions. No boundary conditions are required for this problem.

The differential equation and the initial conditions in this case are stated as

$$\frac{\partial^2 u}{\partial x^2} = \frac{1}{c^2}\frac{\partial^2 u}{\partial t^2} \tag{5.267}$$

$$u(x,0) = f(x) \tag{5.268}$$

$$\frac{\partial u}{\partial t}(x,0) = g(x) \tag{5.269}$$

We first like to transform Equation (5.267) into a canonical form involving mixed derivative (see Article 4.3, Equation (4.41)). In this case the characteristic equation (Equation (4.38)) becomes

$$dx^2 - c^2\,dt^2 = 0 \tag{5.270}$$

which splits into

$$(dx - c\,dt)(dx + c\,dt) = 0$$

or

$$dx - c\,dt = 0 \tag{5.271}$$

and

$$dx + c\,dt = 0 \tag{5.272}$$

Integrating Equations (5.271) and (5.272) yields

$$x - ct = \text{constant} = C_1 \tag{5.273}$$

and

$$x + ct = \text{constant} = C_2 \tag{5.274}$$

Hence, Equations (5.273) and (5.274), according to Article 4.3, yield the transformation variables that reduce Equation (5.267) into its canonical form with mixed derivative. Accordingly, we choose the new variables ξ and η as

$$\xi = x + ct \tag{5.275}$$

$$\eta = x - ct \tag{5.276}$$

Now the variables in the differential equation (Equation (5.267)) are changed from x and t to ξ and η as

$$\frac{\partial u}{\partial x} = \frac{\partial u}{\partial \xi}\frac{\partial \xi}{\partial x} + \frac{\partial u}{\partial \eta}\frac{\partial \eta}{\partial x} = \frac{\partial u}{\partial \xi} + \frac{\partial u}{\partial \eta} \tag{5.277}$$

$$\frac{\partial^2 u}{\partial x^2} = \frac{\partial}{\partial x}\left(\frac{\partial u}{\partial x}\right)$$

$$= \frac{\partial}{\partial \xi}\left(\frac{\partial u}{\partial x}\right) + \frac{\partial}{\partial \eta}\left(\frac{\partial u}{\partial x}\right)$$

$$= \frac{\partial}{\partial \xi}\left(\frac{\partial u}{\partial \xi} + \frac{\partial u}{\partial \eta}\right) + \frac{\partial}{\partial \eta}\left(\frac{\partial u}{\partial \xi} + \frac{\partial u}{\partial \eta}\right)$$

$$= \frac{\partial^2 u}{\partial \xi^2} + 2\frac{\partial^2 u}{\partial \xi\,\partial \eta} + \frac{\partial^2 u}{\partial \eta^2} \tag{5.278}$$

In a similar way we find

$$\frac{\partial^2 u}{\partial t^2} = \frac{\partial}{\partial t}\left(\frac{\partial u}{\partial \xi}\frac{\partial \xi}{\partial t} + \frac{\partial u}{\partial \eta}\frac{\partial \eta}{\partial t}\right)$$

$$= \frac{\partial}{\partial t}\left(c\frac{\partial u}{\partial \xi} - c\frac{\partial u}{\partial \eta}\right)$$

$$= c\frac{\partial}{\partial \xi}\left(c\frac{\partial u}{\partial \xi} - c\frac{\partial u}{\partial \eta}\right) - c\frac{\partial}{\partial \eta}\left(c\frac{\partial u}{\partial \xi} - c\frac{\partial u}{\partial \eta}\right)$$

$$= c^2\left\{\frac{\partial^2 u}{\partial \xi^2} - 2\frac{\partial^2 u}{\partial \xi\,\partial \eta} + \frac{\partial^2 u}{\partial \eta^2}\right\} \tag{5.279}$$

Using Equations (5.278) and (5.279) in Equation (5.267) yields

$$\frac{\partial^2 u}{\partial \xi\,\partial \eta} = 0 \tag{5.280}$$

which is the required canonical form. Integrating Equation (5.280) with respect to ξ gives

$$\frac{\partial u}{\partial \eta} = F_1(\eta) \tag{5.281}$$

where $F_1(\eta)$ is some unknown variable of η. Integrating now Equation (5.281) with respect to η results in

$$u(\xi,\eta) = \int F_1(\eta)\,d\eta + G(\xi)$$

or

$$u(\xi,\eta) = F(\eta) + G(\xi) \tag{5.282}$$

where we set

$$\int F_1(\eta)\,d\eta = F(\eta)$$

In terms of the original variables x and t, Equation (5.282) becomes

$$u(x,t) = F(x - ct) + G(x + ct) \tag{5.283}$$

This form of solution is referred to as D'Alembert solution. It represents two waves, namely:

$F(x - ct)$ a wave moving in the positive x direction with a speed c

and

$G(x + ct)$ a wave moving in the negative x direction with a speed c

The specific forms of these waves depend upon the shapes of $F(x - ct)$ and $G(x + ct)$, which are determined from the imposed initial conditions for the problem. We note that in Article 5.8, Equation (5.109), we arrived at a similar form for the solution to the wave equation by the method of separation of variables, and we discussed the meaning of such a solution.

It remains now to satisfy the initial conditions for the problem as given by Equations (5.268) and (5.269). Applying these conditions we obtain

$$u(x,0) = \overset{\text{known}}{f(x)} = F(x) + G(x) \tag{5.284}$$

$$\frac{\partial u}{\partial t}(x,0) = \overset{\text{known}}{g(x)} = -c\,F'(x) + c\,G'(x) \tag{5.285}$$

where the prime designates differentiation with respect to time. Integrating Equation (5.285) once and dividing by c yields

$$G(x) - F(x) = \frac{1}{c}\int_{x_0}^{x} g(z)\,dz + \text{constant} \tag{5.286}$$

where x_0 is a constant. Equations (5.284) and (5.286) when solved simultaneously give the expressions for $G(x)$ and $F(x)$ as

$$G(x) = \frac{1}{2}f(x) + \frac{1}{2c}\int_{x_0}^{x} g(z)\,dz + \frac{\text{constant}}{2} \tag{5.287}$$

$$F(x) = \frac{1}{2}f(x) - \frac{1}{2c}\int_{x_0}^{x} g(z)\,dz - \frac{\text{constant}}{2} \tag{5.288}$$

Equations (5.287) and (5.288) hold for any value of the argument. Therefore, the solution for $u(x,t)$ as given by Equation (5.283) becomes

$$u(x,t) = \frac{1}{2}[f(x+ct) + f(x-ct)]$$

$$+ \frac{1}{2c}\left[\int_{x_0}^{x+ct} g(z)\,dz - \int_{x_0}^{x-ct} g(z)\,dz\right] \qquad (5.289a)$$

or

$$u(x,t) = \frac{1}{2}[f(x+ct) + f(x-ct)] + \frac{1}{2c}\int_{x-ct}^{x+ct} g(z)\,dz \qquad (5.289b)$$

Equation (5.289) is referred to as D'Alembert Formula or D'Alembert solution to the wave equation.

Physical Interpretation of the Solution

When we examine the solution as given by Equation (5.289), we see that it is made up of two functions, namely:

$$u_D(x,t) = \frac{1}{2}[f(x+ct) + f(x-ct)] \qquad (5.290)$$

and

$$u_V(x,t) = \frac{1}{2c}\int_{x-ct}^{x+ct} g(z)\,dz \qquad (5.291)$$

$u_D(x,t)$ represents waves propagating in the medium due only to the known initial displacement $f(x)$ with zero initial velocity.

$u_V(x,t)$ represents waves propagating in the medium due only to the known initial velocity $g(x)$ with zero initial displacement, such as a blow or an impulse imparted to a string.

The displacement u_D is made up of half the sum of the two waves propagating in opposite directions.

The displacement u_V is made up of the difference of the integral of the initial velocity $g(z)$ divided by $2c$. This is illustrated in the following example.

Example 5.9 A very long string (considered to be infinite) is hit by a hammer of width $2b$ with an initial velocity V_0 and no initial displacement. Study the subsequent vibration of the string (Figure 5-23).

Solution: From Equation (5.289b) the displacement $u(x,t)$ is given by

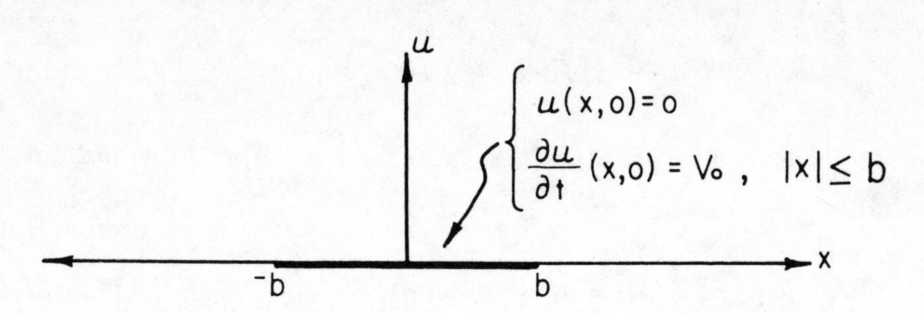

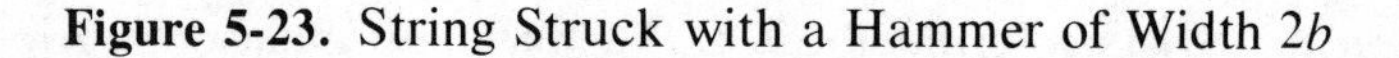

Figure 5-23. String Struck with a Hammer of Width $2b$

$$u(x,t) = u_V(x,t) = \frac{1}{2c}\int_{x-ct}^{x+ct} g(z)\,dz \tag{5.292}$$

or

$$u(x,t) = \frac{1}{2}\,[G(x+ct) - G(x-ct)] \tag{5.293}$$

where

$$G(x) = \frac{1}{c}\int_{x_0}^{x} g(z)\,dz \tag{5.294}$$

For this problem we have:

$$g(x) = \begin{cases} 0 & |x| > b \\ V_0 & |x| \le b \end{cases} \tag{5.295}$$

Therefore, $G(x)$ becomes (Figure 5-24)

$$G(x) = \begin{cases} \dfrac{1}{c}\displaystyle\int_{-b}^{x} V_0\,dz = \dfrac{V_0}{c}(x+b) & -b \le x \le b \qquad (5.296) \\[2ex] \dfrac{1}{c}\displaystyle\int_{-b}^{x} V_0\,dz = \dfrac{V_0}{c}(2b) & x > b \qquad (5.297) \end{cases}$$

Here, the two waves travel in opposite directions on the x-axis and the displacement $u(x,t)$ is equal to half the difference between the local amplitudes of the waves as given by Equation (5.293); (in the case of the propagation of initial displacement, $u(x,t)$ as given by Equation (5.290) is equal to half the sum of the local amplitudes of the waves). The graph of $u(x,t)$ at various times is obtained by summing the two propagating waves as shown in Figure 5-25. One might ask as to what happens when t approaches infinity? The answer is that for very large time "$t \to \infty$" the string

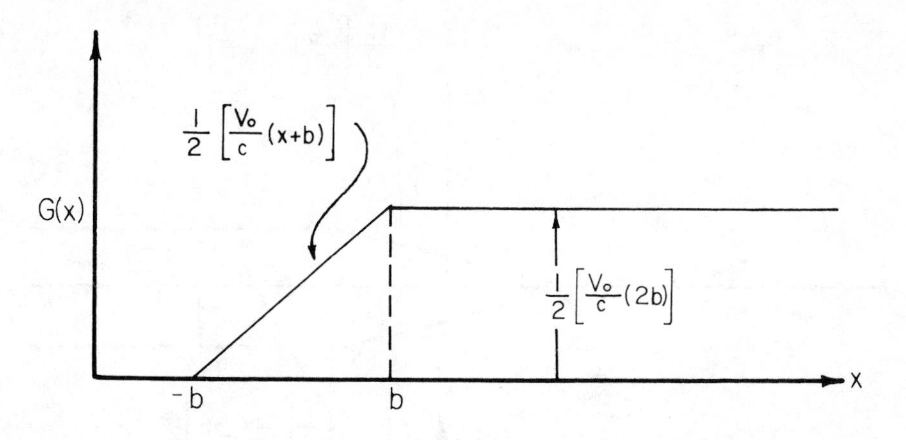

Figure 5-24. Form of Deflection $G(x)$ in Example 5.9

reaches a state of rest "no traveling waves." However, it will not in general assume its original position but it will exhibit what is referred to as a residual displacement.

Example 5.10 Consider two semi-infinite bars as shown in Figure 5-26. Both bars have constant, but different, Young's modulus of elasticity E and constant density ρ, and they have the same shape and cross-sectional area. The bars are welded together at $x = 0$. Let a right-running incident wave generated in bar I by some blow be represented by

$$u(x,t) = f_i(t - (x/c_1)) \tag{5.298}$$

approaches the junction of the two bars. Find: (a) the expression for the transmitted wave in bar II, (b) the expression for the reflected wave in bar I that might be produced by the presence of the interface, and (c) determine under what conditions the reflected wave will be absent.

Solution: Let the displacement in the two bars be represented by $u_1(x,t)$ and $u_2(x,t)$. Because the two bars are rigidly welded together, the displacement of the two bars at the interface should be the same, that is,

$$u_1(0^-,t) = u_2(0^+,t) \tag{5.299}$$

where 0^- represents the limit of x approaching the origin $x = 0$ from the left and always remains to the left of $x = 0$. 0^+ represents the limit of x approaching the origin $x = 0$ from the right and always remains to the right of $x = 0$.

Another condition that should be satisfied at the interface is that the force must be continuous. Hence, because the cross-sectional areas of the two bars are equal, we can write

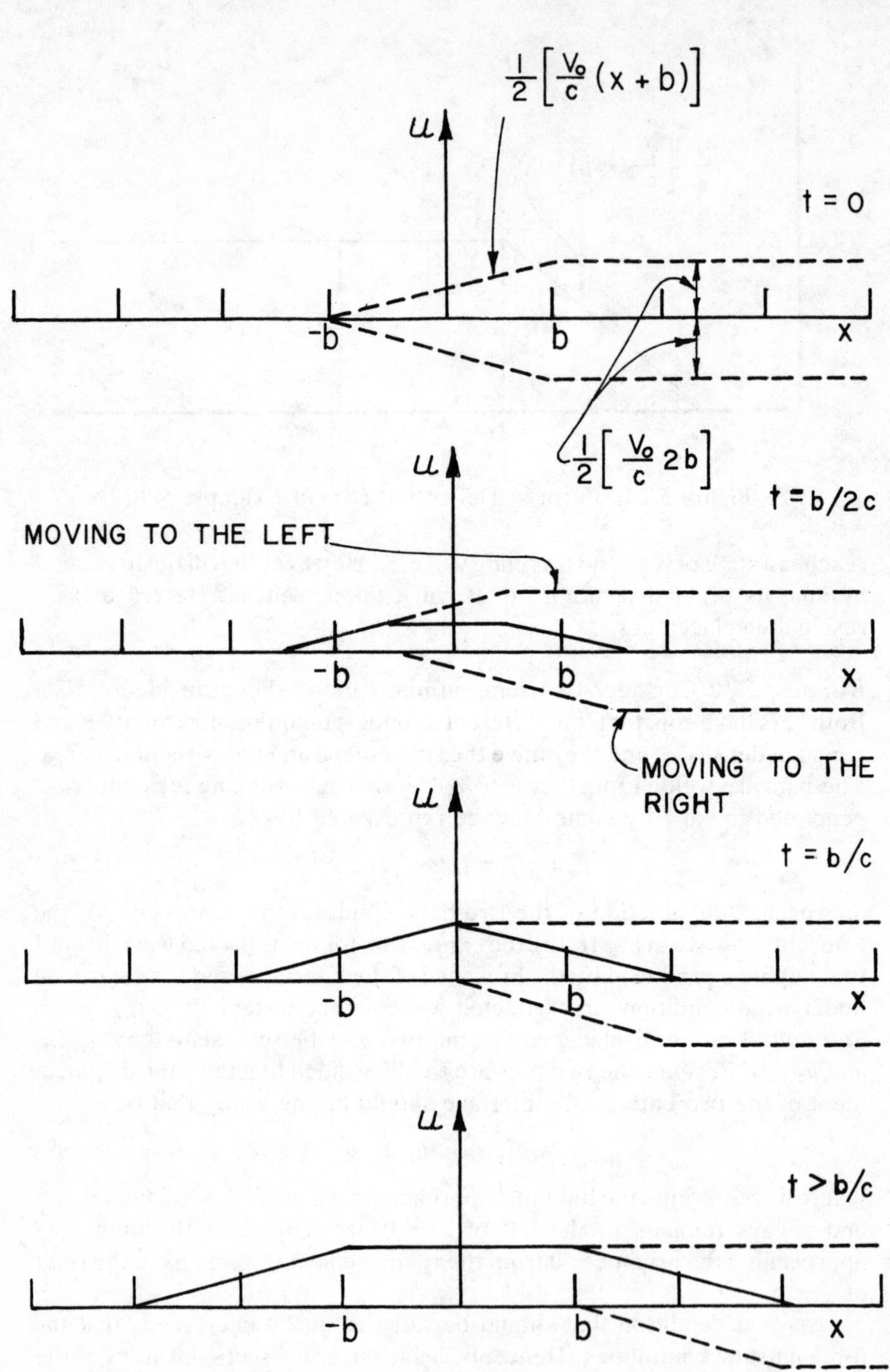

Figure 5-25. Graphical Interpretation of Deflection $u(x,t)$ in Example 5.9

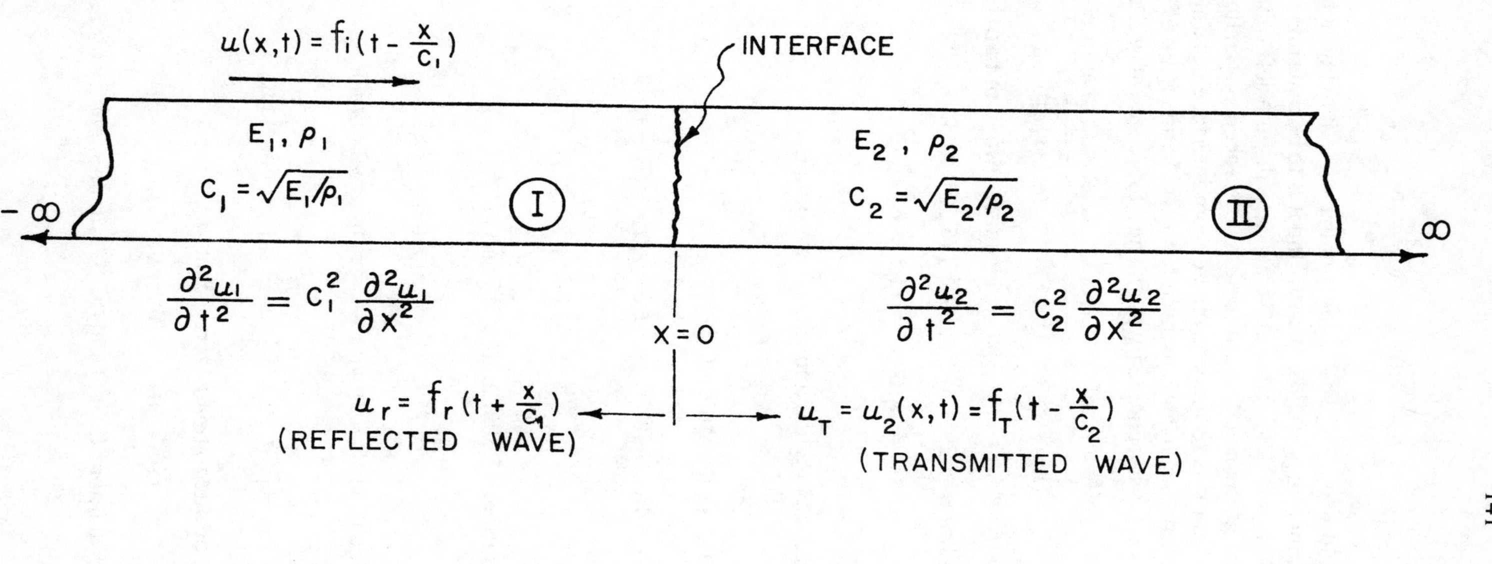

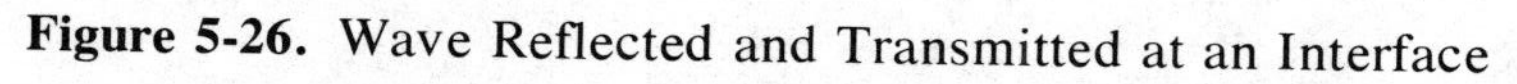

Figure 5-26. Wave Reflected and Transmitted at an Interface

$$E_1 \frac{\partial u_1}{\partial x}(0^-,t) = E_2 \frac{\partial u_2}{\partial x}(0^+,t) \tag{5.300}$$

The transmitted wave will be a right-running wave moving in bar II with speed c_2; the function f_T is still unknown and it is to be determined from the interface conditions. The reflected wave u_r will be a left-running wave moving in bar I in the negative x direction with speed c_1; the function f_r is still unknown and it is to be determined from interface conditions.

The instantaneous displacement in bar I will be made up of the sum of the incident wave and reflected wave. Therefore, we can write

$$u_1(x,t) = \overbrace{f_i(t - (x/c_1))}^{\text{known incident wave}} + \overbrace{f_r(t + (x/c_1))}^{\text{reflected wave in bar I}} \tag{5.301}$$

$$u_2(x,t) = f_T(t - (x/c_2)) \quad \text{transmitted wave in bar II} \tag{5.302}$$

Applying condition (5.299) to Equations (5.301) and (5.302) gives

$$u_1(0^-,t) = f_i(t) + f_r(t) \tag{5.303}$$

$$u_2(0^+,t) = f_T(t) \tag{5.304}$$

Equating Equations (5.303) and (5.304) we obtain

$$f_i(t) + f_r(t) = f_T(t) \tag{5.305}$$

Applying condition (5.300) now yields

$$E_1 \frac{\partial u_1}{\partial x}(0^-,t) = -E_1 \frac{1}{c_1} f_i'(t) + E_1 \frac{1}{c_1} f_r'(t) \tag{5.306}$$

$$E_2 \frac{\partial u_2}{\partial x}(0^+,t) = -E_2 \frac{1}{c_2} f_T'(t) \tag{5.307}$$

Equating Equations (5.306) and (5.307) gives

$$\frac{-E_2}{c_2} f_T'(t) = \frac{-E_1}{c_1}[f_i'(t) - f_r'(t)] \tag{5.308}$$

Integrating Equation (5.308) with respect to x we obtain

$$\frac{E_1}{c_1}[f_i(t) - f_r(t)] = \frac{E_2}{c_2} f_T(t) + \overset{0}{\cancel{\text{constant}}} \tag{5.309}$$

The constant of integration is zero because

$$f_i(0) = 0, \quad f_r(0) = 0, \quad f_T(0) = 0$$

Solving now Equations (5.305) and (5.309) simultaneously for $f_r(t)$ and $f_T(t)$ we get

$$f_T(t) = \frac{2c_2E_1}{c_1E_2 + c_2E_1} f_i(t) \tag{5.310}$$

and

$$f_r(t) = f_T - f_i = \frac{c_2E_1 - c_1E_2}{c_1E_2 + c_2E_1} f_i(t) \tag{5.311}$$

If we let $c_1E_2/c_2E_1 \equiv R$, we have then

$$f_T(t) = \frac{2}{1 + R} f_i(t) \tag{5.312}$$

and

$$f_r(t) = \frac{1 - R}{1 + R} f_i(t) \tag{5.313}$$

Therefore,

1. The expression for the transmitted wave is

$$u_2(x,t) = f_T(x,t) = \frac{2}{1 + R} f_i\left(t - \frac{x}{c_2}\right) \tag{5.314}$$

2. The expression for the reflected wave is

$$u_r(x,t) = f_r(x,t) = \frac{1 - R}{1 + R} f_i\left(t + \frac{x}{c_1}\right) \tag{5.315}$$

3. The deflection in bar I becomes

$$\begin{aligned} u_1(x,t) &= f_i(x,t) + f_r(x,t) \\ &= f_i\left(t - \frac{x}{c_1}\right) + \frac{1 - R}{1 + R} f_i\left(t + \frac{x}{c_1}\right) \end{aligned} \tag{5.316}$$

4. The reflected wave will be absent when

$$f_r(x,t) = 0$$

or

$$R = 1$$

which means

$$c_1E_2 = c_2E_1$$

or

$$\sqrt{\frac{E_1}{\rho_1}} E_2 = \sqrt{\frac{E_2}{\rho_2}} E_1$$

or

$$\rho_1 E_1 = \rho_2 E_2 \tag{5.317}$$

When $R = 1$ the transmitted wave will be identical to the incident wave and no reflection occurs in bar I.

If the incident wave is a sine wave such as

$$f_i(x,t) = \sin\left(t - \frac{x}{c_1}\right) \tag{5.318}$$

the transmitted wave and the reflected wave become respectively

$$u_2(x,t) = \frac{2}{1 + R} \sin\left(t - \frac{x}{c_2}\right) \tag{5.319}$$

$$u_r(x,t) = \frac{1 - R}{1 + R} \sin\left(t + \frac{x}{c_1}\right) \tag{5.320}$$

Example 5.11 Suppose the parameters pertaining to a very long transmission line are such that

$$R/L = G/C$$

This relation implies that the line is distortionless. Find the expression for the voltage in the line.

Solution: The governing differential equation for the problem is Equation (5.41) which is

$$\frac{\partial^2 V}{\partial x^2} = LC \frac{\partial^2 V}{\partial t^2} + (LG + RC)\frac{\partial V}{\partial t} + RGV \tag{5.321}$$

With $R/L = G/C$ and $c^2 = 1/LC$, Equation (5.321) becomes

$$\frac{\partial^2 V}{\partial x^2} = \frac{1}{c^2}\frac{\partial^2 V}{\partial t^2} + 2RC\frac{\partial V}{\partial t} + RGV \tag{5.322}$$

To solve Equation (5.322) we introduce a new variable $Z(x,t)$ defined as

$$V(x,t) = e^{-\sqrt{(RG/LC)}t}\, Z(x,t) = e^{-(R/L)t}\, Z(x,t) \tag{5.323}$$

Introducing Equation (5.323) in Equation (5.322) yields

$$\frac{\partial^2 Z}{\partial x^2} = \frac{1}{c^2}\frac{\partial^2 Z}{\partial t^2} \tag{5.324}$$

Equation (5.324) has the familiar solution

$$Z(x,t) = F_1(x - ct) + F_2(x + ct) \tag{5.325}$$

Therefore, from Equation (5.323) the solution for $V(x,t)$ becomes

$$V(x,t) = e^{-(R/L)t}[F_1(x - ct) + F_2(x + ct)] \tag{5.326}$$

In this case the solution for $V(x,t)$ is made up of two waves propagating without distortion in opposite directions with velocity c. The waves are attenuated by the factor $e^{-Rt/L}$. The functions $F_1(x,t)$ and $F_2(x,t)$ are to be determined from the imposed initial conditions.

Example 5.12 In steady supersonic flow and when the flow is considered irrotational, the perturbation velocity in the fluid is related to a potential function ϕ as $u = \nabla\phi$. The function ϕ, for two-dimensional cases satisfies the following wave equation:

$$(M^2 - 1)\frac{\partial^2\phi}{\partial x^2} = \frac{\partial^2\phi}{\partial y^2} \tag{5.327}$$

The fluid then is considered to have a uniform velocity U_0 plus a velocity of perturbation. Find the expression for the fluid velocity in supersonic flow above a corrugated surface defined in the xy plane by (Figure 5.27)

$$y = Z(x) \tag{5.328}$$

and on which the vertical component of the velocity is approximated by

$$\frac{\partial\phi}{\partial y}(x,0) = U_0\frac{dZ}{dx} \tag{5.329}$$

Solution: In this problem we are interested in the region above the surface in which the characteristics in the xy plane have a positive slope. Therefore, the form of the solution for $\phi\ (x,y)$ becomes

$$\phi(x,y) = F(x - cy) \tag{5.330}$$

where

$$c = \sqrt{M^2 - 1} \tag{5.331}$$

Applying condition (5.329) we obtain

$$\frac{\partial\phi}{\partial y}(x,0) = -c\,F'(x) = U_0\frac{dZ}{dx} \tag{5.332}$$

or

$$F'(x) = -\frac{U_0}{c}\frac{dZ}{dx} \tag{5.333}$$

Integrating Equation (5.333) yields

$$F(x) = -\frac{U_0}{c}Z(x) + \text{constant} \tag{5.334}$$

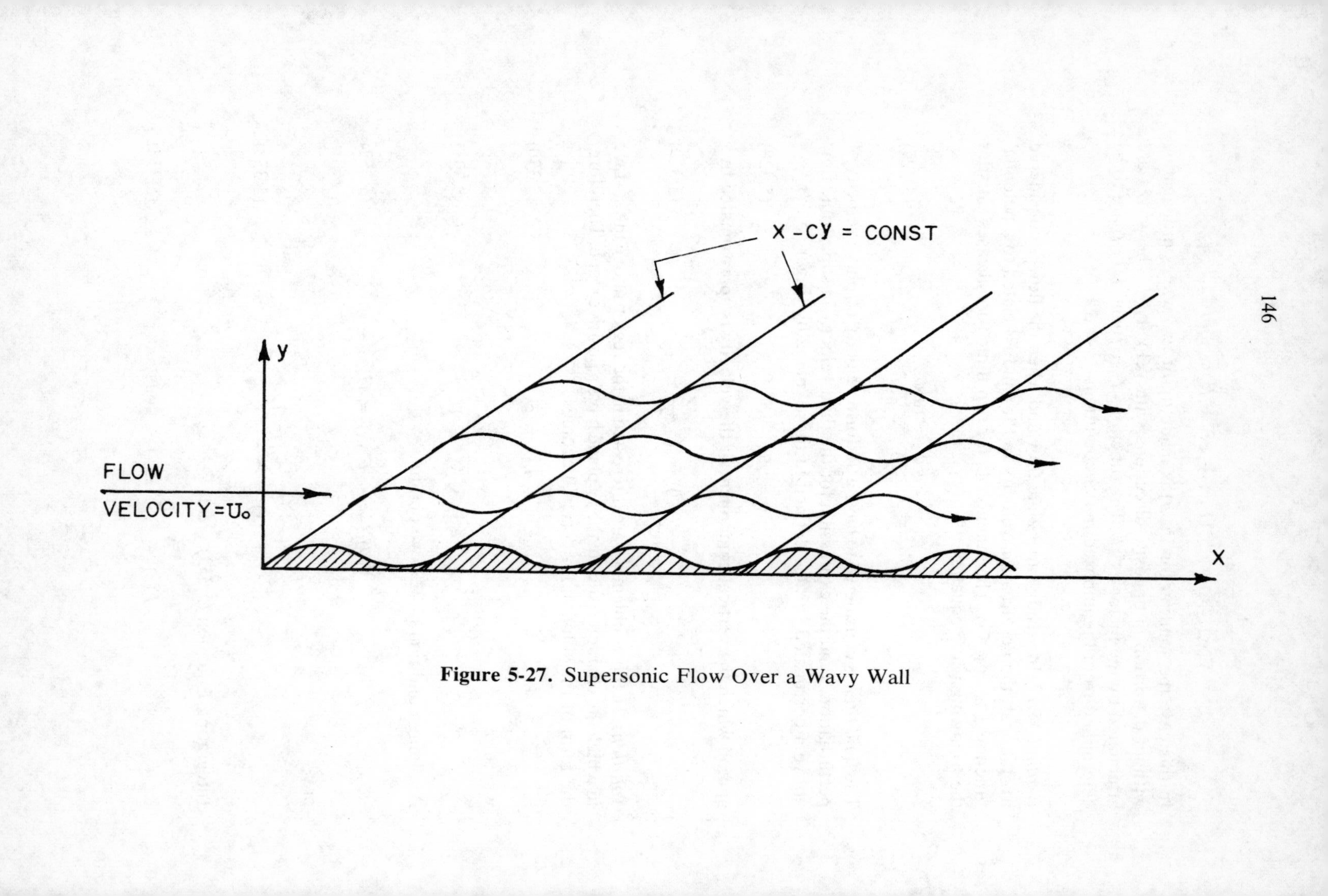

Figure 5-27. Supersonic Flow Over a Wavy Wall

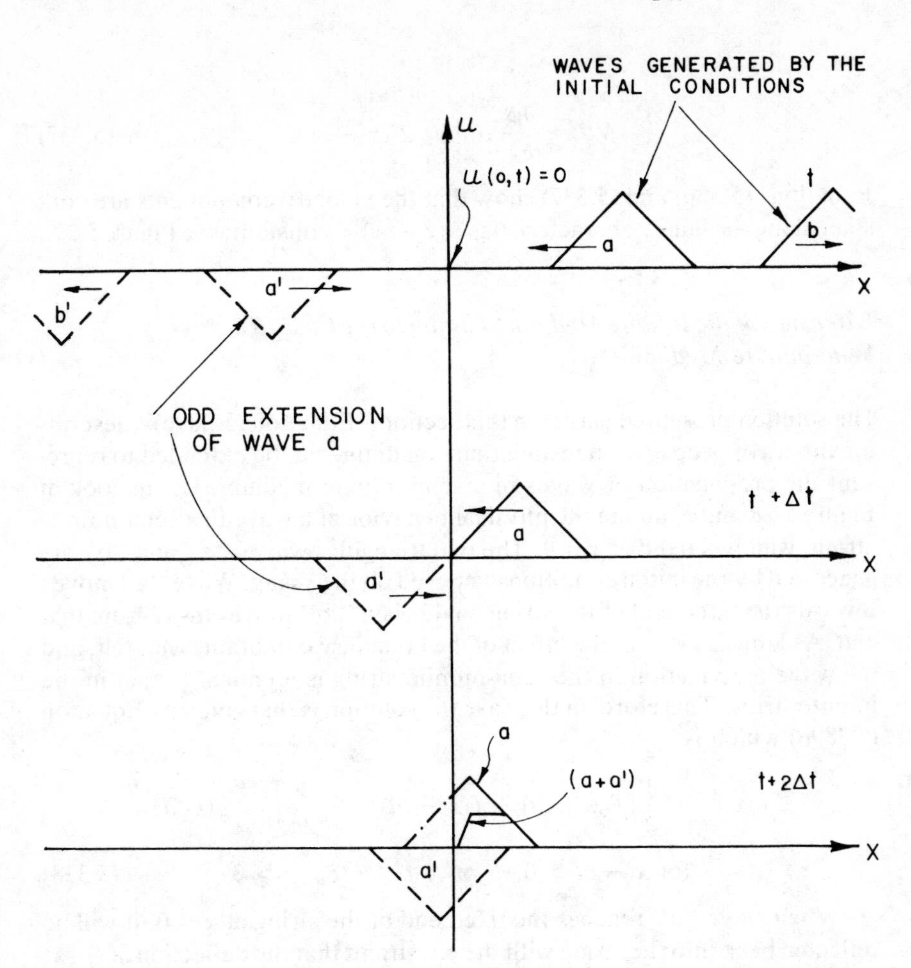

Figure 5-28. Triangular Wave Propagating in a Semi-infinite String Fixed at $x = 0$

Hence,

$$\phi(x,y) = F(x - cy) = -(U_0/c)\, Z(x - cy) + \text{constant} \qquad (5.335)$$

The constant in Equations (5.334) and (5.335) is irrelevant and can be ignored. The velocity components in the x and y directions become

$$V_x = U_0 + \frac{\partial \phi}{\partial x} = U_0 - \frac{U_0}{c} Z'(x - cy)$$

$$= U_0 \left[1 - \frac{Z'}{c}(x - cy) \right] \qquad (5.336)$$

and

$$V_y = \frac{\partial \phi}{\partial y} = U_0 \, Z'(x - cy) \tag{5.337}$$

Equations (5.336) and (5.337) show that the velocity components are constant along the lines "characteristics" $(x - cy) =$ constant; see Figure 5-27.

Extension of the Infinite Medium Solution to the Case of a Semi-infinite Medium

The solution presented earlier in this section "Equation (5.289b)" describing the wave propagation in an infinite medium, can be extended to represent the propagation of waves in a semi-infinite medium. Let us look at Figure 5-28 and examine the physical behavior of a wave in a semi-infinite string, which is fixed at $x = 0$. The two triangular waves "a" and "b" are generated by the initial conditions imposed on the string. Wave "a" moves towards the fixed end of the string while wave "b" moves away from that end. As long as $t < x/c$, the effect of the boundary condition is not felt, and the wave propagation in the semi-infinite string is identical to that in the infinite string. Therefore, in this case the solution is that given by Equation (5.289b) which is

$$u(x,t) = \frac{1}{2}[f(x + ct) + f(x - ct)] + \frac{1}{2c}\int_{x-ct}^{x+ct} g(z)\, dz$$

$$\text{for } x - ct > 0, \quad \text{or} \quad t < x/c; \;\; x > 0 \tag{5.338}$$

When wave "a" reaches the fixed end of the string at $x = 0$, it will be reflected back into the string with the constraint that the deflection $u(x,t)$ at $x = 0$ should remain zero. An odd extension of waves "a" and "b" as shown in Figure 5-28 by the dashed lines accomplishes that. This can be demonstrated as follows:

Let the initial conditions be

$$u(x,0) = f(x), \qquad \frac{\partial u}{\partial t}(x,0) = g(x)$$

The odd extension of the initial conditions is

$$\left. u(x,0) \right]_{\text{odd extension}} = -f(-x), \qquad \left. \frac{\partial u}{\partial t}(x,0) \right]_{\text{odd extension}} = -g(-x)$$

Therefore, if we evaluate the deflection $u(x,t)$ at $x = 0$ as the sum of the waves propagating in an infinite string "a and b" and their odd extension "a′ and b′", we obtain from Equation (5.338) the following

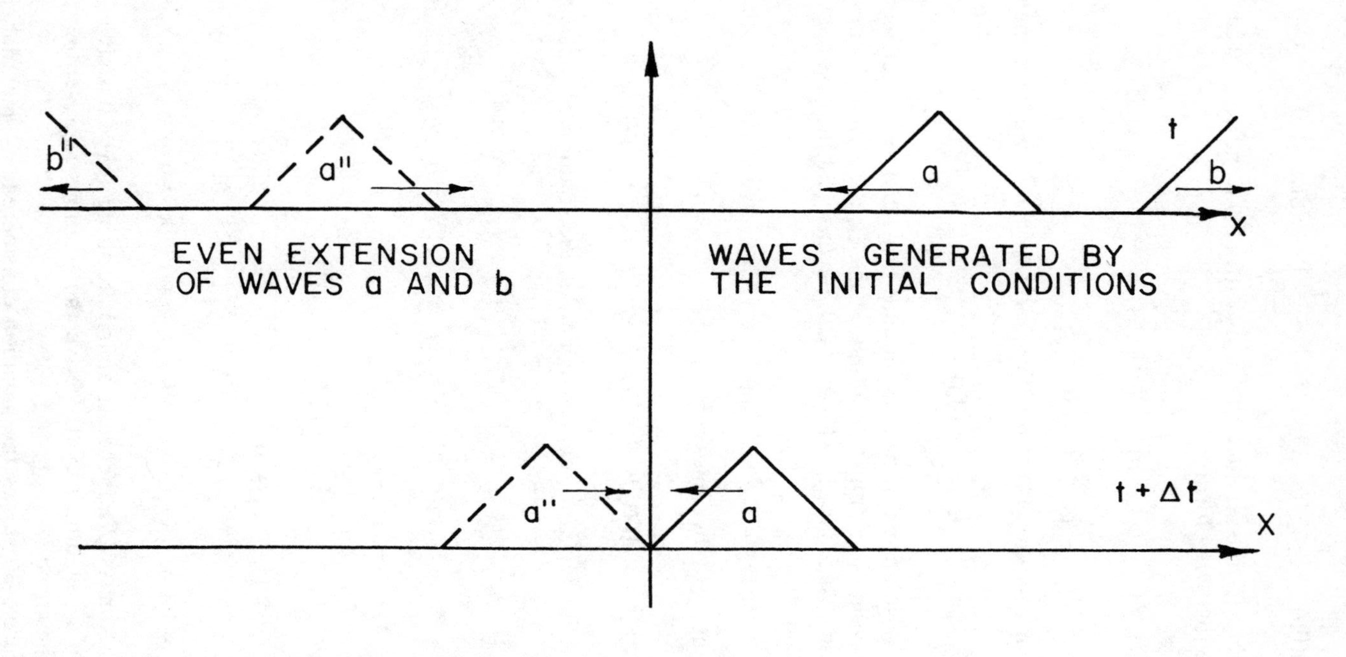

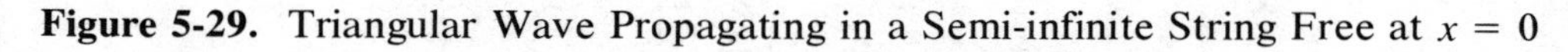

Figure 5-29. Triangular Wave Propagating in a Semi-infinite String Free at $x = 0$

$$u(0,t) = \frac{1}{2}[f(ct) + f(-ct)] + \frac{1}{2c}\int_{-ct}^{ct} g(z)\,dz = 0$$

The solution then for the semi-infinite string with a fixed end at $x = 0$ can be written in the following form

$$u(x,t) = \frac{1}{2}[f(x + ct) - f(ct - x)] + \frac{1}{2c}\int_{ct-x}^{x+ct} g(z)\,dz$$

$$\text{for } x - ct < 0, \quad \text{or} \quad t > x/c; \; x > 0 \tag{5.339}$$

If the initial condition is even with respect to $x = 0$, we have then

$$\frac{\partial u}{\partial x}(0,t) = 0$$

In this case, the solution for the semi-infinite string can be written as the sum of the waves propagating in an infinite string "a and b" and their even extension (a″ and b″) as shown in Figure 5-29. Because, for the case of even extension we have

$$f(x) = f(-x) \quad \text{and} \quad g(x) = g(-x)$$

the condition at $x = 0$ becomes

$$\frac{\partial u}{\partial x}(0,t) = \frac{1}{2}[f'(ct) + f'(-ct)] + \frac{1}{2c}[g(ct) - g(-ct)] = 0$$

Therefore, for this type of boundary condition the solution becomes

$$u(x,t) = \frac{1}{2}[f(x + ct) + f(x - ct)] + \frac{1}{2c}\int_{x-ct}^{x+ct} g(z)\,dz$$

$$\text{for } x - ct > 0, \quad \text{or} \quad t < x/c; \; x > 0 \tag{5.340}$$

and

$$u(x,t) = \frac{1}{2}[f(x + ct) + f(ct - x)] + \frac{1}{2c}\left[\int_{0}^{x+ct} g(z)\,dz + \int_{0}^{ct-x} g(z)\,dz\right]$$

$$\text{for } x - ct < 0, \quad \text{or} \quad t > x/c; \; x > 0 \tag{5.341}$$

Example 5.13 A mass M attached to an infinite string at $x = 0$ is given an initial velocity V_0 at time $t > 0$. Derive the relations representing the vibration of the string.

Solution: Figure 5-30 shows the situation considered; we have labeled $u_1(x,t)$ and $u_2(x,t)$ to be the displacements on either side of the mass. When

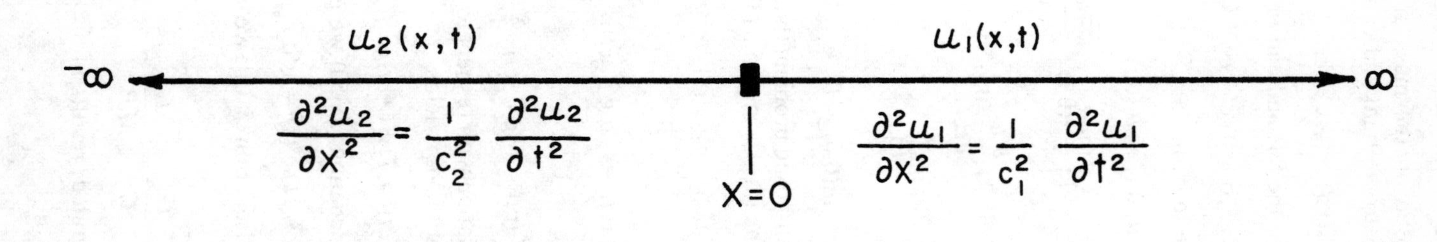

Figure 5-30. Infinite String with Mass Attached at $x = 0$

an initial condition is imposed on the mass at $x = 0$, two waves are generated. One wave moves to the right and can be represented by

$$u_1(x,t) = g(x - ct) \tag{5.342}$$

and the other wave moves to the left and can be represented by

$$u_2(x,t) = f(x + ct) \tag{5.343}$$

The following conditions are to be met:

$$u_2(0,t) = u_1(0,t) \tag{5.344}$$

$$M\frac{\partial^2 u_1}{\partial t^2}(0,t) = T\left[\frac{\partial u_1}{\partial x}(0,t) - \frac{\partial u_2}{\partial x}(0,t)\right]$$

$$= M\frac{\partial^2 u_2}{\partial t^2}(0,t) \tag{5.345}$$

$$\frac{\partial u_2}{\partial t}(0,t) = \frac{\partial u_1}{\partial t}(0,t) = V_0 \tag{5.346}$$

$$u_2(0,0) = u_1(0,0) = 0 \tag{5.347}$$

Therefore, we have now a well-posed problem that we can solve. Applying condition (5.344) to Equations (5.342) and (5.343) gives

$$f(ct) = g(-ct) \tag{5.348}$$

and

$$c\,f'(ct) = -c\,g'(-ct) \tag{5.349a}$$

or

$$f'(ct) = -g'(-ct) \tag{5.349b}$$

where prime means differentiation with respect to the argument. Applying condition (5.345) to Equation (5.342) yields

$$Mc^2 g''(-ct) = T[g'(-ct) - f'(ct)] \tag{5.350}$$

Using Equation (5.349b) in Equation (5.350) we get

$$(Mc^2/2T)\,g''(z) = g'(z) \tag{5.351}$$

where we set $z = ct$. To solve Equation (5.351) we let $G = g'$; then Equation (5.351) becomes

$$\frac{G'}{G} = \frac{2T}{Mc^2} \tag{5.352}$$

Equation (5.352), when integrated, results in

$$G = C_1 e^{(2T/Mc^2)z} \tag{5.353}$$

The constant of integration C_1 is determined from condition (5.346) as

$$G(0) = g'(0) = -\frac{V_0}{c} = C_1 \tag{5.354}$$

Therefore

$$G(z) = g'(z) = -\frac{V_0}{c} e^{(2T/Mc^2)z} \tag{5.355}$$

Integrating Equation (5.355) results in

$$g(z) = -\frac{V_0}{c} \int e^{(2T/Mc^2)z}\, dz + C_2$$

or

$$g(z) = -\frac{V_0 Mc}{2T} e^{(2T/Mc^2)z} + C_2 \tag{5.356}$$

To evaluate the constant of integration C_2 we use condition (5.347) and obtain

$$g(0,0) = 0 = -\frac{V_0\, Mc}{2T} + C_2$$

or

$$C_2 = \frac{V_0 Mc}{2T} \tag{5.357}$$

Therefore, $g(x,t)$ can be obtained from Equations (5.356) and (5.357) and is written as

$$u_1(x,t) = g(x - ct) = \frac{McV_0}{2T}[1 - e^{(2T/Mc^2)(x-ct)}] \tag{5.358}$$

From Equation (5.348) we have

$$g(-z) = f(z)$$

and

$$g(-x - ct) = f(x + ct) \tag{5.359}$$

Therefore, the solution for $u_2(x,t)$ becomes

$$u_2(x,t) = f(x + ct) = \frac{McV_0}{2T}[1 - e^{-(2T/Mc^2)(x+ct)}] \tag{5.360}$$

The displacement in the string is shown in Figure 5-31 as sketched according to Equations (5.358) and (5.360). The locations of x_1 and x_2 are obtained by setting Equations (5.358) and (5.360) respectively equal to zero. Hence, from Equation (5.358) we obtain

$$0 = \frac{McV_0}{2T}[1 - e^{(2T/Mc^2)(x_1-ct)}]$$

or

$$x_1 - ct = 0$$

or

$$x_1 = ct \tag{5.361}$$

Similarly for x_2, from Equation (5.360) we obtain

$$x_2 = -ct \tag{5.362}$$

Equations (5.361) and (5.362) show that the wave fronts x_1 and x_2 move to the right and to the left respectively with velocity c.

The variation with time of the deflection at $x = 0$ is obtained by setting $x = 0$ in either Equation (5.358) or (5.360). The result is

$$u_1(0,t) = u_2(0,t) = \frac{McV_0}{2T}\left[1 - e^{-(2T/Mc)t}\right] \tag{5.363}$$

As $t \to \infty$, the deflection at $x = 0$ becomes

$$u_1(0,\infty) = u_2(0,\infty) = \frac{McV_0}{2T} \tag{5.364}$$

Figure 5-32 is a sketch of the deflection at $x = 0$ as a function of time as obtained from Equation (5.363).

5.13 The Two-dimensional Wave Equation—The Vibrating Rectangular Membrane

The vibration of a stretched membrane is another important problem in the field of vibration and we will analyze it in the present article. Assumptions similar to those cited earlier for the vibrating string in Article 5.2 will be invoked. We assume that: the membrane has a constant mass per unit area and offers no resistance to bending; the weight of the membrane is negligible relative to the tension T, which is considered to remain constant during vibration; the deflection of the membrane is much less than its size such that all the angles of inclination are small. The problem then is to find the transverse deflection of the membrane $u(x,y,t)$ at any location and at any

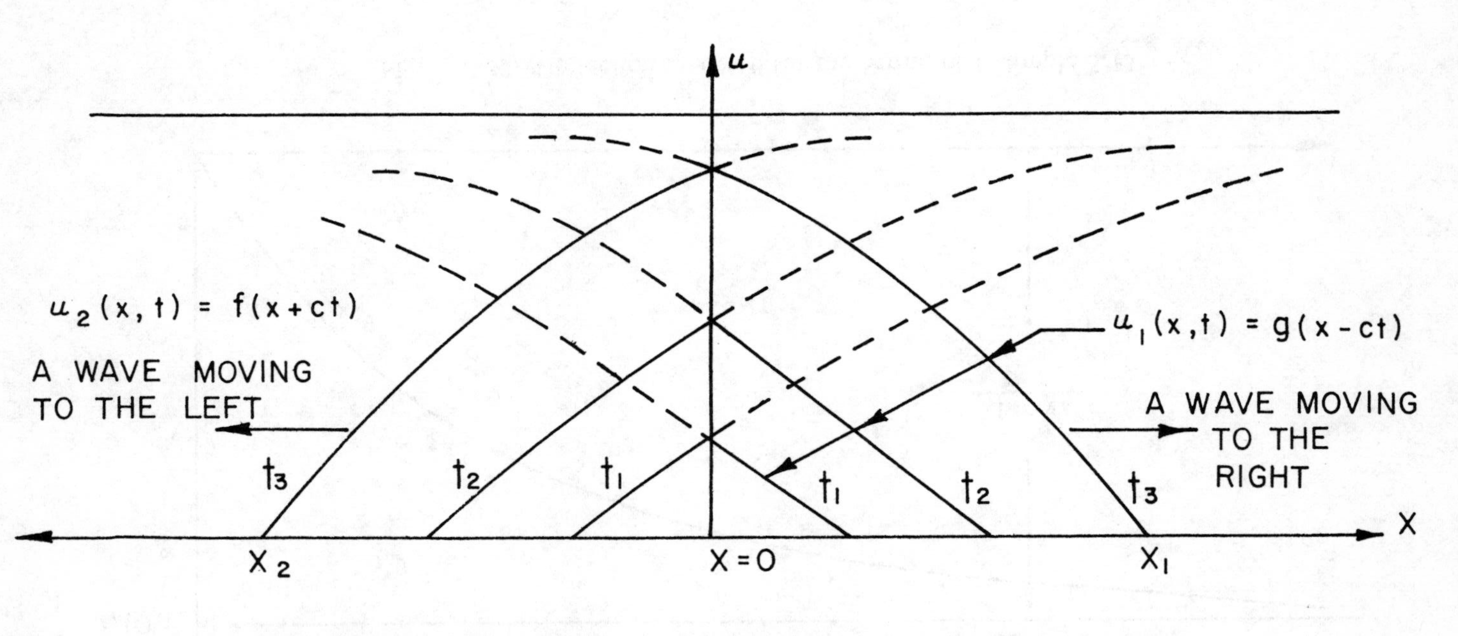

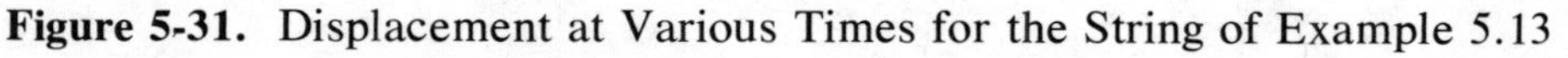
Figure 5-31. Displacement at Various Times for the String of Example 5.13

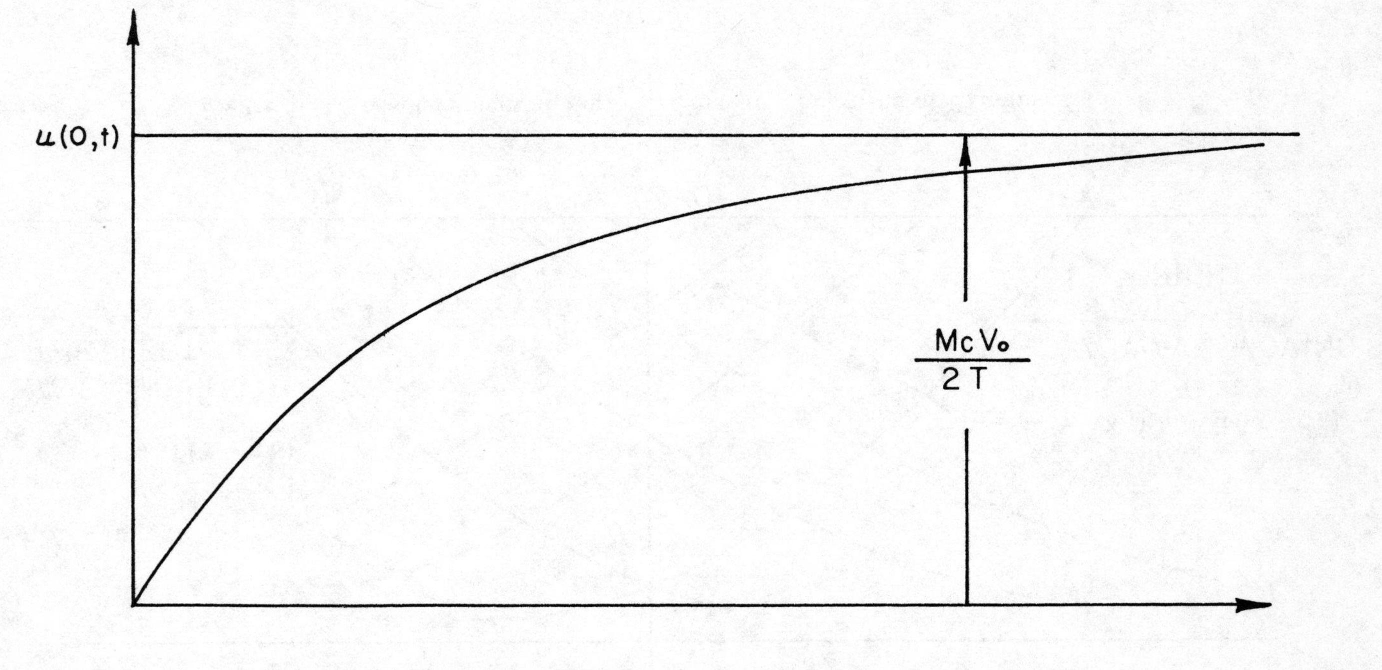

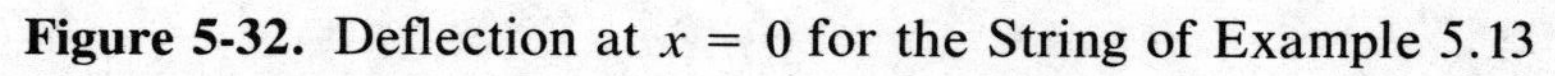
Figure 5-32. Deflection at $x = 0$ for the String of Example 5.13

time once the membrane is disturbed from its equilibrium position. Let us consider a small rectangular element of the membrane ABCD as shown in Figure 5-33. In the displaced position of the element we have the following components of forces:

Net vertical force parallel to yu plane "edges AB and CD" is

$$T\Delta y\,[\sin\theta_1 - \sin\theta_2]$$

For small oscillations, we can make the following approximations

$$\sin\theta_1 \approx \tan\theta_1 = -\frac{\partial u}{\partial x}(x,y_1)$$

$$\sin\theta_2 \approx \tan\theta_2 = -\frac{\partial u}{\partial x}(x + \Delta x,y_2)$$

where y_1 and y_2 are values for y between y and $y + \Delta y$.
Therefore, the net vertical force parallel to yu becomes

$$-T\Delta y\left[\frac{\partial u}{\partial x}(x,y_1) - \frac{\partial u}{\partial x}(x + \Delta x,y_2)\right]$$

Similarly, the net force acting on edges BC and AD is

$$-T\Delta x\left[\frac{\partial u}{\partial y}(x_1,y) - \frac{\partial u}{\partial y}(x_2,y + \Delta y)\right]$$

Where x_1 and x_2 are values for x between x and $x + \Delta x$.
Applying now Newton's Second Law of Motion to the element ABCD of the membrane we obtain

$$m\,\Delta x\,\Delta y\,\frac{\partial^2 u}{\partial t^2} = T\Delta y\left[\frac{\partial u}{\partial x}(x + \Delta x,y_2) - \frac{\partial u}{\partial x}(x,y_1)\right] + T\Delta x\left[\frac{\partial u}{\partial y}(x_2,y + \Delta y) - \frac{\partial u}{\partial y}(x_1,y)\right] \quad (5.365)$$

Dividing Equation (5.365) by $\Delta x\,\Delta y$ and letting Δ approach zero we obtain

$$\frac{1}{c^2}\frac{\partial^2 u}{\partial t^2} = \frac{\partial^2 u}{\partial x^2} + \frac{\partial^2 u}{\partial y^2} \quad (5.366)$$

where $c^2 = T/\rho$, and ρ is mass of the membrane per unit area.

Solution of the Two-dimensional Wave Equation

We pose now the following problem. A rectangular membrane (see Figure

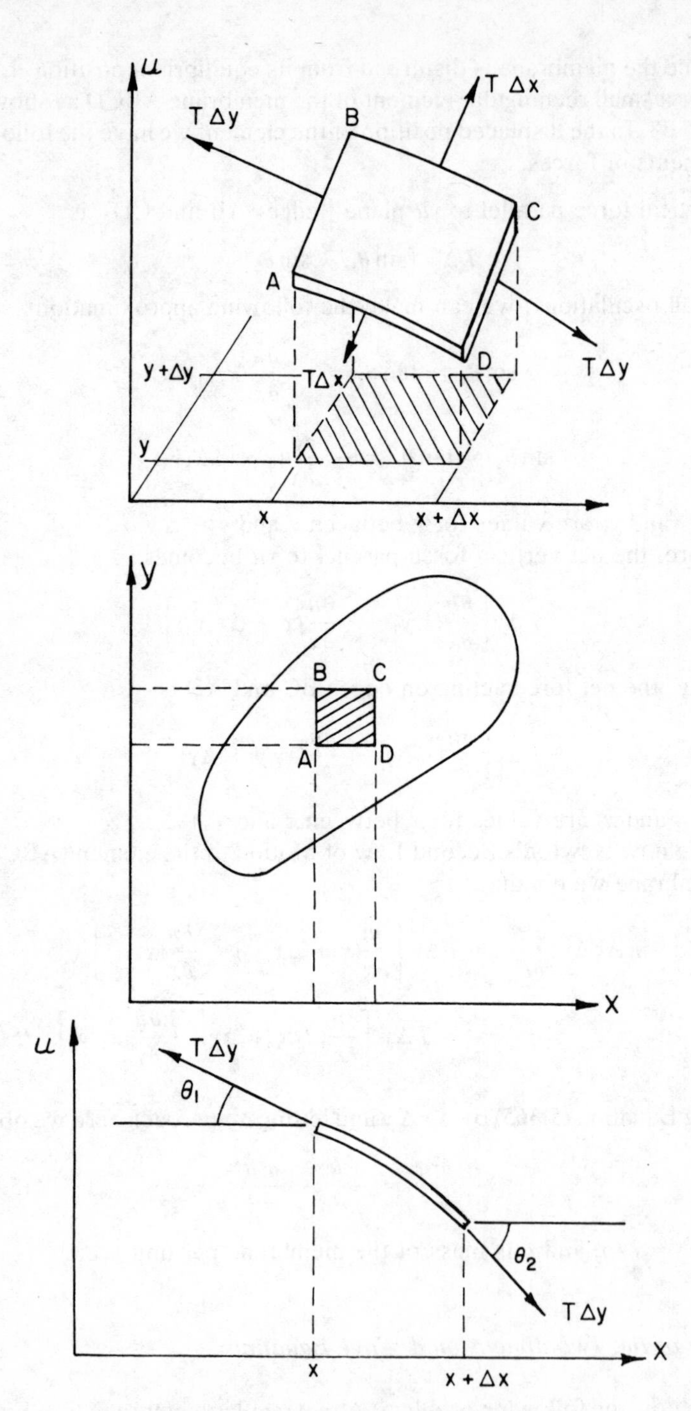

Figure 5-33. Vibrating Membrane

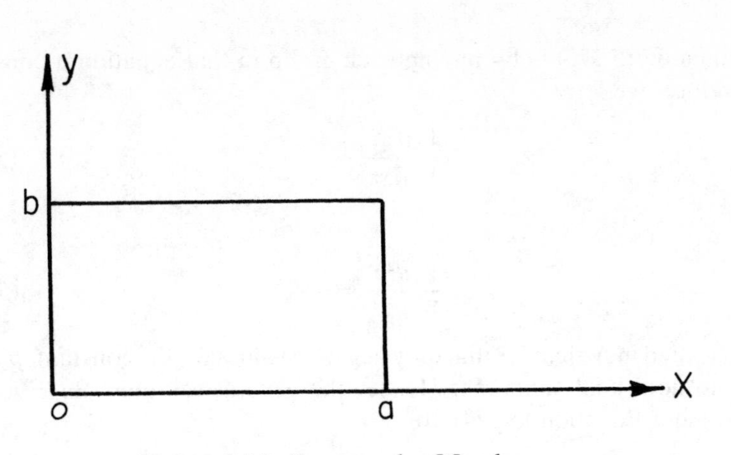

Figure 5-34. Rectangular Membrane

5-34), fixed round its edges has sides of lengths a and b, is subjected to an initial displacement $f(x,y)$ and an initial velocity $g(x,y)$. It is required to find the expression describing the displacement $u(x,y,t)$ of the membrane.

Therefore, the complete problem is now mathematically posed as follows:

$$\frac{1}{c^2}\frac{\partial^2 u}{\partial t^2} = \frac{\partial^2 u}{\partial x^2} + \frac{\partial^2 u}{\partial y^2} \tag{5.366}$$

$$u = 0 \quad \text{for } x = 0,a \quad 0 \leq y \leq b \tag{5.367}$$

$$u = 0 \quad \text{for } y = 0,b \quad 0 \leq x \leq a \tag{5.368}$$

$$u(x,y,0) = f(x,y) \tag{5.369}$$

$$\frac{\partial u}{\partial t}(x,y,0) = g(x,y) \tag{5.370}$$

The method of separation of variables is used and hence we let

$$u(x,y,t) = X(x)\,Y(y)\,G(t) \tag{5.371}$$

Substituting Equation (5.371) in the differential equation "Equation (5.366)" gives

$$\frac{1}{X}\frac{d^2X}{dx^2} + \frac{1}{Y}\frac{d^2Y}{dy^2} = \frac{1}{c^2G}\frac{d^2G}{dt^2} \tag{5.372}$$

In Equation (5.372) the right-hand side contains quantities that are strictly functions of t while the left-hand side involves quantities that are grouped as functions of x and y respectively. In fixing x and y and varying t, or in fixing y and t and varying x, it is deduced that the only way this can be true in Equation (5.372) is by having each group in that equation a constant. Therefore, we let

$$\frac{1}{X}\frac{d^2X}{dx^2} = -p^2 \tag{5.373}$$

and

$$\frac{1}{Y}\frac{d^2Y}{dy^2} = -k^2 \tag{5.374}$$

It was noted in Article 5.8 that only negative values for the constants p and k lead to nontrivial solutions. Hence, the general solutions for Equation (5.373) and Equation (5.374) are

$$X = A\cos px + B\sin px \tag{5.375}$$

and

$$Y = C\cos ky + D\sin ky \tag{5.376}$$

where A, B, C, and D are constants.

The differential equation for $G(t)$ becomes

$$\frac{1}{G}\frac{d^2G}{dt^2} = -c^2(p^2 + k^2) \equiv -\lambda^2 \tag{5.377}$$

and its solution is

$$G(t) = E\cos\lambda t + F\sin\lambda t \tag{5.378}$$

Applying now Equation (5.367) to Equation (5.375) yields

$$X(0) = 0 = A + 0$$

Therefore

$$X = B\sin px$$

with

$$X(a) = 0 = B\sin pa$$

there results

$$\sin pa = 0 \quad \text{and} \quad pa = n\pi \quad \text{or} \quad p = n\pi/a, \quad n = 1, 2, 3, \ldots$$

Hence

$$X_n(x) = B\sin\frac{n\pi}{a}x \tag{5.379}$$

Applying now Equation (5.368) to Equation (5.376) gives

$$Y(0) = 0 = C + 0$$

and

$$Y = D \sin ky$$

with

$$Y(b) = 0 = D \sin kb$$

we have then

$$\sin kb = 0$$

and

$$kb = m\pi \quad \text{or} \quad k = \frac{m\pi}{b}, \qquad m = 1, 2, 3, \ldots$$

Hence

$$Y_m(y) = D \sin \frac{m\pi}{b} y \tag{5.380}$$

From Equation (5.377), λ is $\lambda_{n,m}$ and is given now as

$$\lambda_{n,m}^2 = c^2(p^2 + k^2) = c^2 \left[\left(\frac{n\pi}{a} \right)^2 + \left(\frac{m\pi}{b} \right)^2 \right] \tag{5.381}$$

and therefore,

$$G_{n,m} = E \cos \lambda_{n,m} t + F \sin \lambda_{n,m} t \tag{5.382}$$

In combining Equations (5.379) and (5.380) with Equation (5.382), the solution for $u(x,y,t)$ may now be written as

$$\begin{aligned} u_{n,m}(x,y,t) &= X_n Y_m G_{n,m} \\ &= \sin \frac{n\pi x}{a} \sin \frac{m\pi y}{b} [E' \cos \lambda_{n,m} t + F' \sin \lambda_{n,m} t] \\ &\qquad n = 1, 2, \ldots \qquad m = 1, 2, \ldots \end{aligned} \tag{5.383}$$

where $E' = BDE$ and $F' = BDF$ and $\lambda_{n,m}$ are given by Equation (5.281). $\lambda_{n,m}$ are the eigenvalues of the vibrating membrane. The corresponding eigenfunctions are

$$Z_{n,m}(x,y) \equiv X_n(x)\, Y_m(y) = \sin \frac{n\pi x}{a} \sin \frac{m\pi y}{b}$$

The functions given by Equation (5.383) satisfy the differential equation (Equation (5.366)) and the boundary conditions (Equations (5.367) and

(5.368)). To satisfy the initial conditions, we sum "or superimpose" all the possible solutions from Equation (5.383) for all the eigenfunctions. As we indicated for the case of a vibrating string, for multiple eigenvalues, a linear combination of eigenfunctions constitutes another eigenfunction. Therefore, the general solution becomes

$$u(x,y,t) = \sum_{n=1}^{\infty}\sum_{m=1}^{\infty} \sin\frac{n\pi x}{a} \sin\frac{m\pi y}{b}$$

$$(E_{n,m} \cos\lambda_{n,m}t + F_{n,m} \sin\lambda_{n,m}t) \qquad (5.384)$$

where $E_{n,m}$ and $F_{n,m}$ are to be determined using a double Fourier series as follows:

When we apply the initial condition given by Equation (5.369) representing the initial displacement we obtain

$$u(x,y,0) = f(x,y) = \sum_{n=1}^{\infty}\sum_{m=1}^{\infty} E_{n,m} \sin\frac{n\pi x}{a} \sin\frac{m\pi y}{b} \qquad (5.385)$$

This series is referred to as a double Fourier series representation for $f(x,y)$. To determine the coefficient $E_{n,m}$ we set

$$A_m(y) = \sum_{m=1}^{\infty} E_{n,m} \sin\frac{m\pi y}{b} \qquad (5.386)$$

Hence, substituting Equation (5.386) into Equation (5.385) yields

$$f(x,y) = \sum_{n=1}^{\infty} A_m(y) \sin\frac{n\pi x}{a} \qquad (5.387)$$

Equation (5.386) is a Fourier sine series representation for $A_m(y)$. From chapter 3, section 3.4, the coefficient $E_{n,m}$ becomes

$$E_{n,m} = \frac{2}{b}\int_0^b A_m(y) \sin\frac{m\pi y}{b} dy \qquad (5.388)$$

Similarly, Equation (5.387) is a Fourier sine representation for $f(x,y)$. The coefficient $A_m(y)$, then, is given by

$$A_m(y) = \frac{2}{a}\int_0^a f(x,y) \sin\frac{n\pi x}{a} dx \qquad (5.389)$$

Using Equation (5.389) in Equation (5.388) gives

$$E_{n,m} = \frac{4}{ab}\int_0^b\int_0^a f(x,y) \sin\frac{n\pi x}{a} \sin\frac{m\pi y}{b} dx\, dy$$

$$n = 1, 2, \ldots \quad m = 1, 2, \ldots \qquad (5.390)$$

To determine $F_{n,m}$ we apply the initial condition given by Equation (5.370), which represents the initial velocity. Differentiating Equation (5.384) with respect to time and evaluating the result at $t = 0$ yields

$$\frac{\partial u}{\partial t}(x,y,0) = g(x,y) = \sum_{n=1}^{\infty} \sum_{m=1}^{\infty} (F_{n,m}\, \lambda_{n,m}) \sin \frac{n\pi x}{a} \sin \frac{m\pi y}{b} \tag{5.391}$$

Here again $g(x,y)$ is represented by a double Fourier series. The coefficients of expansion are now $(F_{n,m}\, \lambda_{n,m})$ and can be determined in a way very much similar to that used for $E_{n,m}$. The result is

$$F_{n,m} = \frac{4}{ab\, \lambda_{n,m}} \int_0^b \int_0^a g(x,y) \sin \frac{n\pi x}{a} \sin \frac{m\pi y}{b}\, dx\, dy$$

$$n = 1, 2, \ldots \qquad m = 1, 2, \ldots \tag{5.392}$$

Therefore, the general solution as given by Equation (5.384) will satisfy the initial conditions imposed on the problem when the coefficients $E_{n,m}$ and $F_{n,m}$ satisfy Equations (5.390) and (5.392).

To study the various modes of vibration for the membrane, let us examine again the expression for the eigenvalues as given by Equation (5.381), that is,

$$\lambda_{n,m} = c\pi \left[\left(\frac{n}{a} \right)^2 + \left(\frac{m}{b} \right)^2 \right]^{1/2} \tag{5.393}$$

From the solution given by Equation (5.384), we see that the general term of the series is a periodic function with period $2\pi/\lambda_{n,m}$. The corresponding frequencies are

$$f_{n,m} = \frac{\lambda_{n,m}}{2\pi} = \frac{c}{2} \left[\left(\frac{n}{a} \right)^2 + \left(\frac{m}{b} \right)^2 \right]^{1/2} cps \tag{5.394}$$

and they are called characteristic frequencies. The associated oscillations (given by Equation (5.383)) are called modes. When $n = m = 1$ we have the fundamental mode which is the lowest frequency. This terminology has been presented before in Article 5.8 when we analyzed the vibrating string.

When we dealt with the vibrating string, we found that for every characteristic mode of vibration (whether fundamental or overtones) there were points on the string that did not move and they were designated as nodes or nodal points. These nodes divide the string into equal vibrating segments. In the case of a vibrating membrane, for each characteristic frequency, there are points on the membrane that do not move. Such points are actually lines and designated as nodal lines. It is to be noted though that the position and shape of the nodal lines may not be the same for a given frequency, which implies that for one frequency there will be more than one mode of vibration. When fine powder is sprinkled on a vibrating membrane, nodal lines can then be observed. We take the following example to further interpret the solution from the physical point of view.

Example 5.14 Consider a square membrane with sides $a = b$. Analyze the vibration of the membrane when it is subjected to certain initial displacement and initial velocity.
Solution: The frequency of vibration is given by Equation (5.394) and yields

$$\lambda_{n,m} = \frac{\pi c}{a}(n^2 + m^2)^{1/2} = \beta(n^2 + m^2)^{1/2} \tag{5.395}$$

where we set $\beta = \pi c/a$.
For $n = m = 1$ we have the fundamental mode. Equation (5.384) then gives

$$u_{11} = \sin\frac{\pi x}{a}\sin\frac{\pi y}{a}\left[E_{1,1}\cos\lambda_{1,1}t + F_{1,1}\sin\lambda_{1,1}t\right] \tag{5.396}$$

Here $\lambda_{1,1} = \beta\sqrt{2}$.

If Equation (5.396) is examined, it is found that u_{11} is zero only for $x = 0$, $x = a$, $y = 0$, $y = a$ and therefore, there are no places in the interior of the membrane which have $u_{11} = 0$. Hence, there are no nodal lines. When $n = 1$ and $m = 2$, there results

$$u_{12} = \sin\frac{\pi x}{a}\sin\frac{2\pi y}{a}\left[E_{1,2}\cos\lambda_{1,2}t + F_{1,2}\sin\lambda_{1,2}t\right] \tag{5.397}$$

For $n = 2$ and $m = 1$ we have

$$u_{21} = \sin\frac{2\pi x}{a}\sin\frac{\pi y}{a}\left[E_{2,1}\cos\lambda_{2,1}t + F_{2,1}\sin\lambda_{2,1}t\right] \tag{5.398}$$

In Equations (5.397) and (5.398) $\lambda_{1,2}$ and $\lambda_{2,1}$ are given as

$$\lambda_{1,2} = \lambda_{2,1} = \beta[(1)^2 + (2)^2]^{1/2} = \beta[(2)^2 + (1)^2]^{1/2}$$

$$= \beta\sqrt{5} \tag{5.399}$$

Now, from Equations (5.397) and (5.398) we see respectively that

$$u_{12} = 0 \quad \text{when} \quad y = a/2 \tag{5.400}$$

and

$$u_{21} = 0 \quad \text{when} \quad x = a/2 \tag{5.401}$$

Therefore u_{12} and u_{21} have nodal lines and they are shown in Figure 5-35.

A linear combination of modes u_{12} and u_{21} results in vibration having also a frequency equal to that of u_{12} or u_{21} but with different nodal lines. For example, if we take $E_{12} = E_{21} = 0$ and combine u_{12} and u_{21} there results

$$u_{12} + u_{21} = F_{1,2} \sin \frac{\pi x}{a} \sin \frac{2\pi y}{a} \sin \lambda_{1,2} t$$

$$+ F_{2,1} \sin \frac{2\pi x}{a} \sin \frac{\pi y}{a} \sin \lambda_{2,1} t$$

$$= (\sin \lambda_{1,2} t)\, 2 \sin \frac{\pi x}{a} \sin \frac{\pi y}{a}$$

$$\left(F_{1,2} \cos \frac{\pi y}{a} + F_{2,1} \cos \frac{\pi x}{a} \right) \tag{5.402}$$

In Equation (5.402) the nodal lines are obtained when

$$F_{1,2} \cos \frac{\pi y}{a} + F_{2,1} \cos \frac{\pi x}{a} = 0 \tag{5.403}$$

From Equation (5.403), it is seen that if

$$F_{1,2} = F_{2,1} \tag{5.404}$$

the nodal lines are represented as shown in Figure 5-35 by

$$x + y = a \tag{5.405}$$

However, if

$$F_{1,2} = -F_{2,1} \tag{5.406}$$

the nodal lines are represented by

$$x - y = a \tag{5.407}$$

For u_{22} we have

$$\lambda_{2,2} = \beta(4 + 4)^{1/2} = \beta\sqrt{8}$$

and the nodal lines occur at $x = a/2$ and $y = a/2$. Other nodal lines are shown in Figure 5-35.

The nodal lines can be considered to be boundaries for membranes of different shapes situated within the original one and, hence, they provide a way to study the process of oscillation in such differently shaped membranes.

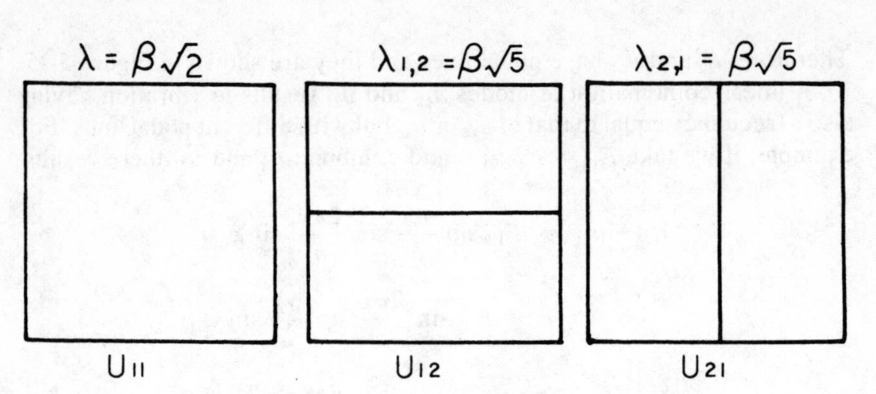

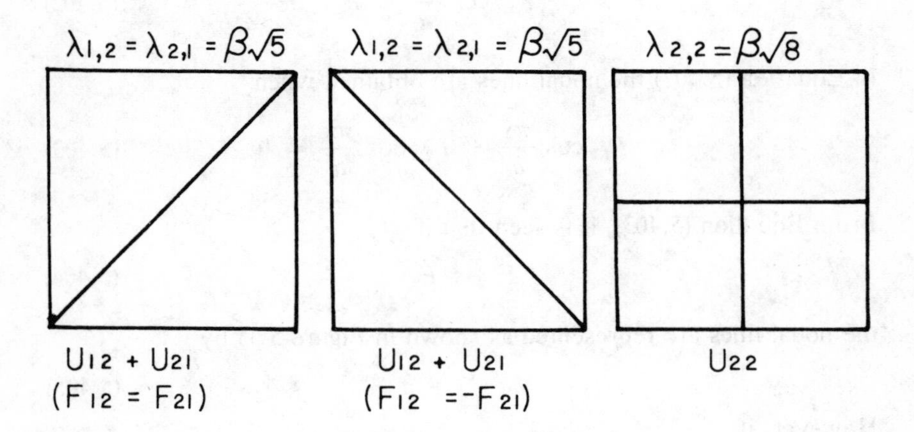

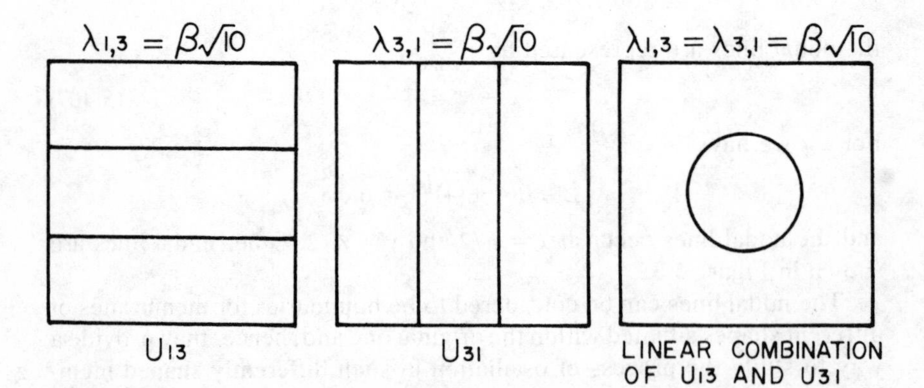

Figure 5-35. Nodal Lines of the Vibrating Square Membrane of Example 5.14

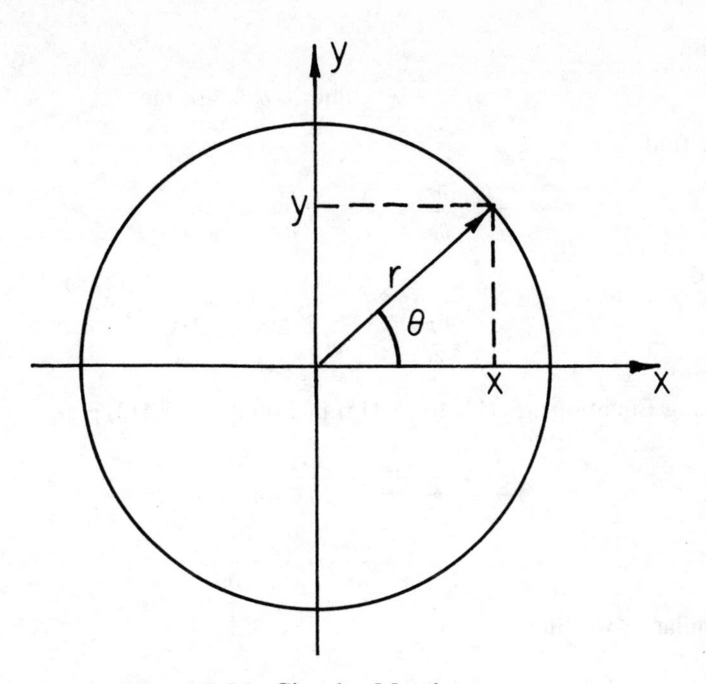

Figure 5-36. Circular Membrane

5.14 Vibration of a Circular Membrane

In the study of oscillations of circular membranes, we have to introduce circular coordinates so that the deflection u is represented as (Figure 5-36)

$$u = u(r,\theta,t) \tag{5.408}$$

and

$$x = r\cos\theta \tag{5.409}$$

$$y = r\sin\theta \tag{5.410}$$

The differential equation for the vibration of a rectangular membrane, Equation (5.366), can be transformed into a cylindrical coordinate system by using Equations (5.409) and (5.410). Therefore, we want a transformation from $u(x,y,t)$ into $u(r,\theta,t)$. To do that we use the chain rule as follows:

$$\frac{\partial u}{\partial x} = \frac{\partial u}{\partial r}\frac{\partial r}{\partial x} + \frac{\partial u}{\partial \theta}\frac{\partial \theta}{\partial x} \tag{5.411}$$

$$\frac{\partial^2 u}{\partial x^2} = \frac{\partial}{\partial x}\left(\frac{\partial u}{\partial x}\right) = \frac{\partial}{\partial x}\left(\frac{\partial u}{\partial r}\frac{\partial r}{\partial x}\right) + \frac{\partial}{\partial x}\left(\frac{\partial u}{\partial \theta}\frac{\partial \theta}{\partial x}\right) \tag{5.412}$$

with

$$r = \sqrt{x^2 + y^2} \quad \text{and} \quad \theta = \arc \tan y/x \tag{5.413}$$

we find

$$\frac{\partial r}{\partial x} = \frac{x}{r}, \quad \frac{\partial \theta}{\partial x} = -\frac{y}{r^2} \tag{5.414}$$

and

$$\frac{\partial^2 r}{\partial x^2} = \frac{y^2}{r^3}, \quad \frac{\partial^2 \theta}{\partial x^2} = \frac{2xy}{r^4} \tag{5.415}$$

Using Equations (5.413) to (5.415) in Equation (5.412) gives

$$\frac{\partial^2 u}{\partial x^2} = \frac{x^2}{r^2}\frac{\partial^2 u}{\partial r^2} - \frac{2xy}{r^3}\frac{\partial^2 u}{\partial r\,\partial \theta} + \frac{y^2}{r^4}\frac{\partial^2 u}{\partial \theta^2} + \frac{y^2}{r^3}\frac{\partial u}{\partial r} + \frac{2xy}{r^4}\frac{\partial u}{\partial \theta} \tag{5.416}$$

Similarly, we find

$$\frac{\partial^2 u}{\partial y^2} = \frac{y^2}{r^2}\frac{\partial^2 u}{\partial r^2} + \frac{2xy}{r^3}\frac{\partial^2 u}{\partial r\,\partial \theta} + \frac{x^2}{r^4}\frac{\partial^2 u}{\partial \theta^2} + \frac{x^2}{r^3}\frac{\partial u}{\partial r} - \frac{2xy}{r^4}\frac{\partial u}{\partial \theta} \tag{5.417}$$

Using Equations (5.416) and (5.417) in Equation (5.366) yields

$$\frac{\partial^2 u}{\partial t^2} = c^2\left(\frac{\partial^2 u}{\partial r^2} + \frac{1}{r}\frac{\partial u}{\partial r} + \frac{1}{r^2}\frac{\partial^2 u}{\partial \theta^2}\right) \tag{5.418}$$

We treat first the case of a circular membrane in which the initial shape is a given function of the radius r so that the vibration is independent of θ. The differential equation then is reduced to

$$\frac{\partial^2 u}{\partial t^2} = c^2\left(\frac{\partial^2 u}{\partial r^2} + \frac{1}{r}\frac{\partial u}{\partial r}\right) \tag{5.419}$$

Boundary condition:

$$u(R,t) = 0 \quad \text{membrane is fixed along the boundary} \tag{5.420}$$

Initial conditions:

The initial conditions, namely the initial deflection and the initial velocity, can be represented respectively as

$$u(r,0) = f(r) \tag{5.421}$$

$$\frac{\partial u}{\partial t}(r,0) = g(r) \tag{5.422}$$

The method of separation of variables is used to solve this problem. Since $u = u(r,t)$ we let

$$u(r,t) = Z(r)\,G(t) \tag{5.423}$$

Introducing Equation (5.423) into Equation (5.419) and dividing by ZG we obtain

$$\frac{1}{c^2G}\frac{d^2G}{dt^2} = \frac{1}{Z}\left(\frac{d^2Z}{dr^2} + \frac{1}{r}\frac{dZ}{dr}\right) \tag{5.424}$$

Again here, it is seen that in Equation (5.424) the left-hand side involves quantities that are strictly functions of t while the right-hand side involves quantities that are functions of r only. Because t and r are independent, the only way for Equation (5.424) to be true is when each side is equal to a constant. As indicated earlier in this chapter, the constant should be taken negative in order to satisfy the boundary conditions and obtain a nontrivial solution. Therefore, we let

$$\frac{1}{c^2G}\frac{d^2G}{dt^2} = -\mu^2 \tag{5.425}$$

so that

$$\frac{d^2G}{dt^2} + \omega^2 G = 0 \quad \text{where } \omega = \mu c \tag{5.426}$$

and

$$\frac{d^2Z}{dr^2} + \frac{1}{r}\frac{dZ}{dr} + \mu^2 Z = 0 \tag{5.427}$$

Equation (5.427) is Bessel's equation of order zero and according to chapter 2 the solution is given as

$$Z(r) = A_1\,J_0(\mu r) + A_2\,Y_0(\mu r) \tag{5.428}$$

where J_0 and Y_0 are Bessel functions of the first and second kind respectively of zero order. Because we are dealing with a membrane that is continuous from $r = 0$ to $r = R$, and because the deflection of the membrane $u(r,t)$ is always finite, the function Y_0 cannot be used for this solution. This is so because Y_0 approaches infinity as its argument approaches zero.

Therefore, we must choose the constant $A_2 = 0$, and the solution for $Z(r)$ becomes

$$Z(r) = A_1 J_0(\mu r) \tag{5.429}$$

Applying the boundary condition as given by Equation (5.420) yields

$$u(R,t) = Z(R)\, G(t) = 0$$

or

$$Z(R) = A_1 J_0(\mu R) = 0 \tag{5.430}$$

Because A_1 cannot be zero or else we have then the trivial solution $u(r,t) = 0$ at all times, we must have

$$J_0(\mu R) = 0 \tag{5.431}$$

The roots of this equation are the values of μR that make $J_0(\mu R) = 0$ and can be obtained from tables of Bessel functions. Denoting by γ_m the mth root of $J_0(\mu R)$, there result then the following eigenvalues to the problem

$$\mu R = \gamma_m \quad \text{or} \quad \mu = \mu_m = \frac{\gamma_m}{R}, \qquad m = 1, 2, \ldots \tag{5.432}$$

and the corresponding eigenfunctions, which vanish at $r = R$, become

$$Z_m(r) = A_1 J_0(\gamma_m(r/R)) \tag{5.433a}$$

or

$$Z_m(r) = A_1 J_0\left(\frac{\omega_m r}{c}\right) \tag{5.433b}$$

Figure 5-37 shows a sketch of $J_0(\mu r)$ and the location of its roots. The first five roots are: 2.404, 5.520, 8.653, 11.791, 14.93.

We turn now to the solution of Equation (5.426), which can now readily be written as

$$G_m(t) = B'_m \cos\omega_m t + C'_m \sin\omega_m t \tag{5.434}$$

Therefore, the solution for $u(r,t)$, which satisfies the boundary conditions, takes the following form

$$\begin{aligned} u_m(r,t) &= Z_m(r)\, G_m(t) \\ &= J_0\left(\frac{\omega_m r}{c}\right)[B_m \cos\omega_m t + C_m \sin\omega_m t] \end{aligned} \tag{5.435}$$

where in Equation (5.435) we set $A_1 B'_m = B_m$ and $A_1 C'_m = C_m$. Equation (5.435) represents the deflection of a vibrating membrane corresponding to

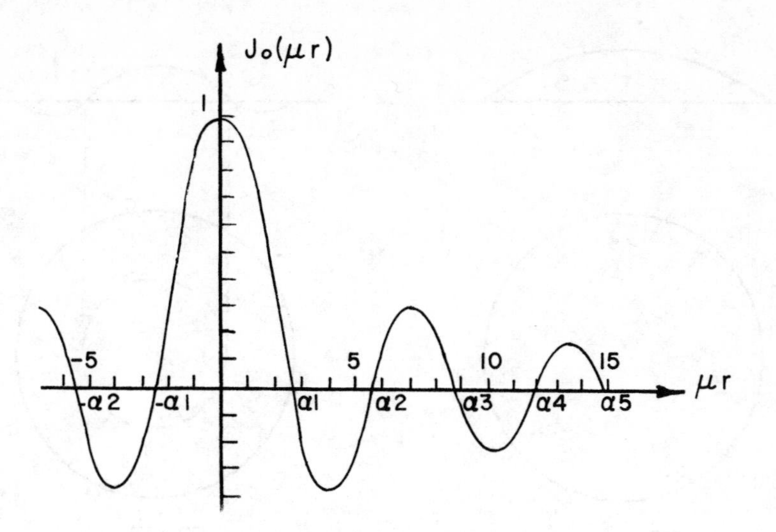

Figure 5-37. Bessel Function $J_0(\mu r)$

the mth normal mode with a frequency ω_m. The first normal mode corresponds to $m = 1$. In this case we have from Equation (5.433)

$$Z_1(r) = A_1 J_0(\gamma_1 (r/R)) \tag{5.436}$$

This equation shows that the only points on the membrane that do not move are the points on the boundary "$r = R$". Therefore, in this case no nodal lines exist within the membrane (Figure 5-38). The second normal mode corresponds to $m = 2$, and it is

$$Z_2(r) = A_2 J_0(\gamma_2(r/R)) \tag{5.437}$$

Equation (5.437) shows that $Z_2(r)$ is zero when

$$\gamma_2(r/R) = \gamma_2, \quad \text{points on the boundary "} r = R \text{"}$$

and

$$\gamma_2(r/R) = \gamma_1$$

or

$$r = \gamma_1/\gamma_2 \, R \tag{5.438}$$

This equation shows that the circle on the membrane that has a radius equal to $\gamma_1/\gamma_2 \, R$ represents a nodal line. During this mode of oscillation, the segments of the membrane that are on either side of this nodal line move in opposite directions (Figure 5-38). The nodal lines in the case of a vibrating

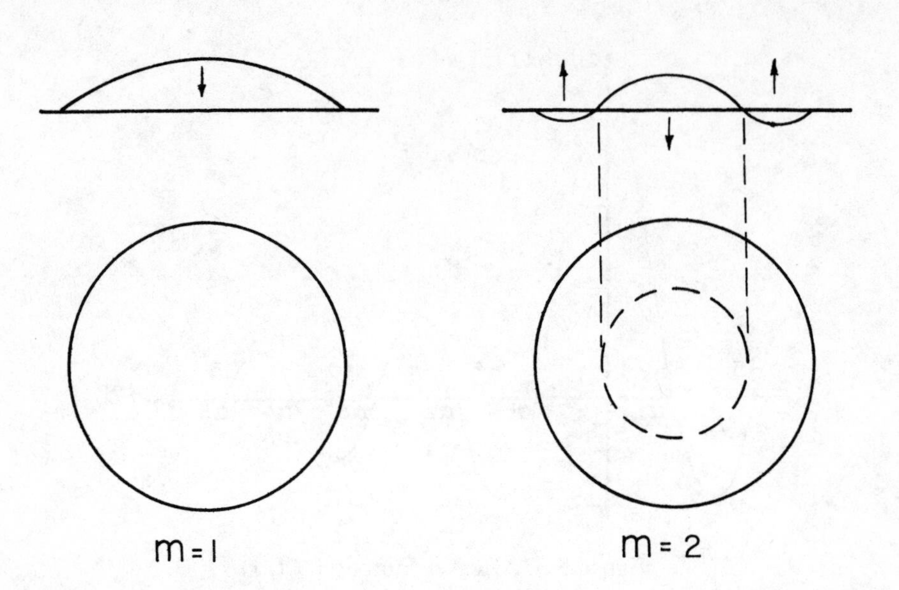

Figure 5-38. Nodal Lines of a Circular Membrane for Vibration Independent of the Angle

circular membrane are concentric circles, and the solution $u_m(r,t)$ has $m - 1$ nodal lines.

We need now to satisfy the initial conditions and, therefore, superimpose all the possible solutions and obtain the infinite series

$$u(r,t) = \sum_{m=1}^{\infty} J_0\left(\frac{\omega_m r}{c}\right) [B_m \cos \omega_m t + C_m \sin \omega_m t] \tag{5.439}$$

Applying the initial deflection now (Equation (5.421)) yields

$$u(r,0) = f(r) = \sum_{m=1}^{\infty} B_m J_0\left(\frac{\omega_m r}{c}\right) \tag{5.440}$$

This equation is a Fourier-Bessel series representation for $f(r)$ in terms of $J_0(\omega_m r/c)$. From chapter 3, the coefficients B_m are given by

$$B_m = \frac{2}{R^2 J_1^2[\omega_m R/c]} \int_0^R r J_0\left(\frac{\omega_m r}{c}\right) f(r)\, dr \tag{5.441a}$$

or

$$B_m = \frac{2}{R^2 J_1^2(\gamma_m)} \int_0^R r J_0\left(\frac{\gamma_m r}{R}\right) f(r)\, dr \tag{5.441b}$$

To satisfy the initial velocity (Equation (5.422)) we differentiate Equation (5.439) with respect to t and obtain

$$\frac{\partial u}{\partial t}(r,t) = \sum_{m=1}^{\infty} J_0\left(\frac{\omega_m r}{c}\right)[-\omega_m B_m \sin \omega_m t + \omega_m C_m \cos \omega_m t]$$

Applying now Equation (5.422) yields

$$\frac{\partial u}{\partial t}(r,0) = g(r) = \sum_{m=1}^{\infty} \omega_m C_m J_0\left(\frac{\omega_m r}{c}\right) \tag{5.442}$$

This equation is also a Fourier-Bessel representation for $g(r)$ in terms of $J_0(\omega_m r/c)$ with coefficients $\omega_m C_m$. Therefore, C_m is given by

$$C_m = \frac{2}{R^2 \omega_m J_1^2[\omega_m R/c]} \int_0^R r J_0\left(\frac{\omega_m r}{c}\right) g(r)\, dr \tag{5.443a}$$

or

$$C_m = \frac{2}{R c \gamma_m J_1^2(\gamma_m)} \int_0^R r J_0\left(\gamma_m \frac{r}{R}\right) g(r)\, dr \tag{5.443b}$$

Hence, in order for the solution $u(r,t)$ to satisfy the initial conditions imposed on the problem, the coefficients B_m and C_m should be represented by Equations (5.441) and (5.443) respectively.

Example 5.15 Consider a circular membrane bounded by two concentric circular boundaries (Figure 5-39) of radii R_1 and R_2 and fastened along all its edges. The membrane is initially deflected into a form symmetrical with respect to the center and then allowed to vibrate. Determine the expression representing the transverse vibration of the membrane $u(r,t)$.

Solution: The governing differential equation, boundary and initial conditions are:

$$\frac{\partial^2 u}{\partial t^2} = c^2\left(\frac{\partial^2 u}{\partial r^2} + \frac{1}{r}\frac{\partial u}{\partial r}\right)$$

$$u(R_1,t) = 0$$

$$u(R_2,t) = 0$$

$$u(r,0) = f(r)$$

$$\frac{\partial u}{\partial t}(r,0) = 0$$

From the method of separation of variables discussed in this article, we found that if we set $u(r,t) = Z(r)\, G(t)$, the expressions for $Z(r)$ and $G(t)$ are then given respectively by

$$Z(r) = A_1 J_0(\mu r) + A_2 Y_0(\mu r) \tag{5.444}$$

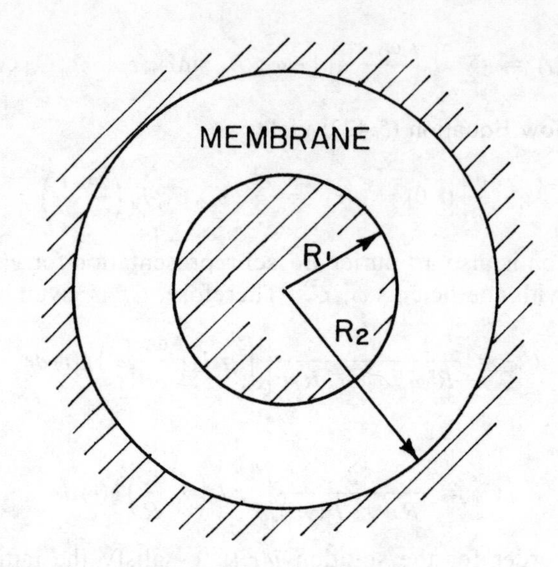

Figure 5-39. Membrane Bounded by Two Concentric Circular Boundaries

and

$$G(t) = B\cos\omega t + C\sin\omega t \tag{5.445}$$

Because the membrane starts at $r = R_1$, the arguments of $J_0(\mu r)$ and $Y_0(\mu r)$ are always greater than zero. Hence, here both $J_0(\mu r)$ and $Y_o(\mu r)$ will appear in the solution. Applying now the boundary conditions at $r = R_1$ and $r = R_2$ gives

$$Z(R_1) = 0 = A_1 J_0(\mu R_1) + A_2 Y_0(\mu R_1) \tag{5.446}$$

$$Z(R_2) = 0 = A_1 J_0(\mu R_2) + A_2 Y_0(\mu R_2) \tag{5.447}$$

Equations (5.446) and (5.447), when solved simultaneously, yield

$$\frac{-A_2\, J_0(\mu R_1)\, Y_0(\mu R_2)}{J_0(\mu R_2)} + A_2 Y_0(\mu R_1) = 0$$

or

$$J_0(\mu R_1) - \frac{J_0(\mu R_2)}{Y_0(\mu R_2)} Y_0(\mu R_1) = 0 \tag{5.448}$$

Equation (5.448) yields the values of $\mu = \mu_m$ for which a solution exists. The solution to this equation is obtained either numerically or graphically.

Because the initial velocity is zero, the constant C in Equation (5.445) is zero (see Equation (5.443b)). Therefore, the general form of the solution to the present problem becomes

$$u(r,t) = \sum_{m=1}^{\infty} B'_m \cos\omega_m t[A_1 J_0(\mu r) + A_2 Y_0(\mu r)] \tag{5.449}$$

A_1 and A_2 are related by Equation (5.446) or (5.447) as

$$A_2 = -A_1 \frac{J_0(\mu R_2)}{Y_0(\mu R_2)} \tag{5.450}$$

Therefore, Equation (5.449) takes the following form

$$u(r,t) = \sum_{m=1}^{\infty} B_m \cos\omega_m t \left[J_0(\mu_m r) - \frac{J_0(\mu_m R_2)}{Y_0(\mu_m R_2)} Y_0(\mu_m r) \right] \tag{5.451}$$

Applying now the initial conditions gives

$$u(r,0) = f(r) = \sum_{m=1}^{\infty} B_m \left[J_0(\mu_m r) - \frac{J_0(\mu_m R_2)}{Y_0(\mu_m R_2)} Y_0(\mu_m r) \right] \tag{5.452}$$

In Equation (5.452), if we set

$$X_0(\mu_m r) = J_0(\mu_m r) - \frac{J_0(\mu_m R_2)}{Y_0(\mu_m R_2)} Y_0(\mu_m r) \tag{5.453}$$

then, (see Byerley, Fourier's Series and Spherical Harmonics)

$$\int_{R_1}^{R_2} r\, X_0(\mu_m r)\, X_0(\mu_n r)\, dr = 0 \tag{5.454}$$

which means that the functions $X_0(\mu_m r)$ and $X_0(\mu_n r)$ are orthogonal on the interval $R_1 < r < R_2$ and that

$$\int_{R_1}^{R_2} r[X_0(\mu_m r)]^2\, dr = \frac{1}{2} \{R_2^2[X_0'(\mu_m R_2)]^2 - R_1^2[X_0'(\mu_m R_1)]^2\} \tag{5.455}$$

Therefore, the expansion coefficients B_m in Equation (5.452) are obtained from the following relations

$$B_m = \frac{2\int_{R_1}^{R_2} r f(r) \left[J_0(\mu_m r) - \frac{J_0(\mu_m R_2)}{Y_0(\mu_m R_2)} Y_0(\mu_m r) \right] dr}{F(R_2) - F(R_1)} \tag{5.456}$$

where

$$F(R_2) = R_2^2 \left[J_0'(\mu_m R_2) - \frac{J_0(\mu_m R_2)}{Y_0(\mu_m R_2)} Y_0'(\mu_m R_2) \right]^2 \tag{5.457}$$

and

$$F(R_1) = R_1^2 \left[J_0'(\mu_m R_1) - \frac{J_0(\mu_m R_2)}{Y_0(\mu_m R_2)} Y_0'(\mu_m R_1) \right]^2 \tag{5.458}$$

The solution given by Equation (5.451) satisfies the boundary conditions and will satisfy the initial conditions if the expansion coefficients B_m are given by Equation (5.456).

5.15 Arbitrary Nonhomogeneity in the Differential Equation—Green's Function

We treat in this section the solution to the one-dimensional wave equation in which there is a nonhomogeneity that is a function of the space variable x and time t; see Equation (5.12). First we consider that the boundary conditions are homogeneous. The problem is then mathematically represented for a vibrating medium of length l as (Figure 5-40)

$$\frac{\partial^2 u}{\partial t^2} = c^2 \frac{\partial^2 u}{\partial x^2} + q(x,t) \tag{5.459}$$

$$u(x,0) = f(x) \tag{5.460}$$

$$\frac{\partial u}{\partial t}(x,0) = g(x) \tag{5.461}$$

$$u(0,t) = 0 \tag{5.462}$$

$$u(l,t) = 0 \tag{5.463}$$

Before going directly to the solution of Equation (5.459), let us recall from chapter 1 the form of the solution to a nonhomogeneous ordinary differential equation such as

$$\frac{d^2 u}{dt^2} + \lambda^2 u = q(t) \tag{5.464}$$

In chapter 1 it was found that this equation has the following solution

$$u(t) = \frac{1}{\lambda}\int_0^t \sin \lambda\,(t - \tau)\, q(\tau)\, d\tau + (C_1 \cos \lambda t + C_2 \sin \lambda t) \tag{5.465}$$

$$= u_1(t) + u_2(t) \tag{5.466}$$

where

$$u_1(t) = \frac{1}{\lambda}\int_0^t \sin \lambda\,(t - \tau)\, q(\tau)\, d\tau \tag{5.467}$$

and

$$u_2(t) = C_1 \cos \lambda t + C_2 \sin \lambda t \tag{5.468}$$

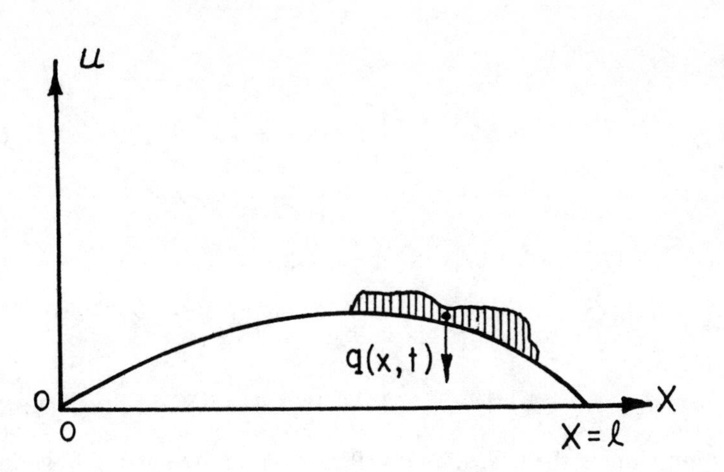

Figure 5-40. Vibrating String Subject to an Arbitrary External Force

$u_1(t)$ accounts for the nonhomogeneity in the differential equation
$u_2(t)$ accounts for the initial conditions

With this brief recall in mind, let us proceed with the solution of Equation (5.459).

We found throughout this chapter that the solution to a vibrating string fixed at $x = 0$ and at $x = l$ leads to a Fourier sine series. Therefore, we like to try a solution to Equation (5.459) in the form of products of functions of time and functions of x represented by

$$u(x,t) = \sum_{n=1}^{\infty} u_n(t) \sin \frac{n\pi x}{l} \tag{5.469}$$

where $u_n(t)$ is to be determined. Let us also represent $q(x,t)$, $f(x)$, and $g(x)$ by Fourier sine series of the form

$$q(x,t) = \sum_{n=1}^{\infty} q_n(t) \sin \frac{n\pi x}{l} \tag{5.470}$$

in which the expression for $q_n(t)$ is obtained from

$$q_n(t) = \frac{2}{l}\int_0^l q(\xi,t) \sin \frac{n\pi\xi}{l} d\xi \tag{5.471}$$

and

$$f(x) = \sum_{n=1}^{\infty} f_n(x) \sin \frac{n\pi x}{l} \tag{5.472}$$

where

$$f_n(x) = \frac{2}{l}\int_0^l f(x)\sin\frac{n\pi x}{l}dx \tag{5.473}$$

and

$$g(x) = \sum_{n=1}^{\infty} g_n(x)\sin\frac{n\pi x}{l} \tag{5.474}$$

where

$$g_n(x) = \frac{2}{l}\int_0^l g(x)\sin\frac{n\pi x}{l}dx \tag{5.475}$$

Substituting now the trial solution (Equations (5.469) and (5.471)) into the differential equation (Equation (5.459)) yields

$$\sum_{n=1}^{\infty} \sin\frac{n\pi x}{l}\left(\frac{d^2u_n(t)}{dt^2}\right) = \sum_{n=1}^{\infty} -\frac{c^2n^2\pi^2}{l^2}u_n(t)\sin\frac{n\pi x}{l} + \sum_{n=1}^{\infty} q_n(t)\sin\frac{n\pi x}{l}$$

or

$$\sum_{n=1}^{\infty} \sin\frac{n\pi x}{l}\left\{\frac{d^2u_n(t)}{dt^2} + \frac{c^2n^2\pi^2}{l^2}u_n(t) - q_n(t)\right\} = 0 \tag{5.476}$$

Equation (5.476) holds if the coefficients of the expansion are equal to zero. Hence, we have then

$$\frac{d^2u_n(t)}{dt^2} + \frac{c^2n^2\pi^2}{l^2}u_n(t) = q_n(t) \tag{5.477}$$

This equation is an ordinary nonhomogeneous differential equation of the second order with constant coefficients. The initial conditions to this equation are obtained as follows

$$u(x,0) = f(x) = \sum_{n=1}^{\infty} f_n\sin\frac{n\pi x}{l} = \sum_{n=1}^{\infty} u_n(0)\sin\frac{n\pi x}{l} \tag{5.478}$$

and

$$\frac{\partial u}{\partial t}(x,0) = g(x) = \sum_{n=1}^{\infty} g_n\sin\frac{n\pi x}{l} = \sum_{n=1}^{\infty}\frac{du_n(0)}{dt}\sin\frac{n\pi x}{l} \tag{5.479}$$

Therefore, Equations (5.478) and (5.479) give

$$u_n(0) = f_n \tag{5.480}$$

$$\frac{du_n(0)}{dt} = g_n \tag{5.481}$$

According to Equation (5.465) we can write the solution to Equation (5.477) with the initial conditions given by Equations (5.480) and (5.481) as

$$u_n(t) = u_{n,1}(t) + u_{n,2}(t) \tag{5.482}$$

where $u_{n,1}(t)$ satisfies the nonhomogeneity in the differential equation, but not the initial conditions, and is given by

$$u_{n,1}(t) = \frac{l}{n\pi c}\int_0^t \sin\frac{n\pi}{l}c\,(t-\tau)\,q_n(\tau)\,d\tau \tag{5.483}$$

and $u_{n,2}(t)$ satisfies the initial conditions only and is given by

$$u_{n,2}(t) = f_n \cos\frac{n\pi}{l}ct + \frac{l}{n\pi c}g_n \sin\frac{n\pi}{l}ct \tag{5.484}$$

Therefore, we can write the solution to the original problem $u(x,t)$ as

$$u(x,t) = \underbrace{\sum_{n=1}^{\infty}\left\{\frac{l}{n\pi c}\int_0^t \sin\frac{n\pi c}{l}(t-\tau)\,q_n(\tau)\,d\tau\right\}\sin\frac{n\pi x}{l}}_{\text{I}}$$

$$+ \underbrace{\sum_{n=1}^{\infty}\left\{f_n\cos\frac{n\pi}{l}ct + \frac{l}{n\pi c}q_n\sin\frac{n\pi}{l}ct\right\}\sin\frac{n\pi x}{l}}_{\text{II}} \tag{5.485}$$

I represents a forced vibration of a string due to the action of an external force $q(x,t)$ and no initial conditions.

II represents the free vibration of a string with imposed initial conditions.

Let the initial conditions now be zero, and let us substitute the expression for $q_n(\tau)$ from Equation (5.471) into Equation (5.485). The result is

$$u(x,t) = \int_0^t\int_0^l G(x,\xi,t-\tau)\,q(\xi,\tau)\,d\xi\,d\tau \tag{5.486}$$

where

$$G(x,\xi,t-\tau) = \frac{2}{\pi c}\sum_{n=1}^{\infty}\frac{1}{n}\sin\frac{n\pi}{l}c(t-\tau)$$

$$\sin\frac{n\pi x}{l}\sin\frac{n\pi}{l}\xi \tag{5.487}$$

This expression is referred to sometimes as Green's function for this class of problem. To study the physical significance of this function, we take the following example:

Example 5.16 A string of length l is acted upon instantaneously by a concentrated force at the location $x = \xi_0$ (Figure 5-41). A force acting in this fashion is referred to as an instantaneous impulse. Find the expression for the deflection of the string.
Solution: Let the force acting on the string have the magnitude $f(x,t)$ per unit length so that the force acting on $\Delta\xi$ becomes

$$f(t) = m \int_{\xi_0}^{\xi_0+\Delta\xi} q(\xi,t)\, d\xi \tag{5.488}$$

The impulse of this force will then be the integral of Equation (5.488) over an infinitesimal time $\Delta\tau$, that is

$$\text{Impulse} = J = \int_{\tau_0}^{\tau_0+\Delta\tau} f(\tau)\, d\tau = m \int_{\tau_0}^{\tau_0+\Delta\tau} \int_{\xi_0}^{\xi_0+\Delta\xi} q(\xi,\tau)\, d\xi\, d\tau \tag{5.489}$$

According to Equation (5.486), $u(x,t)$ is written as

$$u(x,t) = \int_0^t \int_0^l G(x,\ \xi,\ t-\tau)\, q(\xi,\tau)\, d\xi\, d\tau$$

which for the present case becomes

$$u(x,t) = \int_{\tau_0}^{\tau_0+\Delta\tau} \int_{\xi_0}^{\xi_0+\Delta\xi} G(x,\ \xi, t-\tau)\, q(\xi,\tau)\, d\xi\, d\tau \tag{5.490}$$

Applying the mean value theorem to Equation (5.490) (Figure 5-42) yields

$$u(x,t) = G(x,\ \bar{\xi},\ t-\bar{\tau}) \int_{\tau_0}^{\tau_0+\Delta\tau} \int_{\xi_0}^{\xi_0+\Delta\xi} q(\xi,\tau)\, d\xi\, d\tau \tag{5.491}$$

where

$G(x,\ \bar{\xi},\ t-\bar{\tau})$ is the value of $G(x,\ \xi,\ t-\tau)$

evaluated at $\xi = \bar{\xi}$ and at $\tau = \bar{\tau}$. When $\Delta\tau$ and $\Delta\xi$ in Equation (5.491) approach zero, and with the use of Equation (5.489), Equation (5.491) then becomes

$$u(x,t) = G(x,\ \xi_0, t-\tau_0)\, (J/m) \tag{5.492}$$

Equation (5.492) represents the deflection of a string subjected to an instantaneous impulse of magnitude J/m applied at location ξ_0 and at time τ_0. Using Equation (5.487), Equation (5.492) becomes

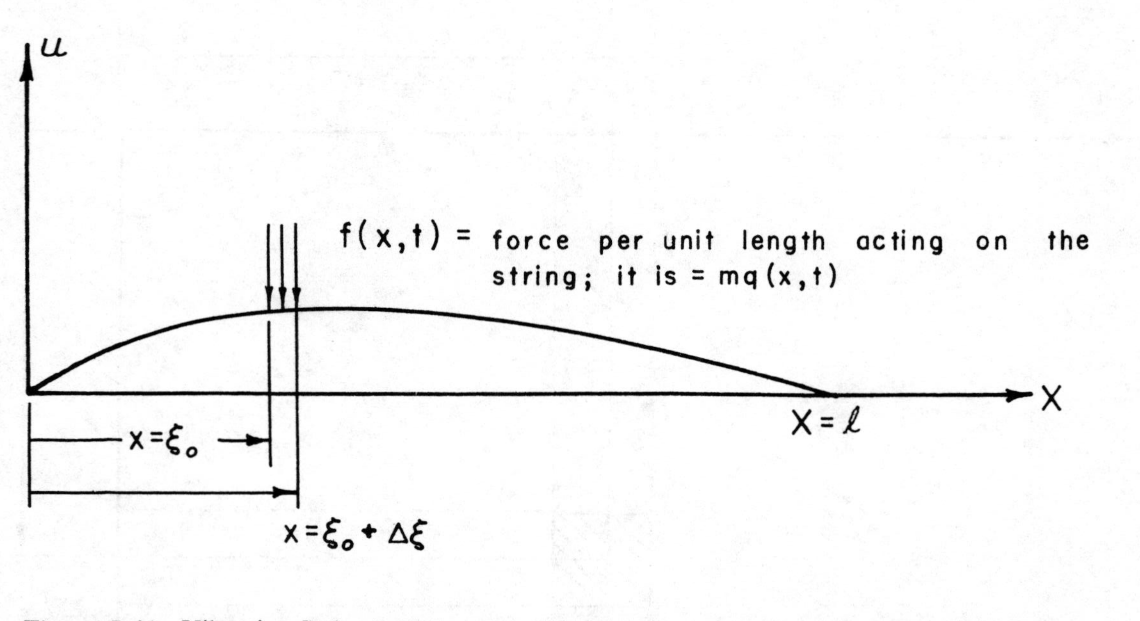

Figure 5-41. Vibrating String in Example 5.16 with an Instantaneous Concentrated Force

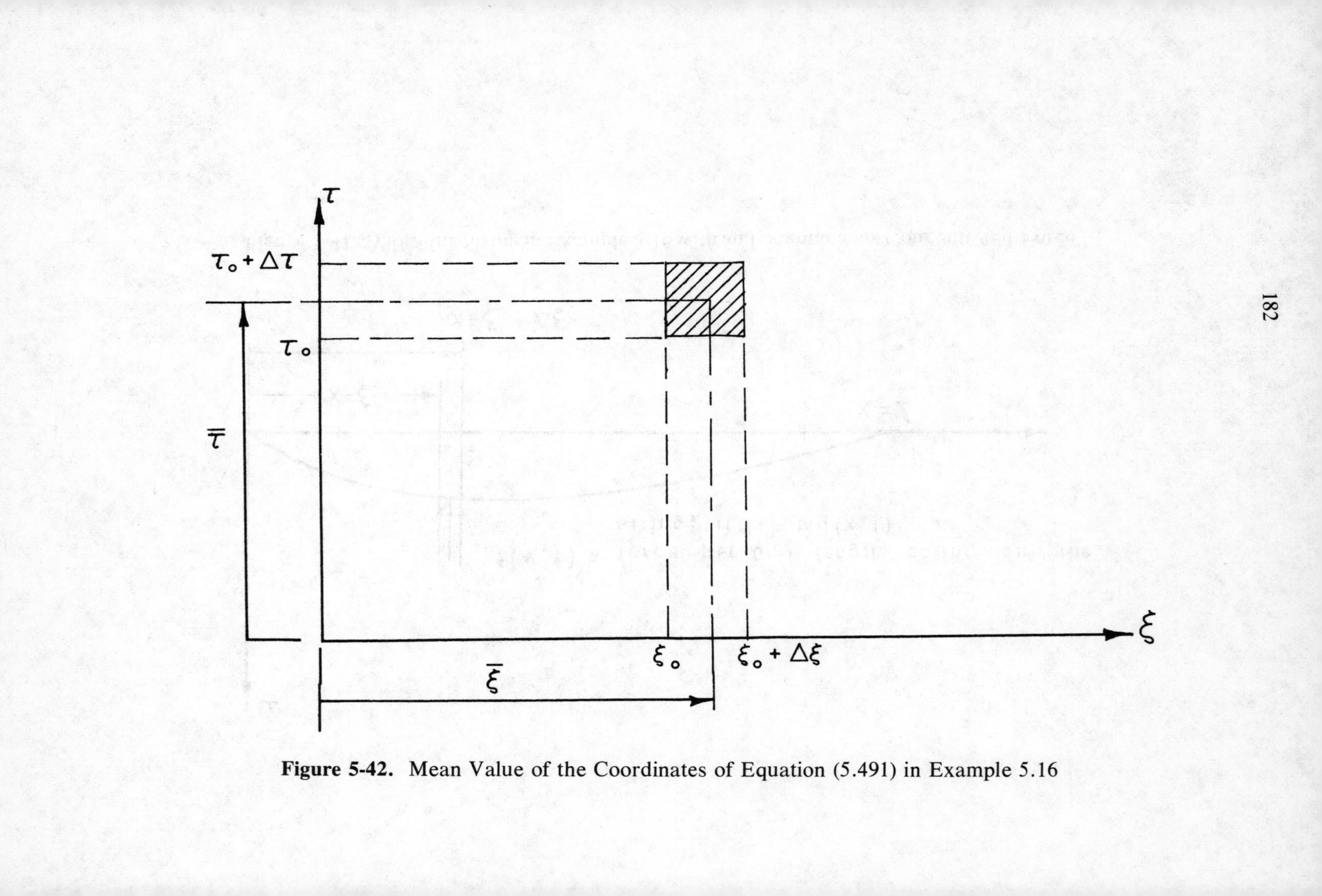

Figure 5-42. Mean Value of the Coordinates of Equation (5.491) in Example 5.16

$$u(x,t) = \frac{2J}{m\pi c} \sum_{n=1}^{\infty} \frac{1}{n} \sin \frac{n\pi}{l} c(t - \tau_0) \sin \frac{n\pi x}{l} \sin \frac{n\pi \xi_0}{l} \tag{5.493}$$

which is applicable for $t > \tau_0$.

Arbitrary Nonhomogeneity in the Differential Equation with Nonhomogeneous Boundary Conditions

The solution presented earlier in this section is extended here to include arbitrary boundary conditions. The problem, therefore, is posed as follows:

$$\frac{\partial^2 u}{\partial t^2} = c^2 \frac{\partial^2 u}{\partial x^2} + q(x,t) \tag{5.494}$$

$$u(x,0) = f(x) \tag{5.495}$$

$$\frac{\partial u}{\partial t}(x,0) = g(x) \tag{5.496}$$

$$u(0,t) = u_1 \tag{5.497}$$

$$u(l,t) = u_2 \tag{5.498}$$

When the boundary conditions were homogeneous, that is, $u_1 = u_2 = 0$, the solution as presented by Equation (5.485) was made up of two solutions. One solution accounts for the nonhomogeneity in the differential equation and the other accounts for the initial conditions of the problem. In the presence of nonhomogeneous boundary conditions, we add another solution to take care of the boundary conditions. This is done as follows:

Let

$$u(x,t) = v(x,t) + w(x) \tag{5.499}$$

so that

$$\frac{\partial^2 u}{\partial t^2} = \frac{\partial^2 v}{\partial t^2} + \cancelto{0}{\frac{\partial^2 w}{\partial t^2}} \tag{5.500}$$

and

$$\frac{\partial^2 u}{\partial x^2} = \frac{\partial^2 v}{\partial x^2} + \frac{d^2 w}{dx^2} \tag{5.501}$$

Introducing Equations (5.500) and (5.501) into Equation (5.494) yields

$$\frac{\partial^2 v}{\partial t^2} = c^2 \left(\frac{\partial^2 v}{\partial x^2} + \frac{d^2 w}{dx^2} \right) + q(x,t) \tag{5.502}$$

or

$$\frac{\partial^2 v}{\partial t^2} = c^2 \frac{\partial^2 v}{\partial x^2} + q(x,t) + c^2 \frac{d^2 w}{dx^2} \tag{5.503}$$

$v(x,t)$ is chosen to satisfy the equation

$$\frac{\partial^2 v}{\partial t^2} = c^2 \frac{\partial^2 v}{\partial x^2} + q(x,t) \tag{5.504}$$

while $w(x)$ accounts for the boundary conditions and satisfies

$$\frac{d^2 w}{dx^2} = 0 \tag{5.505}$$

The initial and boundary conditions become

$$u(x,0) = v(x,0) + w(x) = f(x) \tag{5.506}$$

$$\frac{\partial u}{\partial t}(x,0) = \frac{\partial v}{\partial t}(x,0) + \cancelto{0}{\frac{\partial w}{\partial t}} = g(x) \tag{5.507}$$

$$0 \le x \le l$$

and

$$u(0,t) = v(0,t) + w(0) = u_1 \tag{5.508}$$

$$u(l,t) = v(l,t) + w(l) = u_2 \tag{5.509}$$

$$t > 0$$

From Equations (5.506) and (5.507) we write

$$v(x,0) = f(x) - w(x) \tag{5.510}$$

and

$$\frac{\partial v}{\partial t}(x,0) = g(x) \tag{5.511}$$

Equations (5.508) and (5.509) yield

$$w(0) = u_1 \tag{5.512}$$

$$w(l) = u_2 \tag{5.513}$$

$$v(0,t) = 0 \tag{5.514}$$

$$v(l,t) = 0 \tag{5.515}$$

To summarize: we reduced the original problem into the following two problems

$$\frac{\partial^2 v}{\partial t^2} = c^2 \frac{\partial^2 v}{\partial x^2} + q(x,t) \qquad\qquad \frac{d^2 w}{dx^2} = 0$$

$$\left.\begin{aligned} v(x,0) &= f(x) - w(x) \\ \frac{\partial v}{\partial t}(x,0) &= g(x) \end{aligned}\right\} 0 \le x \le l \qquad\qquad \begin{aligned} w(0) &= u_1 \\ w(l) &= u_2 \end{aligned}$$

$$\left.\begin{aligned} v(0,t) &= 0 \\ v(l,t) &= 0 \end{aligned}\right\} t > 0$$

The problem posed by $v(x,t)$ is the one presented at the beginning of this article by Equations (5.459) to (5.463) where the initial displacement is now given by $f(x) - w(x)$ instead of $f(x)$. Therefore, according to Equation (5.485) the solution for $v(x,t)$ can be written as (see Equations (5.471), (5.473), and (5.475))

$$v(x,t) = \sum_{n=1}^{\infty} \left\{ \frac{l}{n\pi c} \int_0^t \sin \frac{n\pi c}{l}(t - \tau)\, q_n(\tau)\, d\tau \right\} \sin \frac{n\pi x}{l}$$

$$+ \sum_{n=1}^{\infty} \left\{ f_n \cos \frac{n\pi c}{l} t + \frac{l}{n\pi c} g_n \sin \frac{n\pi c}{l} t \right\} \sin \frac{n\pi x}{l} \tag{5.516}$$

where in this case f_n is given by

$$f_n = \frac{2}{l} \int_0^l [f(x) - w(x)] \sin \frac{n\pi x}{l}\, dx \tag{5.517}$$

and q_n and g_n are given respectively by Equations (5.471) and (5.475).

The solution for $w(x)$ is obtained by direct integration and it is

$$w(x) = u_1 + (x/l)(u_2 - u_1) \tag{5.518}$$

Therefore, the solution for the original problem $u(x,t)$ is equal to the sum of Equations (5.516) and (5.518). Schematically, the solution is broken down as shown in the schematic below:

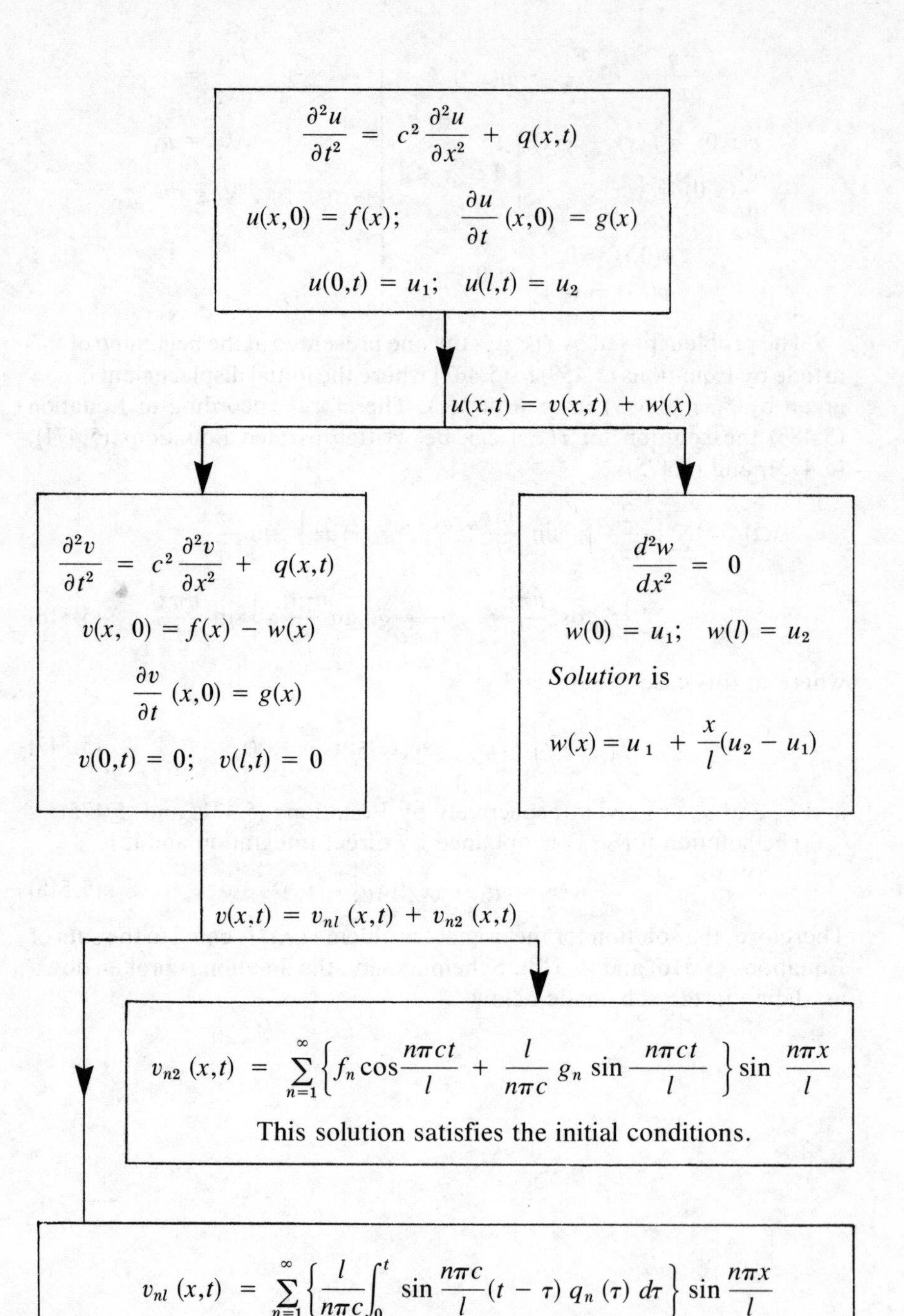

$\frac{\partial^2 u}{\partial t^2} = c^2 \frac{\partial^2 u}{\partial x^2} + q(x,t)$
$u(x,0) = f(x); \quad \frac{\partial u}{\partial t}(x,0) = g(x)$
$u(0,t) = u_1; \quad u(l,t) = u_2$
$u(x,t) = v(x,t) + w(x)$
$\frac{\partial^2 v}{\partial t^2} = c^2 \frac{\partial^2 v}{\partial x^2} + q(x,t)$
$v(x, 0) = f(x) - w(x)$
$\frac{\partial v}{\partial t}(x,0) = g(x)$
$v(0,t) = 0; \quad v(l,t) = 0$
$\frac{d^2 w}{dx^2} = 0$
$w(0) = u_1; \quad w(l) = u_2$
Solution is
$w(x) = u_1 + \frac{x}{l}(u_2 - u_1)$
$v(x,t) = v_{nl}(x,t) + v_{n2}(x,t)$
$v_{n2}(x,t) = \sum_{n=1}^{\infty} \left\{ f_n \cos \frac{n\pi ct}{l} + \frac{l}{n\pi c} g_n \sin \frac{n\pi ct}{l} \right\} \sin \frac{n\pi x}{l}$
This solution satisfies the initial conditions.
$v_{nl}(x,t) = \sum_{n=1}^{\infty} \left\{ \frac{l}{n\pi c} \int_0^t \sin \frac{n\pi c}{l}(t - \tau)\, q_n(\tau)\, d\tau \right\} \sin \frac{n\pi x}{l}$
This solution satisfies the nonhomogeneity in the differential equation.

Partial Differential Equations of the Parabolic Type

6.1 Diffusion Equation

Among the very important differential equations in engineering sciences is the diffusion equation, which is also the heat conduction equation.

Undoubtedly, heat conduction constitutes the most important application of the diffusion equation. There are, however, other important applications such as the mass diffusion in liquids and gases, the propagation of signals in an unloaded submarine or telegraph cable, the study of the skin effect in electrical engineering, the theory of consolidation of soil, the diffusion of momentum, and many others.

The following sections show derivations of the diffusion equation from basic physical principles. Methods of solution follow these derivations.

6.2 Thermal Diffusion—The Heat Conduction Equation

In the derivation of the heat conduction equation we consider a homogeneous slab of width l and of very large extent in the other two dimensions such that the variation of the temperature in the slab occurs only in the x-direction. This condition is also met in an insulated rod in which the temperature can vary only with respect to x and time; see Figure 6-1.

An energy balance on a thin section of width Δx yields

$$A\,q_x = q_{\text{stored}}\,(A\Delta x) + A\,q_{x+\Delta x} \tag{6.1}$$

where q_x = x-direction heat flux into the section by conduction, energy per unit area and per unit time

$q_{x+\Delta x}$ = x-direction heat flux out of the section by conduction, energy per unit area and per unit time

q_{stored} = energy accumulated within the section per unit volume, which acts to change the internal energy of the section.

Equation (6.1) can be rewritten as

$$-\frac{A(q_{x+\Delta x} - q_x)\,\Delta x}{\Delta x} = q_{\text{stored}}\,A\,\Delta x \tag{6.2}$$

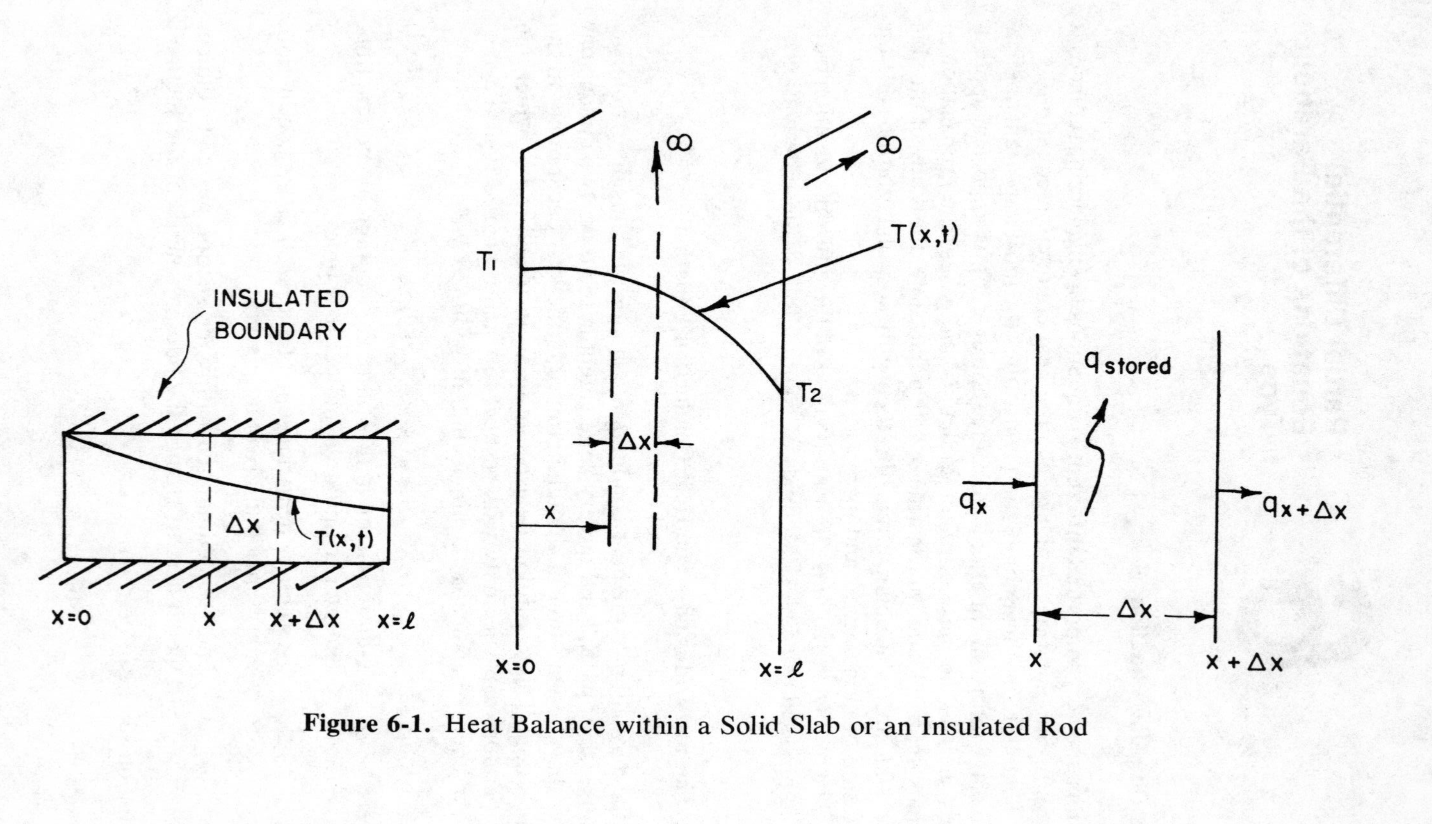

Figure 6-1. Heat Balance within a Solid Slab or an Insulated Rod

The rate of change of internal energy of section Δx is

$$\text{(mass)}\, c\frac{\partial T}{\partial t} = A\,\Delta x\,\rho\, c\frac{\partial T}{\partial t} = q_{\text{stored}}\, A\,\Delta x \tag{6.3}$$

where A is the area of the section, ρ is density, c is specific heat capacity and t is time. Using now Equation (6.3) in Equation (6.2) and having $\Delta x \to 0$ yields

$$-\frac{\partial q}{\partial x} = \rho c\frac{\partial T}{\partial t} \tag{6.4}$$

The heat transfer by conduction in a medium is governed by Fourier's heat conduction law, stated as

$$q_x = -k\frac{\partial T}{\partial x} \tag{6.5}$$

where k is the thermal conductivity of the medium. Introducing Equation (6.5) in Equation (6.40) we obtain

$$\frac{\partial}{\partial x}\left[k\frac{\partial T}{\partial x}\right] = \rho c\frac{\partial T}{\partial t} \tag{6.6}$$

When the slab is considered to be homogeneous with k, ρ and c as constant, Equation (6.6) becomes

$$\alpha\frac{\partial^2 T}{\partial x^2} = \frac{\partial T}{\partial t} \tag{6.7}$$

where

$$\alpha = \frac{k}{\rho c}$$

Equation (6.7) is the diffusion equation representing the one-dimensional transient temperature distribution in a slab. α is referred to as the thermal diffusivity of the medium.

Heat Conduction in the Presence of Heat Generation

When the solid experiences a certain heat generation within it due to such effects as electric current or chemical reaction, the energy balance on the element Δx becomes (Figure 6-2)

$$A\, q_x + A\,\Delta x\, G(x,t) = A\, q_{x+\Delta x} + q_{\text{stored}}\, A\,\Delta x \tag{6.8}$$

or

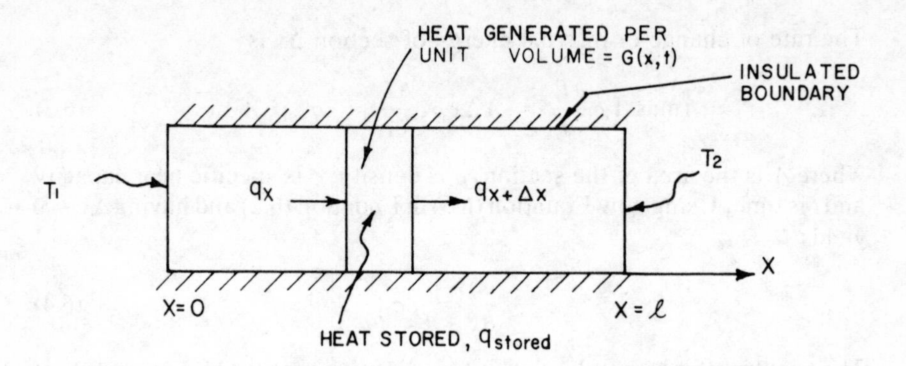

Figure 6-2. Heat Balance in the Presence of Heat Generation

$$-\frac{A[q_{x+\Delta x} - q_x]\Delta x}{\Delta x} + A\,\Delta x\,G(x,t) = A\,\Delta x\,\rho c\,\frac{\partial T}{\partial t} \tag{6.9}$$

We introduce now Fourier's conduction law and let $\Delta x \to 0$ in Equation (6.9); the result becomes

$$k\frac{\partial^2 T}{\partial x^2} + G(x,t) = \rho c\,\frac{\partial T}{\partial t}$$

or

$$\alpha\frac{\partial^2 T}{\partial x^2} + g(x,t) = \frac{\partial T}{\partial t} \tag{6.10}$$

where we set

$$g(x,t) = \frac{G(x,t)}{\rho c}$$

Equation (6.10) is the one-dimensional nonhomogeneous diffusion equation.

Heat Conduction in the Presence of Surface Convection and Radiation

When a solid rod experiences heat generation and loses heat to the surrounding ambient by convection and radiation, the energy balance on element Δx takes the following form (Figure 6-3)

$$A\,q_x + A\,\Delta x\,G(x,t) = A\,q_{x+\Delta x} + q_{\text{stored}}\,A\,\Delta x + (q_c + q_r)A_p \tag{6.11}$$

where A_p = peripheral area of element Δx. The convection and radiative heat transfer processes are given respectively by

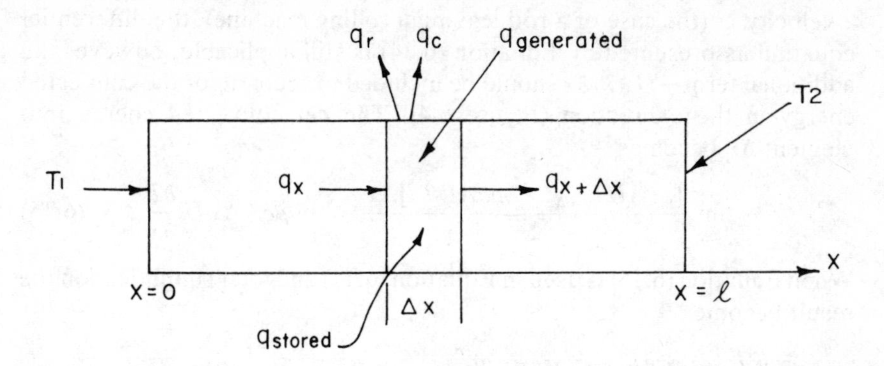

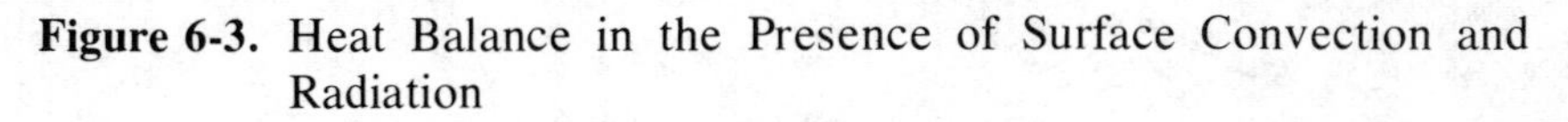

Figure 6-3. Heat Balance in the Presence of Surface Convection and Radiation

$$q_c = A_p \, h \, (T - T_\infty) \tag{6.12}$$

$$q_r = A_p \sigma [\bar{\epsilon} T^4 - \bar{\alpha} T_\infty^4] \tag{6.13}$$

where $A_p = P \, \Delta x$; P = perimeter of element Δx

h = a convective heat transfer coefficient

T_∞ = the ambient temperature surrounding the solid

σ = Stefan-Boltzmann constant

$\bar{\epsilon}$ = surface emissivity of the solid

$\bar{\alpha}$ = surface absorptivity of the solid

Introducing Equations (6.12) and (6.13) along with Fourier's law in Equation (6.11) and having $\Delta x \to 0$ yields

$$k \frac{\partial^2 T}{\partial x^2} + G(x,t) = \rho c \frac{\partial T}{\partial t} + \frac{P}{A} [h(T - T_\infty) + \sigma(\bar{\epsilon} T^4 - \bar{\alpha} T_\infty^4)]$$

or

$$\alpha \frac{\partial^2 T}{\partial x^2} + g(x,t) = \frac{\partial T}{\partial t} + \beta(T - T_\infty) + \gamma(\bar{\epsilon} T^4 - \bar{\alpha} T_\infty^4) \tag{6.14}$$

where we set

$$\beta = \frac{Ph}{\rho c A}; \qquad \gamma = \frac{P\sigma}{\rho c A}$$

Conduction in a Moving Rod

If, in addition to the effects just presented, the rod moves along its axis with a velocity U (the case of a rod leaving a rolling machine), the differential equation as presented by Equation (6.14) is still applicable, however, an additional term $-U\,\partial T/\partial x$ should be included to account for the convected energy in the x-direction (Figure 6-4). The net convected energy into element Δx becomes

$$-\lim_{\Delta x \to 0} \frac{[\rho c A U T_{x+\Delta x} - \rho c A U T_x]\Delta x}{\Delta x} = -\rho c A\,\Delta x\,U \frac{\partial T}{\partial x} \tag{6.15}$$

When Equation (6.15) is used in Equation (6.11) and after simplification the result becomes

$$\alpha \frac{\partial^2 T}{\partial x^2} - U\frac{\partial T}{\partial x} + g(x,t) = \frac{\partial T}{\partial t} + \beta(T - T_\infty) + \gamma(\bar{\epsilon} T^4 - \bar{\alpha} T_\infty^4) \tag{6.16}$$

Heat Conduction in Multidimensional Case

The differential equation representing the heat conduction in more than one dimension can be derived in a similar way to the one-dimensional case. Here, we select a cubic element within the solid of the shape shown in Figure 6-5 and write the energy balance on that element in the following form:

$$\begin{aligned}
&[q_x(\Delta y\,\Delta z) + q_y(\Delta x\,\Delta z) + q_z(\Delta x\,\Delta y)] + G(x,y,z,t)\,\Delta x\,\Delta y\,\Delta z \\
&\qquad = [q_{x+\Delta x}(\Delta y\,\Delta z) + q_{y+\Delta y}(\Delta x\,\Delta z) + q_{z+\Delta z}(\Delta x\,\Delta y)] \\
&\qquad\quad + (\Delta x\,\Delta y\,\Delta z)\rho c \frac{\partial T}{\partial t}
\end{aligned} \tag{6.17}$$

where q_x, q_y, and q_z are respectively the heat fluxes in the x-, y-, and z-directions. Introducing Fourier's conduction law in Equation (6.17) and having $\Delta x \to 0$ yields

$$\alpha\left(\frac{\partial^2 T}{\partial x^2} + \frac{\partial^2 T}{\partial y^2} + \frac{\partial^2 T}{\partial z^2}\right) + g(x,y,z,t) = \frac{\partial T}{\partial t} \tag{6.18}$$

6.3 Equation of Mass Diffusion

Mass transfer by molecular diffusion is quite similar to heat transfer by conduction. When a region is filled nonuniformly with a gas, then diffusion takes place from points of higher concentration of the gas to points having lower concentration. Solutions, also, having gradients in the solutions from one point to another in a volume, experience molecular diffusion in the

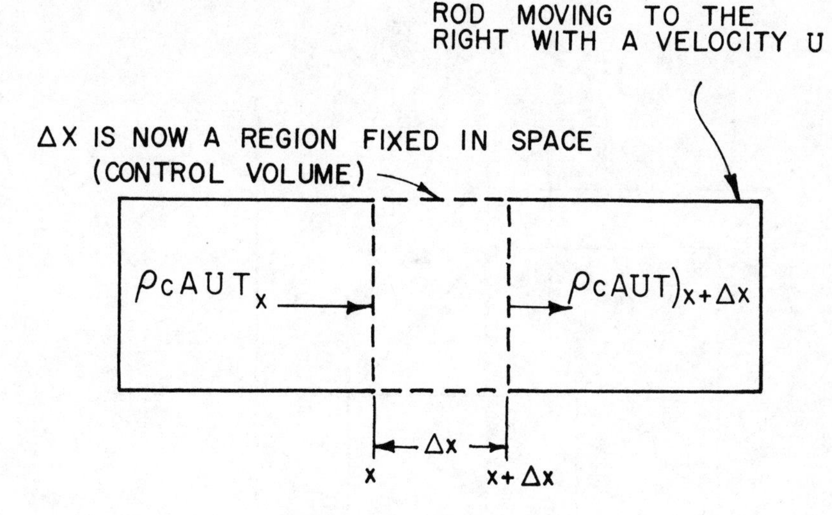

Figure 6-4. Moving Rod

direction of lower concentration of the solute. Consider a volume as shown in Figure 6-6 in which there exists a certain concentration gradient of a substance M. A mass balance on section Δx yields

$$A\, m_x = A\, m_{x+\Delta x} + m_{\text{stored}}$$

or

$$-\frac{A[m_{x+\Delta x} - m_x]\Delta x}{\Delta x} = m_{\text{stored}} \tag{6.19}$$

m, the mass transfer rate, is given by Fick's law of diffusion stated as

$$m = -D\frac{\partial c}{\partial x} \tag{6.20}$$

where D is the mass diffusivity and c is the concentration of substance M. Introducing Equation (6.20) in Equation (6.19) and letting $\Delta x \to 0$ gives

$$A\,\Delta x\frac{\partial}{\partial x}\left[D\frac{\partial c}{\partial x}\right] = A\,\Delta x\frac{\partial c}{\partial t} \tag{6.21}$$

When D is taken as constant, Equation (6.21) becomes

$$D\frac{\partial^2 c}{\partial x^2} = \frac{\partial c}{\partial t} \tag{6.22}$$

Equation (6.22) is identical in form to the heat conduction equation given by Equation (6.7).

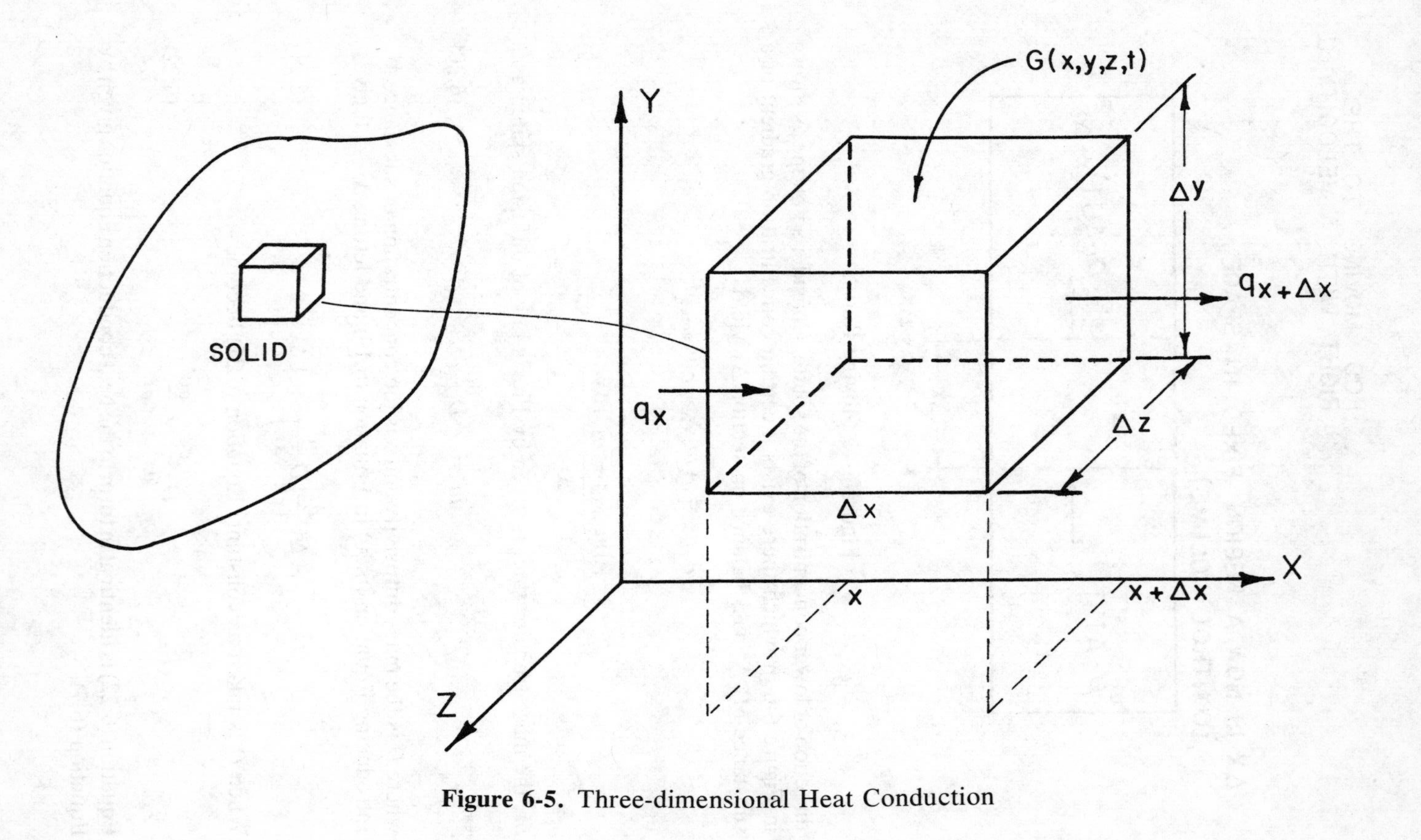

Figure 6-5. Three-dimensional Heat Conduction

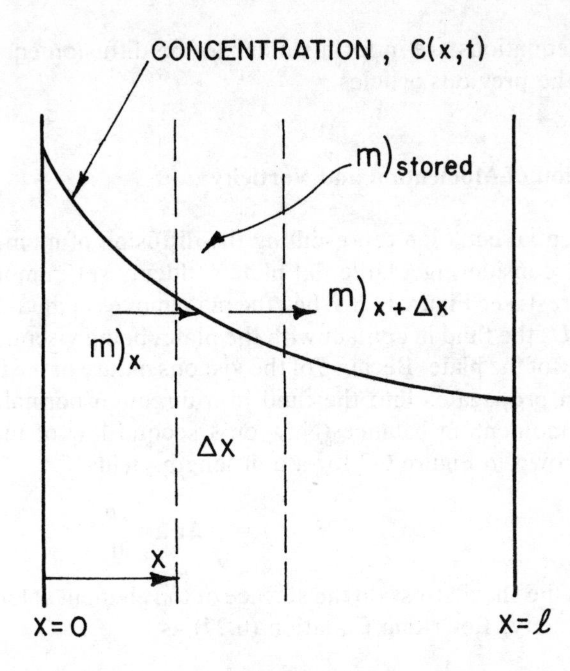

Figure 6-6. Mass Concentration within a Volume of Width l

6.4 Telegraph or Cable Equations

We found in Article 5.5 that the transmission line equations are given by

$$\frac{\partial^2 V}{\partial x^2} = LC\frac{\partial^2 V}{\partial t^2} + (LG + RC)\frac{\partial V}{\partial t} + RGV \tag{6.23}$$

and

$$\frac{\partial^2 I}{\partial x^2} = LC\frac{\partial^2 I}{\partial t^2} + (LG + RC)\frac{\partial I}{\partial t} + RGI \tag{6.24}$$

In many applications to telegraph signaling, the leakage G (conductance per unit length) is quite small and the effect of L (inductance per unit length) is negligible. In this case, Equations (6.23) and (6.24) become

$$\frac{1}{RC}\frac{\partial^2 V}{\partial x^2} = \frac{\partial V}{\partial t} \tag{6.25}$$

and

$$\frac{1}{RC}\frac{\partial^2 I}{\partial x^2} = \frac{\partial I}{\partial t} \tag{6.26}$$

These equations are in form similar to the diffusion equation derived earlier in the previous articles.

6.5 Diffusion of Momentum and Vorticity

The differential equation representing the diffusion of momentum can be derived by considering a large flat plate suddenly set in motion in a fluid initially at rest; see Figure 6-7. When the plate moves in the x-direction with a velocity U, the fluid in contact with the plate, being viscous, moves with the velocity of the plate. Because of the viscous nature of the fluid, the fluid momentum propagates into the fluid in a direction normal to the plate. Writing a momentum balance (Newton's second law of motion) on the element shown in Figure 6-7 for a unit length yields

$$\Delta x(\tau_y - \tau_{y+\Delta y}) = \rho\,\Delta x\,\Delta y\frac{\partial u}{\partial t} \tag{6.27}$$

where τ_y is the shear stress on the surface of the element at location y and u is fluid velocity. Rewriting Equation (6.27) as

$$-\Delta x\,\Delta y\left[\frac{\tau_{y+\Delta y} - \tau_y}{\Delta y}\right] = \rho\,\Delta x\,\Delta y\frac{\partial u}{\partial t} \tag{6.28}$$

and with $\Delta y \to 0$ in Equation (6.28), there results

$$-\frac{\partial \tau}{\partial y} = \rho\frac{\partial u}{\partial t} \tag{6.29}$$

The shear stress for a Newtonian fluid (such as water, air, and many fluids) is related to the viscosity, μ, and fluid velocity by Newton's law of friction stated in the form

$$\tau = -\mu\frac{\partial u}{\partial y} \tag{6.30}$$

Introducing Equation (6.30) in Equation (6.29) we get, for constant viscosity

$$\mu\frac{\partial^2 u}{\partial y^2} = \rho\frac{\partial u}{\partial t}$$

or

$$\nu\frac{\partial^2 u}{\partial y^2} = \frac{\partial u}{\partial t}; \quad \nu = \frac{\mu}{\rho} \tag{6.31}$$

where ν is referred to as the kinematic viscosity or molecular diffusivity for

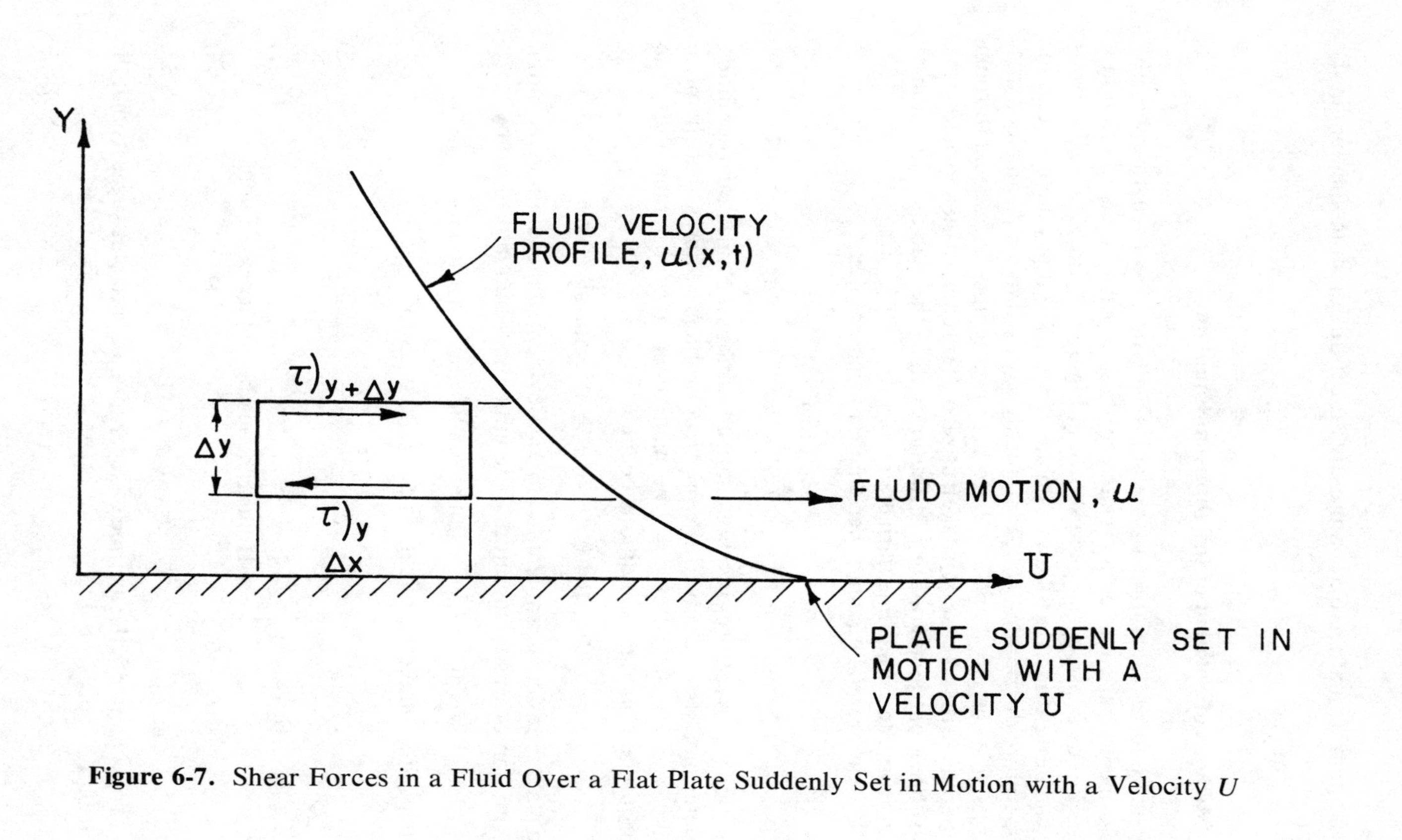

Figure 6-7. Shear Forces in a Fluid Over a Flat Plate Suddenly Set in Motion with a Velocity U

momentum. Equation (6.31) represents the diffusion of momentum in the fluid.

6.6 Methods of Solution of the Diffusion Equation

The method of separation of variables will first be used to solve the one-dimensional equation as related to the temperature distribution in a plate. Let us consider the plate shown in Figure 6-8. Physically, the problem can be described as follows: A plate of width l, initially having a certain temperature distribution $f(x)$, is suddenly exposed to a cold ambient such that its sides at $x = 0$ and $x = l$ are kept at a temperature equal to room temperature; in this case, it is taken as zero. It is required then to determine the subsequent temperature distribution within the slab as a function of space variable x and time variable t.

The differential equation pertaining to this problem is

$$\alpha \frac{\partial^2 T}{\partial x^2} = \frac{\partial T}{\partial t} \tag{6.32}$$

Equation (6.32) requires two boundary conditions and one initial condition to determine its solution. If the physical problem requires that the plate surfaces at $x = 0$ and at $x = l$ be maintained at zero temperature, the appropriate boundary conditions become

$$T(0,t) = 0, \qquad T(l,t) = 0 \qquad \text{for all } t$$

The plate time dependent temperature, as evident, depends upon the initial distribution of this temperature. The initial distribution is given by

$$T(x,0) = f(x) \qquad \text{initial temperature distribution}$$

Hence, the complete problem is posed now as follows:

$$\alpha \frac{\partial^2 T}{\partial x^2} = \frac{\partial T}{\partial t} \tag{6.32}$$

$$\left.\begin{aligned} T(0,t) &= 0 \qquad (6.33) \\ T(l,t) &= 0 \qquad (6.34) \end{aligned}\right\} \quad \text{Homogeneous boundary conditions}$$

$$T(x,0) = f(x) \tag{6.35}$$

From chapter 5, the method of separation of variables suggests that the solution be represented as

$$T(x,t) = X(x)\,G(t) \tag{6.36}$$

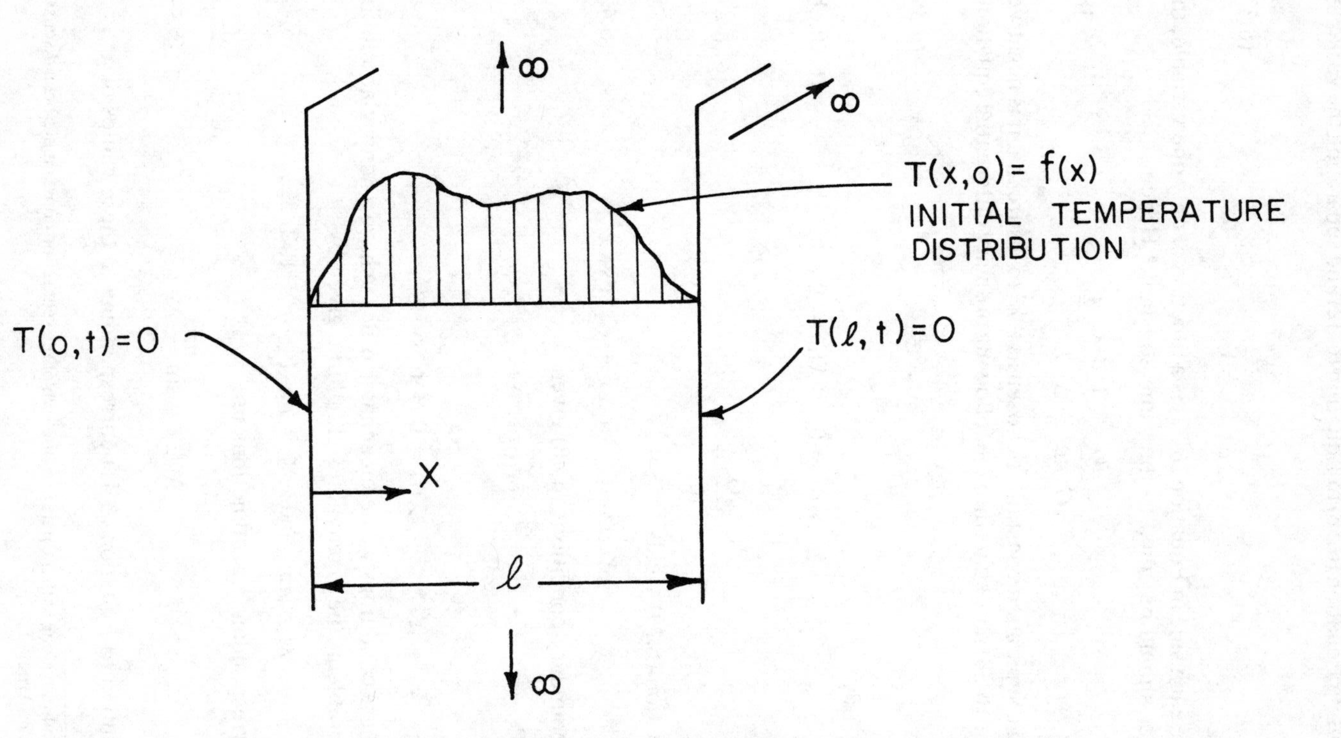

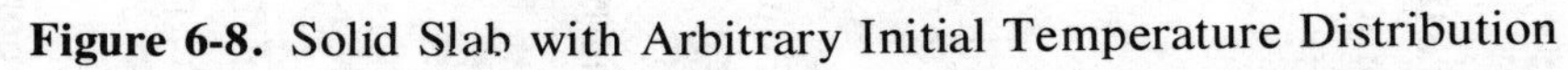
Figure 6-8. Solid Slab with Arbitrary Initial Temperature Distribution

Substituting Equation (6.36) in Equation (6.32) and separating the variables yields

$$\frac{1G'}{\alpha G} = \frac{X''}{X} \tag{6.37}$$

Because each side in Equation (6.37) is a function of only one variable, the two sides should be equal to the same constant.[a] Hence

$$\frac{1}{\alpha G}\frac{dG}{dt} = \frac{1}{X}\frac{d^2X}{dx^2} = -\lambda^2 \tag{6.38}$$

where the separation constant is chosen as $-\lambda^2$ to insure that it is negative. Equation (6.38) now splits into the following ordinary differential equations

$$\frac{d^2X}{dx^2} + \lambda^2 X = 0 \tag{6.39}$$

with

$$X(0) = 0, \quad X(l) = 0 \tag{6.40a,b}$$

and

$$\frac{dG}{dt} + \lambda^2 \alpha\, G = 0 \tag{6.41}$$

Equation (6.39) yields

$$X(x) = C_1 \sin \lambda x + C_2 \cos \lambda x \tag{6.42}$$

Using boundary conditions (6.40) gives

$$X(0) = 0 = C_2$$

and

$$X(l) = 0 = C_1 \sin \lambda l$$

From these conditions we deduce that in order to have a nontrivial solution to the problem, the eigenvalues should be given by

$$\lambda l = n\pi \quad \text{and} \quad \lambda_n = n\pi/l \quad n = 1, 2, 3, \ldots \tag{6.43}$$

The corresponding eigenfunctions are

$$X_n(x) = \sin \frac{n\pi x}{l} \tag{6.44}$$

The solution to Equation (6.41) corresponding to these values of λ_n is

---

[a]This condition along with the nature of the sign of the separation constant has been discussed in detail in Article 5.8.

$$G_n(t) = C_n e^{-\lambda^2 \alpha t} \tag{6.45}$$

where C_n are coefficients to be determined. Therefore, according to Equation (6.36) the solution can be written in the form

$$T_n(x,t) = X_n(x)\, G_n(t) = C_n\, e^{-\lambda^2 \alpha t} \sin \frac{n\pi x}{l} \tag{6.46}$$

$$n = 1, 2, 3, \ldots$$

Equation (6.46) satisfies the differential equation and the boundary conditions imposed on the problem. We can easily see, however, that a single solution from Equation (6.46) does not satisfy the imposed initial conditions. To satisfy the initial conditions, we consider the sum of all the particular solutions given by Equation (6.46) and obtain

$$\begin{aligned} T(x,t) &= \sum_{n=1}^{\infty} T_n(x,t) \\ &= \sum_{n=1}^{\infty} C_n e^{-\lambda^2 \alpha t} \sin \frac{n\pi x}{l} \end{aligned} \tag{6.47}$$

Applying now the initial condition given by Equation (6.35) yields

$$T(x,0) = f(x) = \sum_{n=1}^{\infty} C_n \sin \frac{n\pi x}{l} \tag{6.48}$$

Equation (6.48) represents the expansion of an arbitrary function $f(x)$ in terms of a Fourier sine series. The coefficient, C_n, of this expansion is given by

$$C_n = \frac{2}{l} \int_0^l f(x) \sin \frac{n\pi x}{l} dx \tag{6.49}$$

The expression for $T(x,t)$ as given by Equation (6.47) with the coefficients C_n given by Equation (6.49) satisfies all the conditions of the problem and therefore constitutes the solution to the problem. As was indicated in the previous chapters, the series representation of $T(x,t)$ does converge. The series representation for $\partial^2 T/\partial x^2$ and $\partial T/\partial t$ as obtained by termwise differentiation of Equation (6.47) do converge and are continuous.

Example 6.1 Consider a plate of width l having a constant initial temperature T_0. If the boundaries suddenly are dropped in temperature to a value of zero, find the expression representing the time dependent temperature distribution. (Figure 6-9)

Solution: This problem is basically the one just analyzed in this section and the temperature distribution is represented by Equation (6.47), that is,

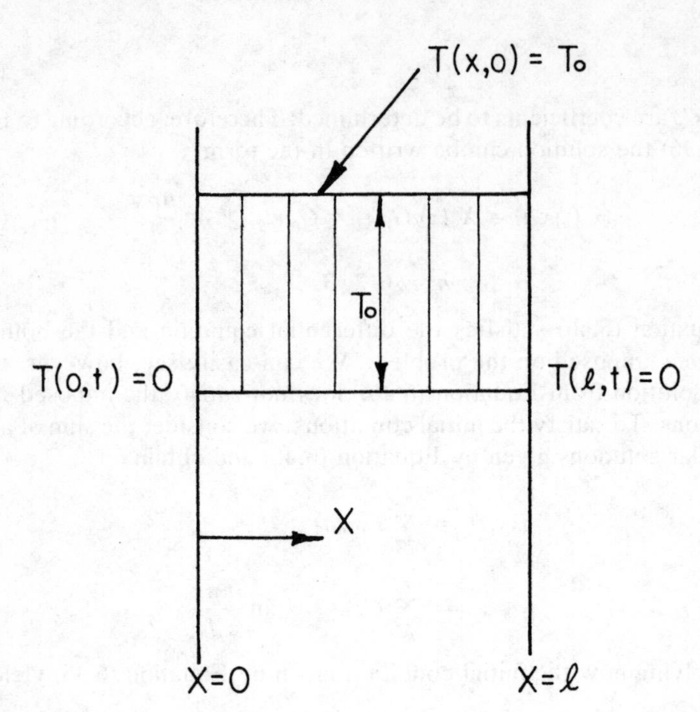

Figure 6-9. Solid Slab with Uniform Initial Temperature Distribution

$$T(x,t) = \sum_{n=1}^{\infty} C_n e^{-\lambda^2 \alpha t} \sin \frac{n\pi x}{l} \tag{6.50}$$

and C_n is given by Equation (6.49) as

$$C_n = \frac{2}{l}\int_0^l f(x) \sin \frac{n\pi x}{l} dx \tag{6.51}$$

$f(x)$ in this problem is equal to T_0. Hence,

$$C_n = \frac{2T_0}{l}\int_0^l \sin \frac{n\pi x}{l} dx \tag{6.52}$$

$$= \frac{2T_0}{n\pi}\left[-\cos \frac{n\pi x}{l}\right]_0^l = \frac{2T_0}{n\pi}[1 - \cos n\pi] \tag{6.53}$$

Therefore, $T(x,t)$ becomes

$$T(x,t) = \frac{2T_0}{\pi}\sum_{n=1}^{\infty} \frac{(1 - \cos n\pi)}{n} e^{-(n\pi/l)^2 \alpha t} \sin \frac{n\pi x}{l} \tag{6.54}$$

Note that in Equation (6.54), the terms with $n = 2, 4, 6, \ldots$ will drop out since $1 - \cos n\pi = 0$ for even values of n. Only terms with odd values for n

will remain, $n = 1, 3, 5, \ldots$ and for these values of n $1 - \cos n\pi = 2$. Therefore, Equation (6.54) can be written as

$$T(x,t) = \frac{4T_0}{\pi} \sum_{m=0}^{\infty} \frac{1}{2m + 1} e^{-\{[(2m+1)\pi]/l\}^2 \alpha t} \sin \frac{(2m + 1)\pi x}{l} \quad (6.55)$$

Figure 6-10 shows the shape of the temperature distribution at various times. As $t \to \infty$, the steady state solution to the problem is attained. In Equation (6.54), as $t \to \infty$, $T(x,t) \to 0$. Physically this is expected.

6.7 Nonhomogeneous Boundary Conditions

Cases in which the boundary conditions of the problem are not homogeneous are many and are of greater physical interest. Physically, one type of nonhomogeneity in the boundary conditions appears when a temperature other than zero is prescribed at one or more than one boundary. This prescribed temperature may be a constant or may be a function of time. The simplest case occurs when both sides of the slab analyzed in Article 6.6 are subjected suddenly to constant temperature different from zero. In this case the problem is modified to the following (Figure 6-11)

$$\alpha \frac{\partial^2 T}{\partial x^2} = \frac{\partial T}{\partial t} \quad (6.56)$$

$$T(0,t) = T_1 \quad (6.57)$$

$$T(l,t) = T_2 \quad (6.58)$$

$$T(x,0) = f(x) \quad (6.59)$$

It can easily be seen that when all the transients die out in this problem, the slab will have a temperature distribution invariant with time but varies in the x-direction between T_1 and T_2. Because the method of separation of variables requires that the boundary conditions be homogeneous in order for it to apply, we can then formulate the problem in the following way, which was discussed thoroughly in chapter 5.
Let us set

$$T(x,t) = v(x,t) + w(x) \quad (6.60)$$

Correspondingly we have

$$\frac{\partial^2 T}{\partial x^2} = \frac{\partial^2 v}{\partial x^2} + \frac{d^2 w}{dx^2} \quad (6.61)$$

and

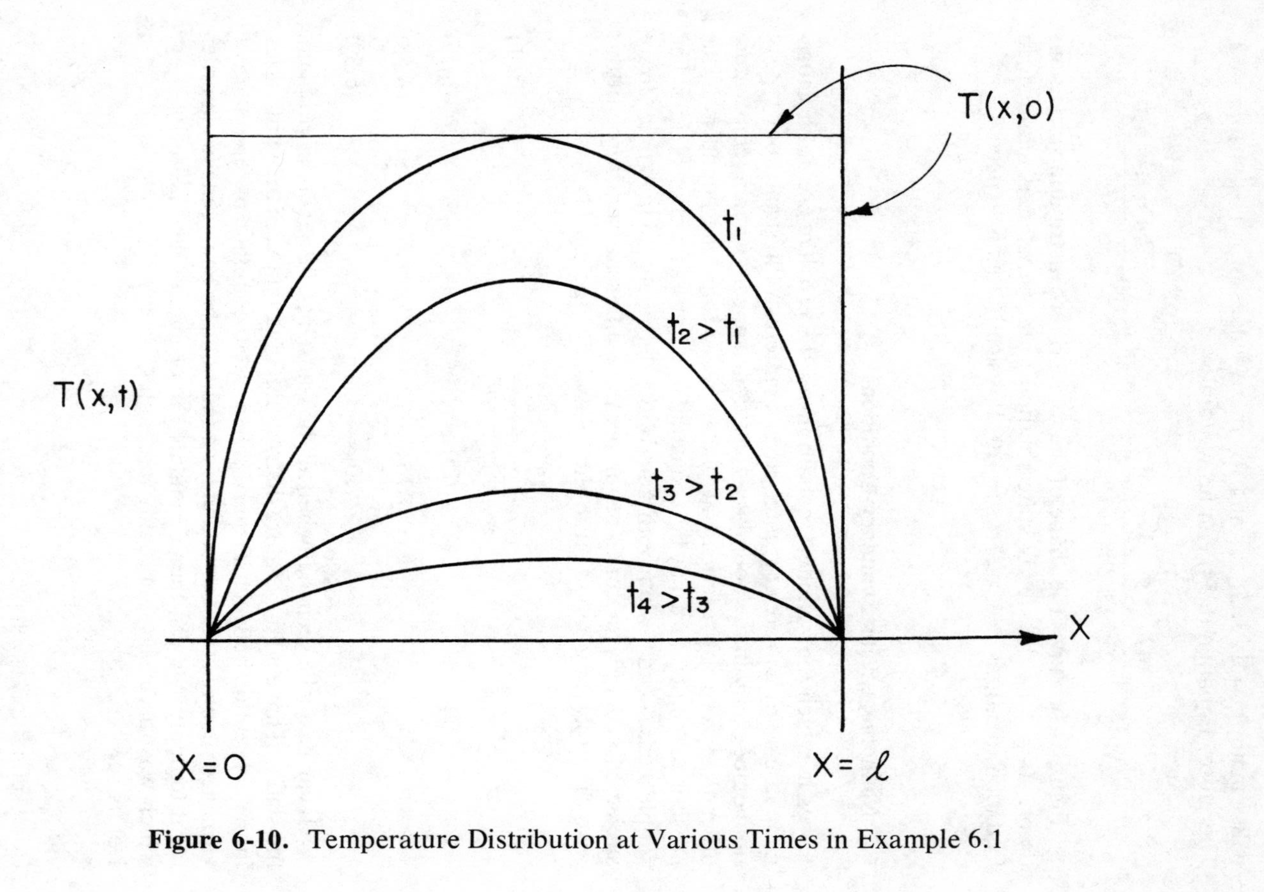

Figure 6-10. Temperature Distribution at Various Times in Example 6.1

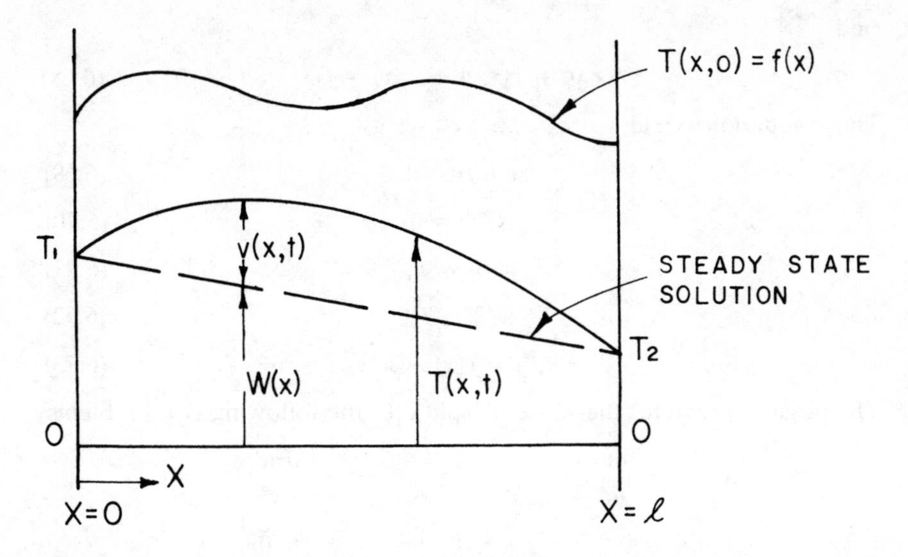

Figure 6-11. Nonhomogeneous Boundary Condition and Steady State Solution for a Slab

$$\frac{\partial T}{\partial t} = \frac{\partial v}{\partial t} + \cancelto{0}{\frac{\partial w}{\partial t}} \tag{6.62}$$

Using Equations (6.61) and (6.62) in Equation (6.56) yields

$$\alpha\left[\frac{\partial^2 v}{\partial x^2} + \frac{d^2 w}{dx^2}\right] = \frac{\partial v}{\partial t} \tag{6.63}$$

In order for $v(x,t)$ to satisfy the diffusion equation, we should have

$$\alpha \frac{\partial^2 v}{\partial x^2} = \frac{\partial v}{\partial t} \tag{6.64}$$

Hence,

$$\frac{d^2 w}{dx^2} = 0 \tag{6.65}$$

Equation (6.65) can be solved by direct integration. However, to solve Equation (6.64) by the method of separation of variables, homogeneous boundary conditions should be imposed on it. The boundary and initial conditions can, therefore, be written as follows:

$$T(0,t) = v(0,t) + w(0) = T_1 \tag{6.66}$$

$$T(l,t) = v(l,t) + w(l) = T_2 \tag{6.67}$$

and

$$T(x,0) = v(x,0) + w(x) = f(x) \tag{6.68}$$

These equations yield

$$v(0,t) = 0 \tag{6.69}$$

$$v(l,t) = 0 \tag{6.70}$$

$$w(0) = T_1 \tag{6.71}$$

$$w(l) = T_2 \tag{6.72}$$

$$v(x,0) = f(x) - w(x) \tag{6.73}$$

The original problem, therefore, is split into the following two problems

$\alpha \dfrac{\partial^2 v}{\partial x^2} = \dfrac{\partial v}{\partial t}$	$\dfrac{d^2 w}{dx^2} = 0$
$v(0,t) = 0$	$w(0) = T_1$
$v(l,t) = 0$	$w(l) = T_2$
$v(x,0) = f(x) - w(x)$	which yields
	$w(x) = T_1 + (T_2 - T_1)\dfrac{x}{l}$

The new problem posed by $v(x,t)$ can now be solved by the method of separation of variables. It is basically the problem analyzed in Article 6.6 with the solution given by Equation (6.47)

$$v(x,t) = \sum_{n=1}^{\infty} C_n e^{-\lambda^2 \alpha t} \sin \frac{n\pi x}{l} \tag{6.74}$$

where C_n is given now by

$$C_n = \frac{2}{l}\int_0^l [f(x) - w(x)] \sin \frac{n\pi x}{l} dx \tag{6.75}$$

The final solution to the original problem becomes

$$T(x,t) = w(x) + v(x,t)$$

or

$$T(x,t) = T_1 + (T_2 - T_1)\frac{x}{l} + \sum_{n=1}^{\infty} C_n e^{-\lambda^2 \alpha t} \sin \frac{n\pi x}{l} \tag{6.76}$$

and C_n is given by Equation (6.75).

The first two terms on the right-hand side of Equation (6.76) represent the steady state solution, whereas the third term gives the initial transients in the slab temperature. When $t \to \infty$ (steady state condition), the third term becomes zero and the solution becomes

$$T(x,t) = T_1 + (T_2 - T_1)\frac{x}{l} \tag{6.77}$$

This solution is the steady state solution to the problem.

Special Cases

Boundary Conditions at Zero. When the boundaries at $x = 0$ and $x = l$ are subjected to the same temperature, that is, $T(0,t) = T_1$, $T(l,t) = T_1$, the solution can be obtained directly from Equation (6.76) as

$$T = T_1 + \sum_{n=1}^{\infty} C_n e^{-\lambda^2 \alpha t} \sin \frac{n\pi x}{l} \tag{6.78}$$

If we set $\theta = T - T_1$, this solution becomes

$$\theta(x,t) = \sum_{n=1}^{\infty} C_n e^{-\lambda^2 \alpha t} \sin \frac{n\pi x}{l} \tag{6.79}$$

Of course, the steady state solution is $T = T_1$ or $\theta = 0$. The expansion coefficients C_n in this case become, from Equation (6.75)

$$C_n = \frac{2}{l}\int_0^l [f(x) - T_1] \sin \frac{n\pi x}{l} dx \tag{6.80}$$

Equation (6.79) basically represents the following boundary value problem

$$\alpha \frac{\partial^2 \theta}{\partial x^2} = \frac{\partial \theta}{\partial t} \tag{6.81}$$

$$\theta(0,t) = 0 \tag{6.82}$$

$$\theta(l,t) = 0 \tag{6.83}$$

$$\theta(x,0) = \bar{f}(x) = f(x) - T_1 \tag{6.84}$$

This problem, therefore, is exactly the one defined by Equations (6.32) to (6.35). Accordingly, if in a certain problem, a boundary is prescribed to be at zero temperature, in effect, this condition includes the case of a constant temperature at the boundary if the solution is interpreted as θ, ($\theta = T - T_1$).

Boundary Conditions at Unity. Let us now take the solution given by Equation (6.78) and define a temperature $\phi = T/T_1$. The solution then takes the following form

$$\phi = 1 + \sum_{n=1}^{\infty} \bar{C}_n \, e^{-\lambda^2 \alpha t} \sin \frac{n\pi x}{l} \tag{6.85}$$

with

$$\bar{C}_n = \frac{2}{l}\int_0^l \left[\frac{f(x)}{T_1} - 1\right] \sin \frac{n\pi x}{l} dx \tag{6.86}$$

This solution represents the following problem

$$\alpha \frac{\partial^2 \phi}{\partial x^2} = \frac{\partial \phi}{\partial t} \tag{6.87}$$

$$\phi(0,t) = 1 \tag{6.88}$$

$$\phi(l,t) = 1 \tag{6.89}$$

$$\phi(x,0) = \bar{f}(x) = \frac{f(x)}{T_1} \tag{6.90}$$

Accordingly, a prescribed surface temperature of unity includes the case of a constant temperature at the surface if the temperature is interpreted as ϕ, ($\phi = T/T_1$).

Nondimensional Temperature Distribution. Let us recall Example 6.1 and define a temperature $\psi(x,t)$ as

$$\psi(x,t) = \frac{T(x,t) - T_1}{T_0 - T_1} \tag{6.91}$$

The differential equation and the corresponding conditions become

$$\alpha \frac{\partial^2 \psi}{\partial x^2} = \frac{\partial \psi}{\partial t} \tag{6.92}$$

$$\psi(0,t) = 0 \tag{6.93}$$

$$\psi(l,t) = 0 \tag{6.94}$$

$$\psi(x,0) = 1 \tag{6.95}$$

Correspondingly, the solution for $\psi(x,t)$ becomes [Example 6.1, Equation (6.55)]

$$\psi(x,t) = \frac{4}{\pi}\sum_{m=0}^{\infty} \frac{1}{2m+1} e^{-\{[(2m+1)\pi]^2/l\}\alpha t} \sin \frac{(2m+1)\pi x}{l}$$

$$\equiv M(x,t) \tag{6.96}$$

Hence,

$$\frac{T - T_1}{T_0 - T_1} = M(x,t) \tag{6.97}$$

If $T_1 = 0$, there results

$$\frac{T}{T_0} = M(x,t) \quad \text{(zero boundary condition)} \tag{6.98}$$

If $T_0 = 0$, we have

$$\frac{T}{T_1} = 1 - M(x,t) \quad \text{(zero initial condition)} \tag{6.99}$$

These findings can be outlined as follows:

$$\bar{\phi} = \frac{T}{T_0},$$

T_0 = initial temperature = constant

$$\alpha\frac{\partial^2\bar{\phi}}{\partial x^2} = \frac{\partial\bar{\phi}}{\partial t}$$

$$\bar{\phi}(0,t) = 0$$

$$\bar{\phi}(l,t) = 0$$

$$\bar{\phi}(x,0) = 1$$

Solution:

$$\bar{\phi}(x,t) = M(x,t)$$

$$\phi \equiv \frac{T}{T_1},$$

T_1 = surface temperature = constant

$$\alpha\frac{\partial^2\phi}{\partial x^2} = \frac{\partial\phi}{\partial t}$$

$$\phi(0,t) = 1$$

$$\phi(l,t) = 1$$

$$\phi(x,0) = 0$$

Solution:

$$\phi(x,t) = 1 - M(x,t)$$

$M(x,t)$ is given by Equation (6.96).

6.8 Time Dependent Boundary Conditions—Duhamel's Superposition Integral

We will restrict for now our analysis to a body initially at zero temperature (Figure 6-12), that is,

$$T(x,0) = 0$$

and suddenly for $t > 0$, the end at $x = 0$ is kept at $T = 0$ while the temperature at $x = l$ is varied with time according to

$$T(0,t) = 0, \quad T(l,t) = F(t)$$

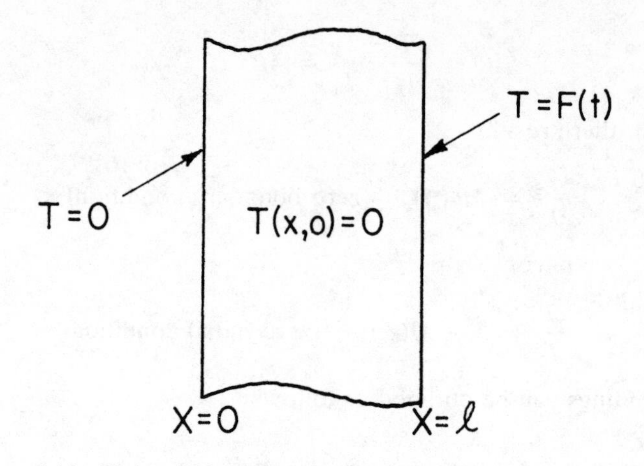

Figure 6-12. Slab with a Time Dependent Boundary Condition

Initially we solve the problem for the case where $F(t) = 1$ and establish a fundamental solution upon which the time dependent solution will depend. From the previous section, the problem in which the following conditions hold

$$T(0,t) = 0$$

$$T(l,t) = 1$$

$$T(x,0) = 0$$

can be readily solved with the solution defined as $N(x,t)$ and determined to be

$$\begin{aligned} T(x,t) &= N(x,t) \\ &= \frac{x}{l} + \frac{2}{\pi} \sum_{n=1}^{\infty} \frac{(-1)^n}{n} e^{-(n\pi/l)^2 \alpha t} \sin \frac{n\pi x}{l} \end{aligned} \tag{6.100}$$

If at $t = 0$, the surface temperature at $x = l$ is raised to a constant value $F(0) > 1$ and held at that level, the solution then becomes

$$T(x,t) = F(0)\, N(x,t)$$

Suppose now that at $t = \tau_1$ the surface temperature is raised to $F(\tau_1)$ and held there; the solution now becomes

$$T(x,t) = F(0)\, N(x,t) + [F(\tau_1) - F(0)]\, N(x,t - \tau_1)$$

where the new contribution is acting for a time $t - \tau_1$, Figure 6-13. The addition of contribution is valid because of the linearity of the problem.

Suppose the surface temperature at $x = l$ undergoes such step changes in temperature as shown in Figure 6-14. The solution then can be written as

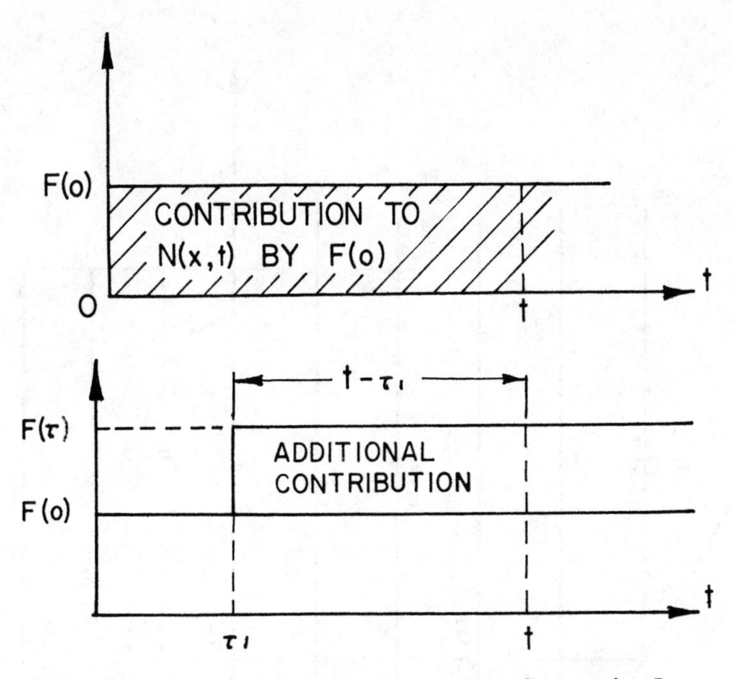

Figure 6-13. Superimposed Contributions at Two Successive Increments of Time

$$T(x,t) = F(0)\ N(x,t) + [F(\tau_1) - F(0)]\ N(x,t - \tau_1) + [F(\tau_2) - F(\tau_1)]\ N(x,t - \tau_2) + \ldots + [F(\tau_n) - F(\tau_{n-1})]\ N(x,t - \tau_n) \tag{6.101}$$

Define

$$\Delta F_n = F(\tau_n) - F(\tau_{n-1})$$

$$\Delta\tau_n = \tau_n - \tau_{n-1}$$

Equation (6.101) can then be written in the form

$$T(x,t) = F(0)\ N(x,t) + \sum_{n=1}^{\infty} N(x,t - \tau_n)\left(\frac{\Delta F}{\Delta\tau}\right)_n \Delta\tau_n \tag{6.102}$$

When the number of jumps becomes infinite such that the successive jumps ΔF and $\Delta\tau$ tend to zero, Equation (6.102) becomes

$$T(x,t) = F(0)\ N(x,t) + \int_0^t N(x,t - \tau)\frac{dF(\tau)}{d\tau}d\tau \tag{6.103}$$

This relation is referred to as *Duhamel's superposition integral.* Another

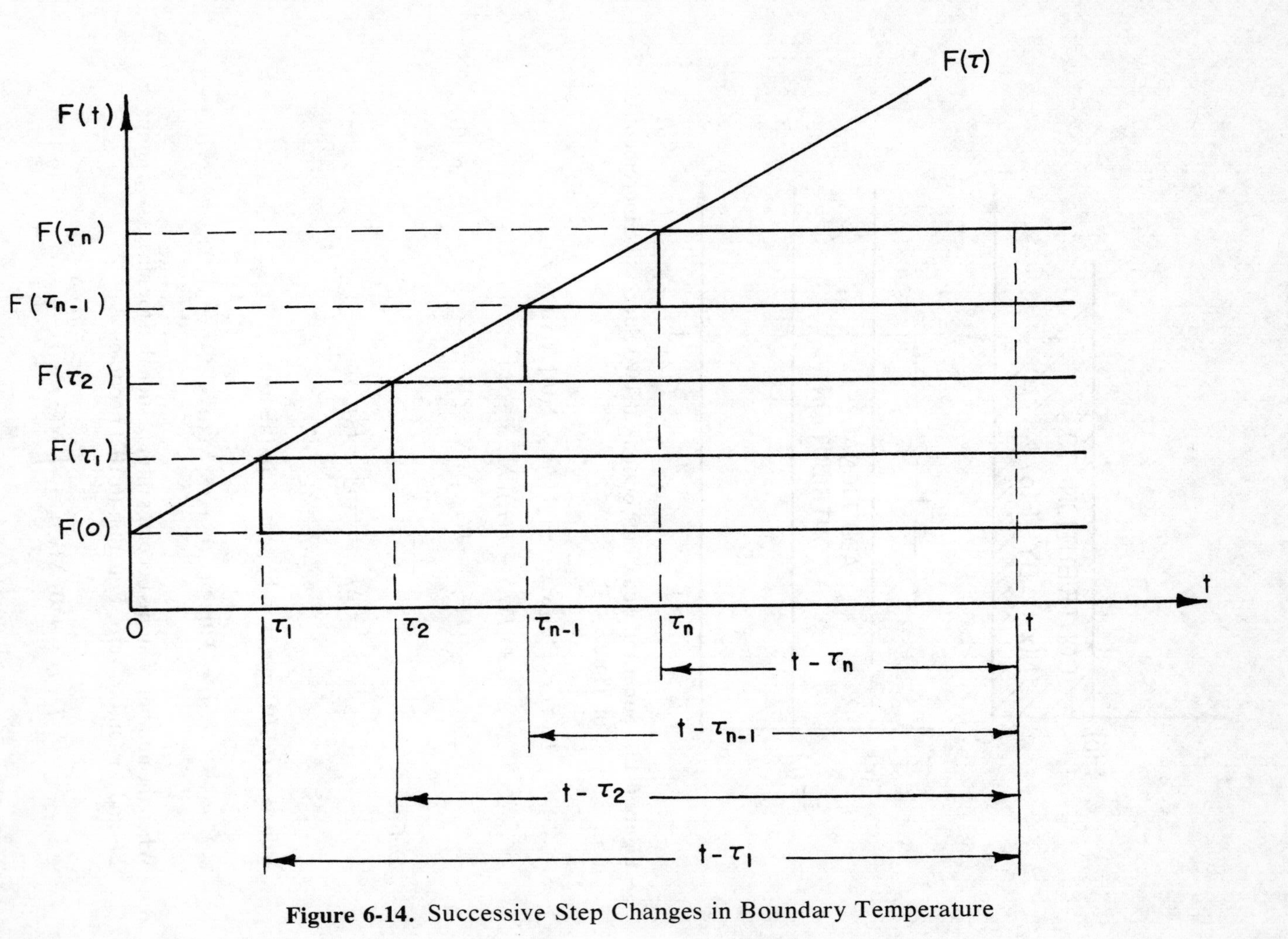

Figure 6-14. Successive Step Changes in Boundary Temperature

form of Equation (6.103) can be obtained and usually it is more convenient. Let us integrate Equation (6.103) by parts

$$T(x,t) = F(0)\,N(x,t) + \Big[N(x,t-\tau)\,F(\tau)\Big]_{\tau=0}^{\tau=t} - \int_0^t F(\tau)\frac{\partial N}{\partial \tau}(x,t-\tau)\,d\tau$$

This equation simplifies to

$$T(x,t) = N(x,0)\,F(t) + \int_0^t F(\tau)\frac{\partial N}{\partial t}(x,t-\tau)\,d\tau \tag{6.104}$$

where we used

$$\frac{\partial N}{\partial \tau}(x,t-\tau) = -\frac{\partial N}{\partial t}(x,t-\tau)$$

From the statement of the initial problem, the initial condition imposed on the problem is zero. Hence, $N(x,0) = 0$, and Equation (6.104) becomes

$$T(x,t) = \int_0^t F(\tau)\frac{\partial N}{\partial t}(x,t-\tau)\,d\tau \tag{6.105}$$

which is the alternative form for Duhamel's superposition integral.

Example 6.2 A slab of thickness l initially at a temperature $T_i = f(x)$ is suddenly exposed to ambient conditions such that the temperature at both surfaces ($x = 0$ and $x = l$) are functions of time (Figure 6-15). Find the temperature distribution within the slab.
Solution: The mathematical statement of the problem is represented as follows:

$$\alpha\frac{\partial^2 T}{\partial x^2} = \frac{\partial T}{\partial t} \tag{6.106}$$

$$T(0,t) = F_1(t) \tag{6.107}$$

$$T(l,t) = F_2(t) \tag{6.108}$$

$$T(x,0) = f(x) \tag{6.109}$$

The following points should be emphasized now:

1. The method of separation of variables requires that the boundary conditions be homogeneous.
2. Duhamel's superposition integral requires that the initial condition be zero.

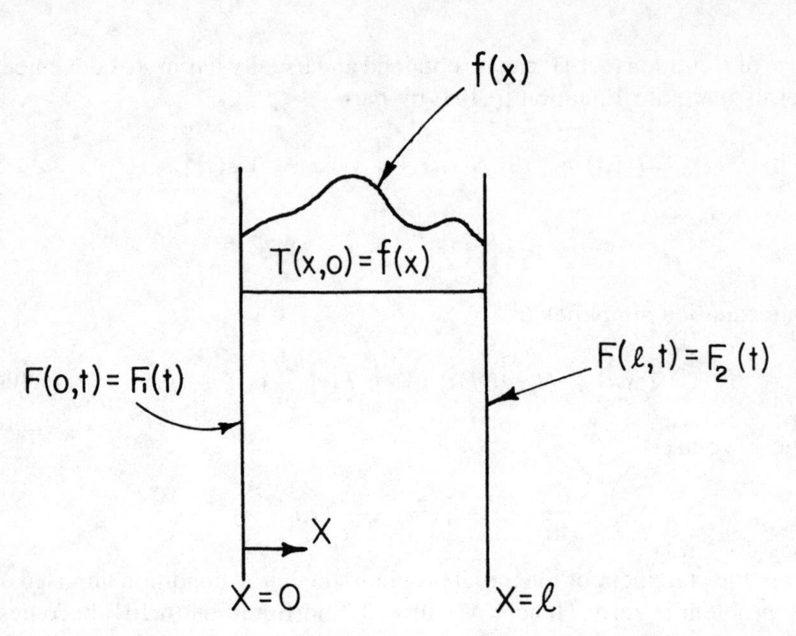

Figure 6-15. Slab with Two Sides Exposed to Time Dependent Boundary Conditions

Let the temperature distribution $T(x,t)$ be represented as

$$T(x,t) = u(x,t) + w(x,t) \tag{6.110}$$

where $u(x,t)$ satisfies the given initial conditions with zero boundary conditions while $w(x,t)$ satisfies the nonhomogeneous boundary conditions with zero initial conditions. Therefore, we obtain the following two problems:

$$\alpha \frac{\partial^2 u}{\partial x^2} = \frac{\partial u}{\partial t} \qquad \alpha \frac{\partial^2 w}{\partial x^2} = \frac{\partial w}{\partial t}$$

$$u(0,t) = 0 \qquad w(0,t) = F_1(t)$$

$$u(l,t) = 0 \qquad w(l,t) = F_2(t)$$

$$u(x,0) = f(x) \qquad w(x,0) = 0$$

The solution for $u(x,t)$ is given by Equations (6.47) and (6.49) as

$$u(x,t) = \frac{2}{l}\sum_{n=1}^{\infty} e^{-(n\pi/l)^2 \alpha t} \sin\frac{n\pi x}{l}\int_0^l f(x) \sin\frac{n\pi x}{l}\, dx \tag{6.111}$$

To obtain the solution for $w(x,t)$ we use Duhamel's integral (Equation (6.105)). $w(x,t)$ can be written as

$$w(x,t) = w_1(x,t) + w_2(x,t) \qquad (6.112)$$

where $w_1(x,t)$ and $w_2(x,t)$ satisfy the following

$\alpha \dfrac{\partial^2 w_1}{\partial x^2} = \dfrac{\partial w_1}{\partial t}$	$\alpha \dfrac{\partial^2 w_2}{\partial x^2} = \dfrac{\partial w_2}{\partial t}$
$w_1(0,t) = F_1(t)$	$w_2(0,t) = 0$
$w_1(l,t) = 0$	$w_2(l,t) = F_2(t)$
$w(x,0) = 0$	$w(x,0) = 0$

Using Article 6.7 and Equations (6.100) and (6.105) the solutions for $w_1(x,t)$ and $w_2(x,t)$ can readily be written as

$$w_1(x,t) = \int_0^t F_1(\tau) \frac{\partial N_1}{\partial t}(x,t-\tau)\, d\tau \qquad (6.113)$$

$$w_2(x,t) = \int_0^t F_2(\tau) \frac{\partial N_2}{\partial t}(x,t-\tau)\, d\tau \qquad (6.114)$$

where

$$N_1(x,t-\tau) = 1 - \frac{x}{l} - \frac{2}{\pi} \sum_{n=1}^{\infty} \frac{1}{n} e^{-(n\pi/l)^2 \alpha (t-\tau)} \sin \frac{n\pi x}{l} \qquad (6.115)$$

$$N_2(x,t-\tau) = \frac{x}{l} + \frac{2}{\pi} \sum_{n=1}^{\infty} \frac{(-1)^n}{n} e^{-(n\pi/l)^2 \alpha (t-\tau)} \sin \frac{n\pi x}{l} \qquad (6.116)$$

Finally, the solution is obtained by taking the sum of Equations (6.111), (6.113) and (6.114), that is,

$$T(x,t) = u(x,t) + w_1(x,t) + w_2(x,t) \qquad (6.117)$$

6.9 Similarity Solution of the Diffusion Equation

In this section, a powerful approach, referred to as similarity[b] transforma-

[b]Excellent coverage of the subject of similarity solutions can be found in the work of T.Y. Na. We cite here the following two: T.Y. Na, "Similarity Solution of Power Law Fluids Near an Oscillating Plate," *AIAA Journal* (Feb. 1965): 378-79. T.Y. Na, and A.G. Hansen, "Similarity Analysis of Differential Equations by Lie Group," *Journal of the Franklin Institute* 292, no. 6 (Dec. 1971).

tion, is presented for solving some partial differential equations of mathematical physics. The idea behind the method is simply to change the given form of a problem into a simpler form that can be solved by some standard solution techniques. In the new and simpler form, the number of the independent variables is reduced. We apply the method to solve the diffusion equation

$$\alpha \frac{\partial^2 \theta}{\partial x^2} = \frac{\partial \theta}{\partial t} \tag{6.118}$$

subject to the conditions

$$\theta(0,t) = \theta_0 \tag{6.119}$$

$$\theta(\infty,t) = 0 \tag{6.120}$$

$$\theta(x,0) = 0 \tag{6.121}$$

Let us introduce now the following transformations

$$t = A^{\beta_1}\bar{t}, \qquad x = A^{\beta_2}\bar{x}, \qquad \theta = A^{\beta_3}\bar{\theta} \tag{6.122}$$

In terms of these new variables, Equation (6.118) becomes

$$\alpha\, A^{\beta_3 - 2\beta_2} \frac{\partial^2 \bar{\theta}}{\partial \bar{x}^2} = A^{\beta_3 - \beta_1} \frac{\partial \bar{\theta}}{\partial \bar{t}} \tag{6.123}$$

Equation (6.123) will be invariant when compared to Equation (6.118) if

$$\beta_3 - 2\beta_2 = \beta_3 - \beta_1 \quad \text{or} \quad \beta_2 = \beta_1/2 \tag{6.124}$$

Because

$$\frac{x}{(t)^{\beta_2/\beta_1}} = \frac{\bar{x}}{(\bar{t})^{\beta_2/\beta_1}} \quad \text{and} \quad \frac{\theta}{(t)^{\beta_3/\beta_1}} = \frac{\bar{\theta}}{(\bar{t})^{\beta_3/\beta_1}} \tag{6.125a,b}$$

we can define new invariant variables and write

$$\eta \equiv \frac{x}{t^{\beta_2/\beta_1}} = \frac{x}{t^{1/2}}\,; \quad f(\eta) \equiv \frac{\theta}{t^{\beta_3/\beta_1}} \tag{6.126a,b}$$

From Equation (6.119) we have

$$A^{\beta_3}\bar{\theta} = \theta_0 \tag{6.127}$$

Therefore, β_3 should be zero and $f(\eta) = \theta/t^0 = \theta(\eta)$. Correspondingly, if the new invariants η and θ are used, there result

$$\frac{d\eta}{dx} = \frac{1}{t^{1/2}}, \qquad \frac{d\eta}{dt} = -\frac{1}{2}\frac{x}{t^{1/2}}\frac{1}{t} = -\frac{1}{2}\frac{\eta}{t}$$

and

$$\frac{\partial \theta}{\partial t} = \frac{\partial \theta}{\partial \eta}\frac{d\eta}{dt} = -\frac{1}{2}\frac{\eta}{t}\frac{d\theta}{d\eta}$$

$$\frac{\partial \theta}{\partial x} = \frac{\partial \theta}{\partial \eta}\frac{d\eta}{dx} = \frac{1}{t^{1/2}}\frac{d\theta}{d\eta}$$

$$\frac{\partial^2 \theta}{\partial x^2} = \frac{1}{t}\frac{d^2\theta}{d\eta^2}$$

When these derivatives are used in Equation (6.119), we get

$$\frac{d^2\theta}{d\eta^2} + \frac{\eta}{2\alpha}\frac{d\theta}{d\eta} = 0 \tag{6.128}$$

The conditions given by Equations (6.119) to (6.121) are reduced to the following two conditions

$$\theta(0) = \theta_0, \qquad \theta(\infty) = 0 \tag{6.129a,b}$$

Accordingly, the original partial differential equation, Equation (6.118), is transformed into an ordinary differential equation in which θ is a function of one variable η. The solution to Equation (6.128) is obtained by direct integration as follows: Let $z = d\theta/d\eta$ in Equation (6.128); we get

$$\frac{dz}{d\eta} = -\frac{\eta}{2\alpha} z$$

Separating the variables and integrating yields

$$\int \frac{dz}{z} = -\frac{1}{2\alpha}\int \eta \, d\eta + C$$

or

$$z = C_1 \, e^{-(\eta^2/4\alpha)} = \frac{d\theta}{d\eta}$$

Hence,

$$\theta = C_1 \int_0^{\eta} e^{-(\bar{\eta}^2/4\alpha)} \, d\bar{\eta} + C_2 \tag{6.130}$$

Applying the conditions given by Equation (6.129a,b), namely

$$\eta = 0, \theta = \theta_0; \qquad \eta = \infty, \theta = 0$$

we get

$$C_2 = \theta_0$$

and

$$0 = C_1 \int_0^{\infty} e^{-(\bar{\eta}^2/4\alpha)} \, d\bar{\eta} + \theta_0 \tag{6.131}$$

If we set $\bar{\xi}^2 = \bar{\eta}^2/4\alpha$ in this equation, there results

$$0 = C_1\sqrt{4\alpha}\int_0^\infty e^{-\bar{\xi}^2}\, d\bar{\xi} + \theta_0 \tag{6.132}$$

with

$$\int_0^\infty e^{-\bar{\xi}^2}\, d\bar{\xi} = \frac{\sqrt{\pi}}{2}$$

the value of C_1 becomes

$$C_1 = -\frac{\theta_0}{\sqrt{\alpha\pi}} \tag{6.133}$$

Introducing the values of C_1 and C_2 in Equation (6.130) gives

$$\frac{\theta}{\theta_0} = \left[1 - \frac{2}{\sqrt{\pi}}\int_0^\xi e^{-\bar{\xi}^2}\, d\bar{\xi}\right], \quad \xi = \frac{\eta}{\sqrt{4\alpha}} = \frac{x}{\sqrt{4\alpha t}} \tag{6.134}$$

The integral expressed by

$$\frac{2}{\sqrt{\pi}}\int_0^\xi e^{-\bar{\xi}^2}\, d\bar{\xi}$$

is referred to as the error integral or error function and is designated by erf ξ. The quantity $1 - \operatorname{erf}\xi$ is the complementary error function and is designated by erfc ξ. The graphs of both erf ξ and erfc ξ are shown in Figure 6-16.

Example 6.3 Find the rate of heat transfer into a semi-infinite solid initially at a zero temperature and suddenly the surface at $x = 0$ is exposed to a temperature T_0.

Solution: Obviously, the temperature distribution for this example is given by Equation (6.134) as

$$\frac{T}{T_0} = \operatorname{erfc}\left(\frac{x}{\sqrt{4\alpha t}}\right) = 1 - \frac{2}{\sqrt{\pi}}\int_0^{x/\sqrt{4\alpha t}} e^{-\xi^2}\, d\xi \tag{6.135}$$

The rate of heat transfer into the slab per unit area is

$$q)_{x=0} = \left(-k\frac{\partial T}{\partial x}\right)_{x=0} \tag{6.136}$$

Using Equation (6.135) in Equation (6.136) we get

$$\frac{\partial T}{\partial x} = \frac{\partial T}{\partial \xi}\frac{d\xi}{dx} \tag{6.137}$$

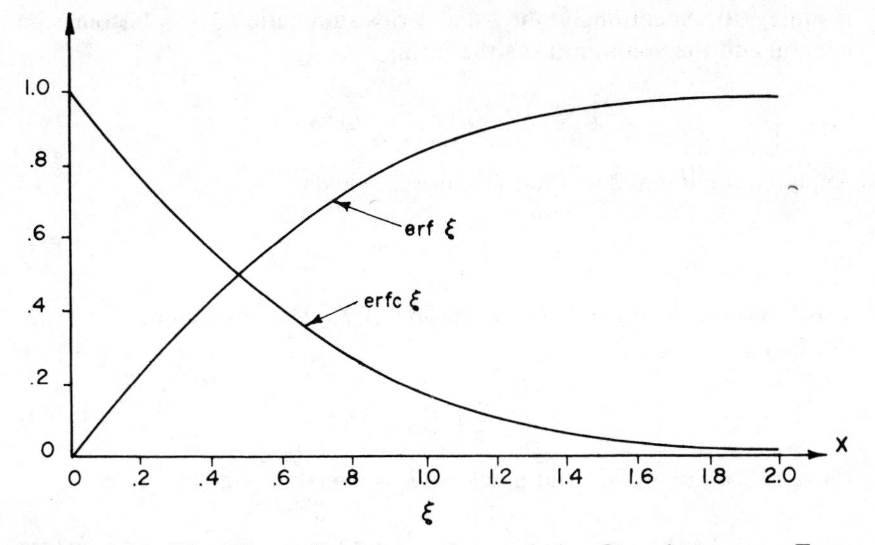

Figure 6-16. The Error Function erf ξ and the Complementary Error Function erfc ξ

$$= -\frac{2T_0}{\sqrt{\pi}}(e^{-(x^2/4\alpha t)})\frac{1}{\sqrt{4\alpha t}} = -\frac{T_0}{\sqrt{\pi\alpha t}}e^{-(x^2/4\alpha t)} \qquad (6.138)$$

Hence,

$$-k\left.\frac{\partial T}{\partial x}\right)_{x=0} = +\frac{kT_0}{\sqrt{\pi\alpha t}} \qquad (6.139)$$

It can readily be shown that if the initial temperature of the solid is T_i and the surface at $x = 0$ is suddenly exposed to a temperature equal to zero, the temperature distribution within the solid will be

$$\frac{T}{T_i} = \text{erf}\left(\frac{x}{\sqrt{4\alpha t}}\right) = \frac{2}{\sqrt{\pi}}\int_0^{x/\sqrt{4\alpha t}} e^{-\xi^2}\, d\xi \qquad (6.140)$$

This case can be solved easily by the method of separation of variables. Let the problem conditions be

$$T(x,0) = T_i; \quad T(0,t) = 0, \quad T(\infty,t) = 0$$

From separation of variables we have

$$T_\lambda = e^{-\lambda^2\alpha t}[c_1 \cos\lambda x + c_2 \sin\lambda x] \qquad (6.141)$$

From $T(0,t) = 0$, $c_1 = 0$. No other finite length exists to evaluate λ. This requires us to consider all possible solutions and values for λ (integral and nonintegral). Accordingly, the usual series summation $0 - \infty$ becomes an integral and the solution takes the form

$$T = \int_{\lambda=0}^{\infty} c(\lambda)\, e^{-\lambda^2 \alpha t} \sin \lambda x \, d\lambda \tag{6.142}$$

Applying the initial condition $T(x,0) = T_i$ yields

$$T_i = \int_{\lambda=0}^{\infty} c(\lambda) \sin \lambda x \, d\lambda \tag{6.143}$$

This is Fourier's integral representation of T_i. The coefficient $c(\lambda)$ is defined by

$$c(\lambda) = \frac{2}{\pi} \int_0^{\infty} T_i \sin \lambda x' \, dx' \tag{6.144}$$

Using (6.144) in (6.142) and invoking T_i = constant we get

$$T(x,t) = \frac{2T_i}{\pi} \int_{x'=0}^{\infty} \left[\int_{\lambda=0}^{\infty} e^{-\lambda^2 \alpha t} \sin \lambda x' \sin \lambda x \, d\lambda \right] dx' \tag{6.145}$$

The integral with respect to λ becomes

$$\int_{\lambda=0}^{\infty} e^{-\lambda^2 \alpha t} \sin \lambda x' \sin \lambda x \, d\lambda$$

$$= \frac{1}{2} \left\{ \int_{\lambda=0}^{\infty} e^{-\lambda^2 \alpha t} [\cos \lambda (x' - x) - \cos \lambda (x' + x)] \, d\lambda \right\} \tag{6.146}$$

$$= \frac{1}{4} \sqrt{\frac{\pi}{\alpha t}} \, [e^{-(x'-x)^2/4\alpha t} - e^{(x'+x)^2/4\alpha t}] \tag{6.147}$$

Therefore, Equation (6.145) takes the form

$$T(x,t) = \frac{T_i}{2\sqrt{\pi \alpha t}} \int_{x'=0}^{\infty} [e^{-(x'-x)^2/4\alpha t} - e^{-(x'+x)^2/4\alpha t}] dx' \tag{6.148}$$

If we introduce the variable $\xi = x/\sqrt{4\alpha t}$ in (6.148) there results

$$T(x,t) = \frac{T_i}{\sqrt{\pi}} \left[\int_{-(x/\sqrt{4\alpha t})}^{\infty} e^{-\xi} \, d\xi - \int_{x/\sqrt{4\alpha t}}^{\infty} e^{-\xi^2} \, d\xi \right] \tag{6.149}$$

or

$$\frac{T(x,t)}{T_i} = \frac{1}{\sqrt{\pi}} \int_{-x/\sqrt{4\alpha t}}^{x/\sqrt{4\alpha t}} e^{-\xi^2} \, d\xi = \frac{2}{\sqrt{\pi}} \int_0^{x/\sqrt{4\alpha t}} e^{-\xi^2} \, d\xi \tag{6.150}$$

$$= \operatorname{erf}\left(\frac{x}{\sqrt{4\alpha t}} \right) \tag{6.151}$$

6.10 Diffusion in Cylindrical and Spherical Media

In a general form, the diffusion equation given by Equation (6.18) can be written as

$$\alpha \nabla^2 T + g(x,y,z,t) = \frac{\partial T}{\partial t} \tag{6.152}$$

In cylindrical coordinates, r, θ, and z, (see Figure 6-17a), we have

$$x = r\cos\theta, \qquad y = r\sin\theta, \qquad z = z$$

and accordingly, in the absence of source effects, Equation (6.152) becomes

$$\alpha\left[\frac{\partial^2 T}{\partial r^2} + \frac{1}{r}\frac{\partial T}{\partial r} + \frac{1}{r^2}\frac{\partial^2 T}{\partial \theta^2} + \frac{\partial^2 T}{\partial z^2}\right] = \frac{\partial T}{\partial t} \tag{6.153}$$

In spherical coordinates (Figure 6-17b) r, θ, and ψ, where

$$x = r\cos\psi\sin\theta, \qquad y = r\sin\psi\sin\theta, \qquad z = r\cos\theta$$

and in the absence of source effects, the equation becomes

$$\alpha\left[\frac{1}{r^2}\frac{\partial}{\partial r}\left(r^2\frac{\partial T}{\partial r}\right) + \frac{1}{r^2\sin\theta}\frac{\partial}{\partial \theta}\left(\sin\theta\frac{\partial T}{\partial \theta}\right) + \frac{1}{r^2\sin^2\theta}\frac{\partial^2 T}{\partial \psi^2}\right] = \frac{\partial T}{\partial t} \tag{6.154}$$

In what follows we specialize Equations (6.153) and (6.154) to problems of common physical applications.

Example 6.4 A very long solid cylinder of radius a initially having a temperature distribution $f(r)$ is suddenly plunged into a very cold bath such that its surface is maintained at zero temperature. Find the expression for the temperature distribution.

Solution: In this problem, because the initial condition is independent of θ and z, the subsequent temperature distribution will be independent of θ and z. Therefore, Equation (6.153) becomes

$$\alpha\left[\frac{\partial^2 T}{\partial r^2} + \frac{1}{r}\frac{\partial T}{\partial r}\right] = \frac{\partial T}{\partial t} \tag{6.155}$$

subject to:

$$T(0,t) = \text{finite} \tag{6.156}$$

$$T(a,t) = 0 \tag{6.157}$$

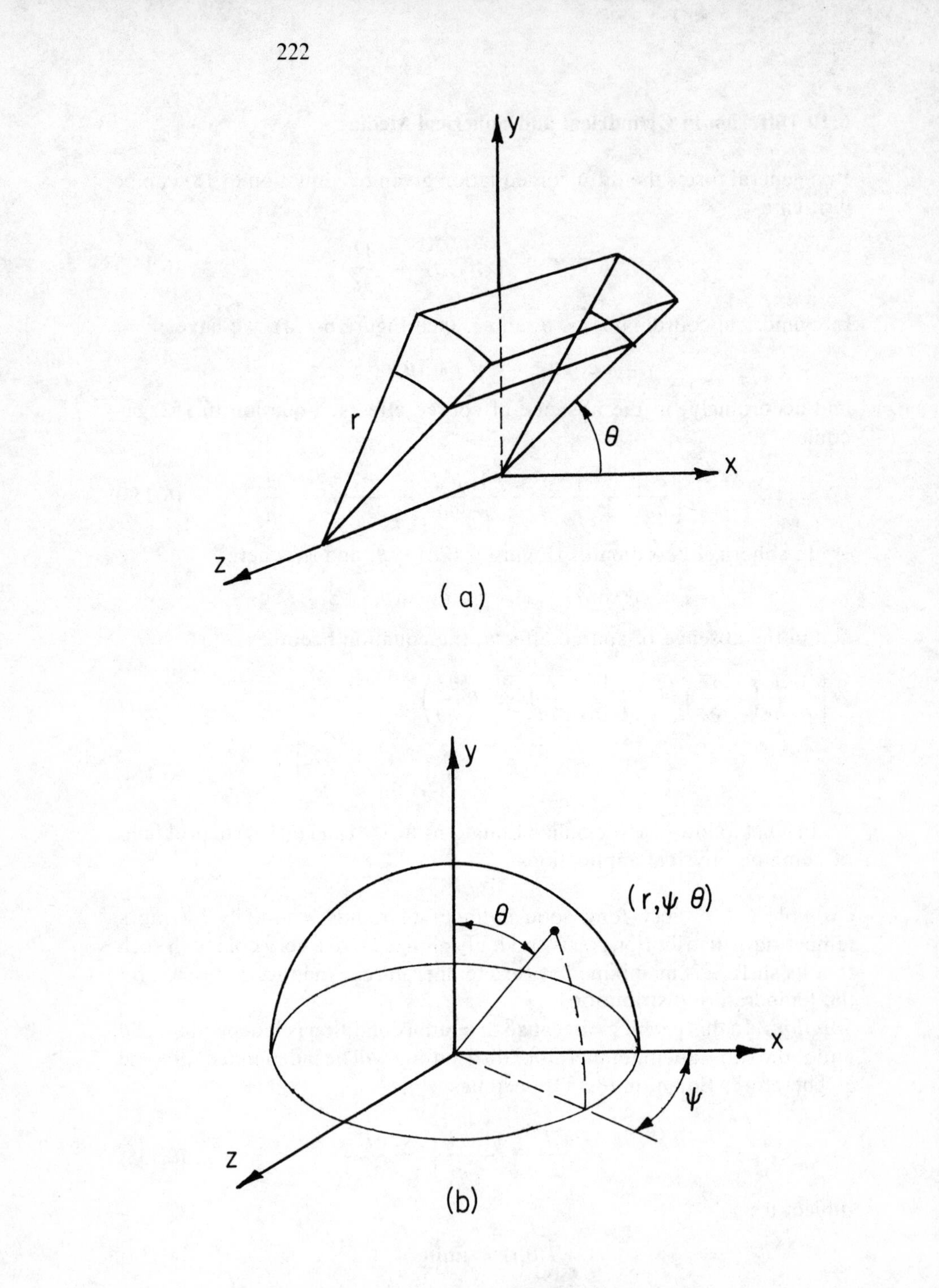

Figure 6-17. Cylindrical and Spherical Coordinates

$$T(r,0) = f(r) \tag{6.158}$$

By assuming $T(r,t) = R(r)\,G(t)$ and following the procedures for the method of separation of variables, Equation (6.155) yields the expressions

$$G(t) = C_1\, e^{-\lambda^2 \alpha t} \tag{6.159}$$

and

$$R(r) = A\, J_0(\lambda r) + B\, Y_0(\lambda r) \tag{6.160}$$

The condition given by Equation (6.156) requires that B should vanish. Equation (6.157) yields

$$J_0(\lambda a) = 0 \quad \text{and}\ \lambda = \lambda_n = \gamma_n / a, \qquad n = 1, 2, 3, \ldots \tag{6.161}$$

where γ_n are the zeros of $J_0(s) = 0$. Hence, the general solution for $T(r,t)$ becomes

$$T(r,t) = \sum_{n=1}^{\infty} a_n\, J_0(\lambda_n r)\, e^{-\lambda_n^2 \alpha t} \tag{6.162}$$

Applying now the initial condition given by Equation (6.158) gives

$$T(r,0) = f(r) = \sum_{n=1}^{\infty} a_n\, J_0(\lambda_n r) \tag{6.163}$$

and

$$a_n = \frac{\int_0^a r\, f(r)\, J_0(\lambda_n r)\, dr}{\int_0^a r\, J_0^2(\lambda r)\, dr}$$

$$= \frac{\int_0^a r f(r)\, J_0(\lambda_n r)\, dr}{\frac{a^2}{2} J_1^2(\lambda_n a)} \tag{6.164}$$

If $f(r)$ is a constant T_i, Equation (6.164) yields

$$a_n = \frac{T_i \int_0^a r\, J_0(\lambda_n r)\, dr}{\frac{a^2}{2} J_1^2(\lambda_n a)} = \frac{T_i\, a\, J_1(\lambda_n a)}{\lambda_n \frac{a^2}{2} J_1^2(\lambda_n a)} \tag{6.165}$$

or

$$a_n = \frac{2\, T_i}{a \lambda_n J_1(\lambda_n a)} \tag{6.166}$$

and the general solution becomes

$$T(r,t) = \frac{2\,T_i}{a}\sum_{n=1}^{\infty}\frac{1}{\lambda_n}\frac{J_0(\lambda_n r)}{J_1(\lambda_n a)}e^{-\lambda_n^2\alpha t}$$

Example 6.5 Consider now the problem of a solid sphere of radius a initially at a temperature $f(r)$ and suddenly its surface is exposed to zero temperature. Determine the temperature distribution within the sphere. *Solution*: For this problem, the temperature distribution will be independent of the angles θ and ψ, and accordingly, Equation (6.154) is reduced to

$$\alpha\left[\frac{\partial^2 T}{\partial r^2} + \frac{2}{r}\frac{\partial T}{\partial r}\right] = \frac{\partial T}{\partial t} \tag{6.168}$$

subject to:

$$T(0,t) = \text{finite} \tag{6.169}$$

$$T(a,t) = 0 \tag{6.170}$$

$$T(r,0) = f(r) \tag{6.171}$$

If we set $u = rT$ in Equation (6.168), the equation is reduced to

$$\alpha\frac{\partial^2 u}{\partial r^2} = \frac{\partial u}{\partial t} \tag{6.172}$$

with the conditions:

$$u(0,t) = 0 \tag{6.173}$$

$$u(a,t) = 0 \tag{6.174}$$

$$u(r,0) = r\,f(r) \tag{6.175}$$

Equations (6.172) to (6.175) represent now a problem similar to that analyzed in Article 6.6. From Equations (6.47) to (6.49) the result can be written directly as

$$T(r,t) = \frac{u(r,t)}{r} = \sum_{n=1}^{\infty}\frac{C_n}{r}\sin(\lambda_n r)e^{-\lambda_n^2\alpha t} \tag{6.176}$$

where

$$\lambda_n = \frac{n\pi}{a}$$

and

$$C_n = \frac{2}{a}\int_0^a r f(r)\sin\frac{n\pi r}{a}\,dr \tag{6.177}$$

If $f(r)$ is a constant T_i, C_n and $T(r,t)$ respectively become

$$C_n = \frac{2\,T_i}{a}\int_0^a r \sin\frac{n\pi r}{a}\,dr$$

$$= \frac{2\,T_i\,a}{n\pi}(-1)^{n+1} \tag{6.178}$$

and

$$T(r,t) = \frac{2\,T_i\,a}{\pi r}\sum_{n=1}^{\infty}\frac{(-1)^{n+1}}{n}\sin(\lambda_n r)e^{-\lambda_n^2 \alpha t} \tag{6.179}$$

Example 6.6 A sphere of permeable wall having a radius R is suspended in a large medium of quiescent air. At $t = 0$ the sphere is filled with a liquid such that at its surface, a vapor concentration u_0 is maintained at all times. The air initially has a zero vapor concentration. Find the concentration distribution $u(r,t)$ of the vapor around the sphere. The sphere and the air medium are at room temperature.

Solution: The concentration distribution in the air surrounding the sphere is governed by the differential equation

$$D\left[\frac{\partial^2 u}{\partial r^2} + \frac{2}{r}\frac{\partial u}{\partial r}\right] = \frac{\partial u}{\partial t} \qquad r > R \tag{6.180}$$

where D is the mass diffusivity. The equation is subject to

$$u(R,t) = u_0 \tag{6.181}$$

$$u(\infty,t) = 0 \tag{6.182}$$

$$u(r,0) = 0 \tag{6.183}$$

Introducing the transformation $v = ur$ and $\xi = r - R$ in Equation (6.180), we get

$$D\frac{\partial^2 v}{\partial \xi^2} = \frac{\partial v}{\partial t} \tag{6.184}$$

with

$$v(0,t) = R\,u_0 \tag{6.185}$$

$$v(\infty,t) = 0 \tag{6.186}$$

$$v(\xi,0) = 0 \tag{6.187}$$

Equations (6.184) to (6.187) now represent the case analyzed in Example 6.3. Therefore, the solution can be written directly as

$$\frac{v}{R\,u_0} = \text{erfc}\left(\frac{\xi}{2\sqrt{Dt}}\right) = \text{erfc}\left(\frac{r-R}{2\sqrt{Dt}}\right) \tag{6.188}$$

and with $v = ur$, we obtain

$$\frac{u(r,t)}{R\,u_0} = \frac{1}{r}\,\text{erfc}\left(\frac{r-R}{2\sqrt{Dt}}\right) \tag{6.189}$$

6.11 Nonhomogeneous Diffusion Equation

In the presence of source effects, the diffusion equation takes the form given by Equation (6.10), that is,

$$\alpha\frac{\partial^2 T}{\partial x^2} + g(x,t) = \frac{\partial T}{\partial t} \tag{6.190}$$

First, the case of a time independent heat source will be treated. Accordingly, we pose the following problem:

$$\alpha\frac{\partial^2 T}{\partial x^2} + g(x) = \frac{\partial T}{\partial t} \tag{6.191}$$

$$T(0,t) = T_1 \tag{6.192}$$

$$T(l,t) = T_2 \tag{6.193}$$

$$T(x,0) = f(x) \tag{6.194}$$

We ask again here the question as to whether there exists a steady state solution to the problem. The answer to that is definitely yes (Figure 6-18). Therefore, we let

$$T(x,t) = v(x,t) + w(x) \tag{6.195}$$

Hence, using this equation in Equations (6.191) to (6.194) yields

$$\alpha\frac{\partial^2 v}{\partial x^2} = \frac{\partial v}{\partial t}$$

$$v(0,t) = 0$$

$$v(l,t) = 0$$

$$v(x,0) = f(x) - w(x)$$

$v(x,t)$ is given by Equations (6.74) and (6.75).

$$\alpha\frac{d^2 w}{dx^2} + g(x) = 0$$

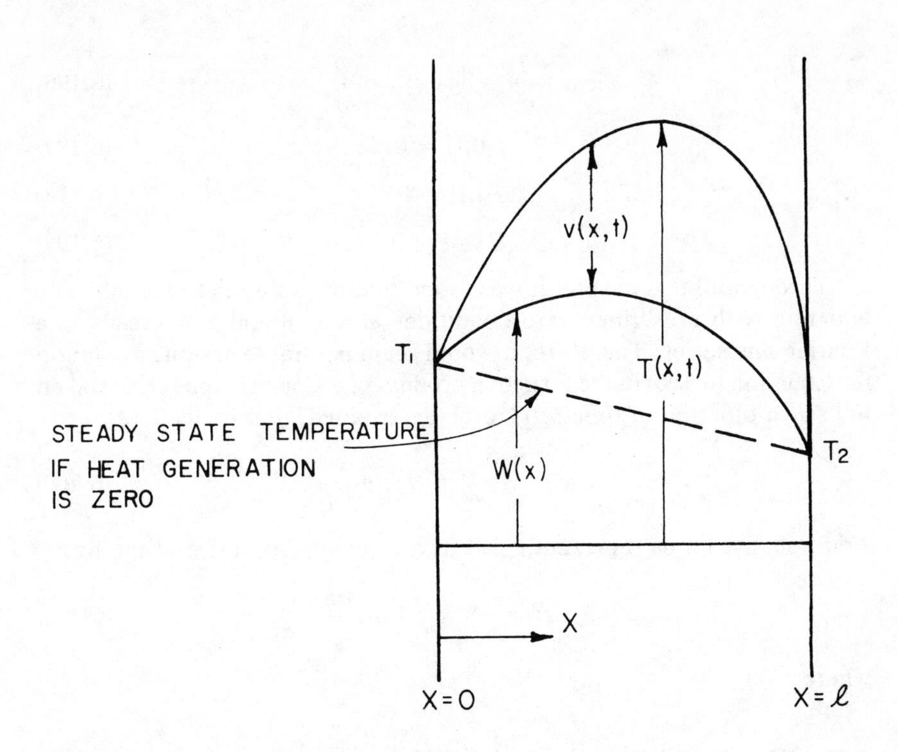

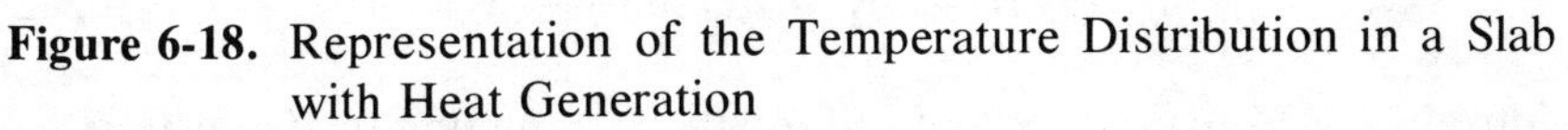

Figure 6-18. Representation of the Temperature Distribution in a Slab with Heat Generation

$$w(0) = T_1$$

$$w(l) = T_2$$

$w(x)$ is obtained by direct integration once $g(x)$ is specified. If $g(x)$ is a constant, $w(x)$ becomes

$$w(x) = \left(\frac{T_2 - T_1}{l}\right)x + T_1 + \frac{gl^2}{2\alpha}\left(\frac{x}{l} - \frac{x^2}{l^2}\right)$$

6.12 Arbitrary Nonhomogeneity in the Differential Equation

This section treats the diffusion equation in which the nonhomogeneous term is a function of both the space variable x and time. Initially, we consider the case of homogeneous boundary conditions and zero initial conditions. The procedure to be followed is similar to that covered in Article 5.15. The problem can then be stated in the following form

$$\alpha \frac{\partial^2 T}{\partial x^2} + g(x,t) = \frac{\partial T}{\partial t} \tag{6.196}$$

$$T(0,t) = 0 \tag{6.197}$$

$$T(l,t) = 0 \tag{6.198}$$

$$T(x,0) = 0 \tag{6.199}$$

Throughout this chapter it was found that the solution to the diffusion equation with prescribed zero conditions at $x = 0$ and $x = l$ leads to a Fourier sine series. Therefore, it would seem natural to assume a solution for Equation (6.196) in the form of a product of a Fourier sine series and an unknown function of time $u_n(t)$ to be determined, that is, we let

$$T(x,t) = \sum_{n=1}^{\infty} u_n(t) \sin \frac{n\pi x}{l} \tag{6.200}$$

Additionally, let us represent $g(x,t)$ as a Fourier sine series of the form

$$g(x,t) = \sum_{n=1}^{\infty} g_n(t) \sin \frac{n\pi x}{l} \tag{6.201}$$

where

$$g_n(t) = \frac{2}{l} \int_0^l g(x,t) \sin \frac{n\pi x}{l} \, dx \tag{6.202}$$

Introducing Equations (6.200) and (6.201) in Equation (6.196) we get

$$\sum_{n=1}^{\infty} \sin \frac{n\pi x}{l} \left[\left(\frac{\pi n}{l} \right)^2 \alpha \, u_n(t) + \frac{du_n}{dt} - g_n(t) \right] = 0 \tag{6.203}$$

Because sin $n\pi x/l$ cannot be zero for all times, the quantity in the bracket should be zero, that is,

$$\frac{du_n}{dt} + \left(\frac{n\pi}{l} \right)^2 \alpha \, u_n(t) = g_n(t) \tag{6.204}$$

From Equations (6.199) and (6.200) we have

$$T(x,0) = \sum_{n=1}^{\infty} u_n(0) \sin \frac{n\pi x}{l} = 0$$

from which we deduce

$$u_n(0) = 0 \tag{6.205}$$

Equation (6.204) is a first order ordinary nonhomogeneous differential equation. Its solution (see Article 1.5), subject to $u_n(0) = 0$, is

$$u_n(t) = \int_0^t e^{-(n\pi/l)^2 \alpha(t-\tau)} g_n(\tau)\, d\tau \tag{6.206}$$

Hence, introducing this expression for $u_n(t)$ in Equation (6.200) results in

$$T(x,t) = \sum_{n=1}^{\infty} \left\{ \int_0^t e^{-(n\pi/l)^2 \alpha(t-\tau)} g_n(\tau)\, d\tau \right\} \sin \frac{n\pi x}{l} \tag{6.207}$$

When Equation (6.202) is used in Equation (6.207) we get

$$T(x,t) = \int_0^t \int_0^l G(x,x',t-\tau)\, g(x',t)\, dx'\, d\tau \tag{6.208}$$

where

$$G(x,x',t-\tau) = \frac{2}{l} \sum_{n=1}^{\infty} e^{-(n\pi/l)^2 \alpha(t-\tau)} \sin \frac{n\pi x}{l} \sin \frac{n\pi x'}{l} \tag{6.209}$$

This function is referred to as the source function or Green's function. It represents the temperature effect in the solid of an instantaneous point source of heat at location x' and at time τ. Correspondingly, Equation (6.208) can be interpreted to represent the effect of heat sources continuously distributed and defined by $g(x,t)$.

When arbitrary nonhomogeneities are included in the differential equation and in the boundary conditions, the procedure of solution is represented schematically below:

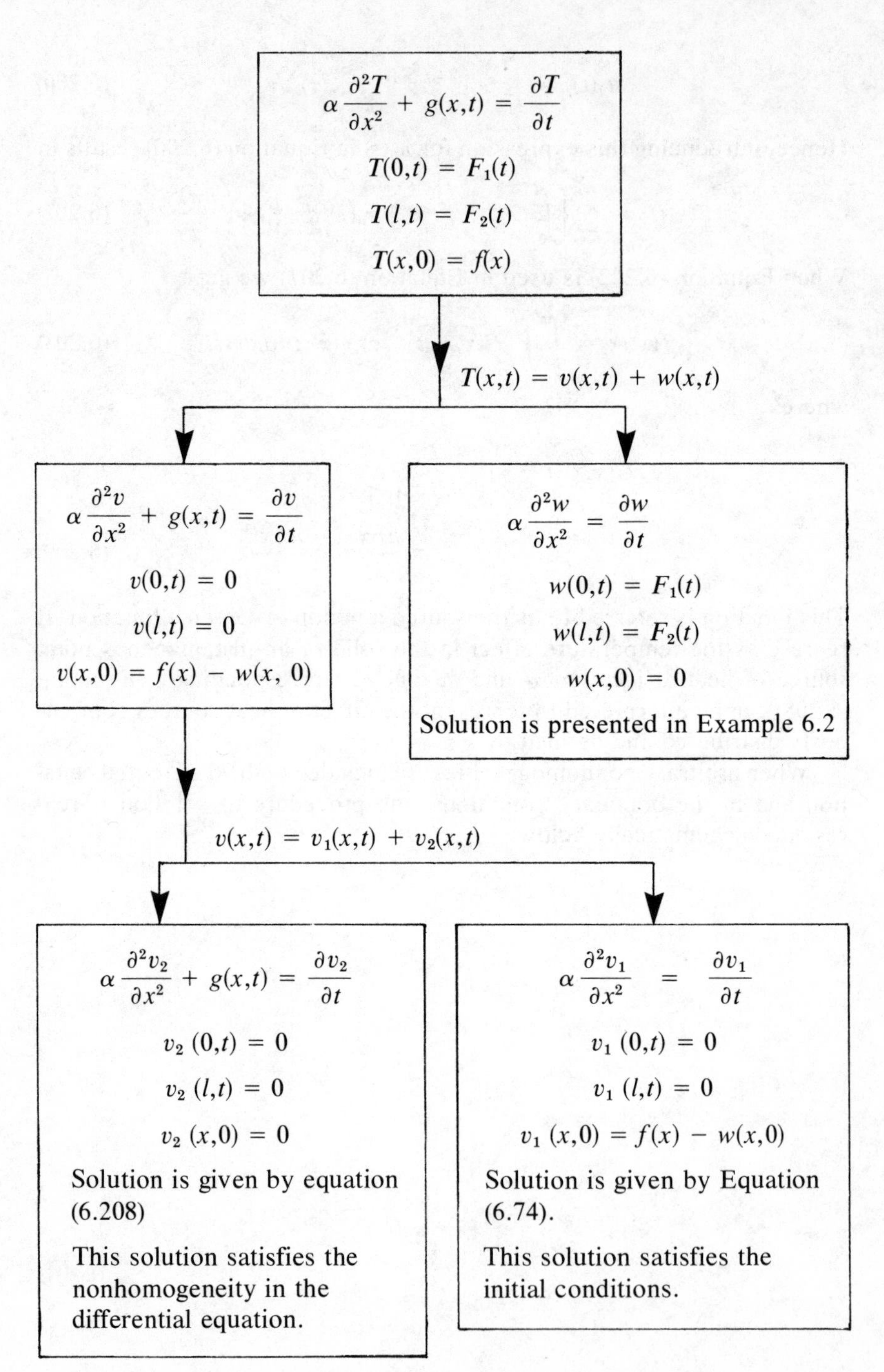

$\alpha \frac{\partial^2 T}{\partial x^2} + g(x,t) = \frac{\partial T}{\partial t}$
$T(0,t) = F_1(t)$
$T(l,t) = F_2(t)$
$T(x,0) = f(x)$
$T(x,t) = v(x,t) + w(x,t)$
$\alpha \frac{\partial^2 v}{\partial x^2} + g(x,t) = \frac{\partial v}{\partial t}$
$v(0,t) = 0$
$v(l,t) = 0$
$v(x,0) = f(x) - w(x, 0)$
$\alpha \frac{\partial^2 w}{\partial x^2} = \frac{\partial w}{\partial t}$
$w(0,t) = F_1(t)$
$w(l,t) = F_2(t)$
$w(x,0) = 0$
Solution is presented in Example 6.2
$v(x,t) = v_1(x,t) + v_2(x,t)$
$\alpha \frac{\partial^2 v_2}{\partial x^2} + g(x,t) = \frac{\partial v_2}{\partial t}$
$v_2\,(0,t) = 0$
$v_2\,(l,t) = 0$
$v_2\,(x,0) = 0$
Solution is given by equation (6.208)
This solution satisfies the nonhomogeneity in the differential equation.
$\alpha \frac{\partial^2 v_1}{\partial x^2} = \frac{\partial v_1}{\partial t}$
$v_1\,(0,t) = 0$
$v_1\,(l,t) = 0$
$v_1\,(x,0) = f(x) - w(x,0)$
Solution is given by Equation (6.74).
This solution satisfies the initial conditions.

Partial Differential Equations of the Elliptic Type

7.1 Laplace's Equation—Harmonic Functions

The most common equation of the elliptic type is *Laplace's equation*

$$\nabla^2 \phi = 0 \tag{7.1}$$

This equation appears in the study of steady state problems in areas such as: (1) electromagnetic phenomena including electrostatic fields, steady current in conductors and magnetostatics, (2) heat and mass transfer, (3) potential flow theory and surface waves, and (4) gravitational potential.

Steady state phenomena with source effects such as the electrostatic potential at those points in space where electric charges occur, and steady temperature distributions in solids in the presence of heat sources satisfy the equation

$$\nabla^2 \phi = -g \tag{7.2}$$

This equation is referred to as *Poisson's equation*; g depends upon the strength of the source.

Solutions of Laplace's equation are designated as harmonic functions. Harmonic functions have the following general properties:

1. A harmonic function ϕ cannot have a maximum or a minimum value within a given region. The maximum and minimum values are attained at the surface.
2. Only harmonic functions that are constant can reach their maximum value (and hence minimum value) at interior points of the region.
3. If the normal derivative of a harmonic function ϕ is zero at all points of a closed boundary σ, the function then must be constant within the region enclosed by the boundary σ (Figure 7-1).

7.2 Steady State Temperature Distribution

In chapter 6, Article 6.2, the differential equation for the temperature distribution in a homogeneous solid was derived in rectangular geometry. For a three-dimensional case and in the presence of heat sources, the resulting equation (Equation (6.18)) is

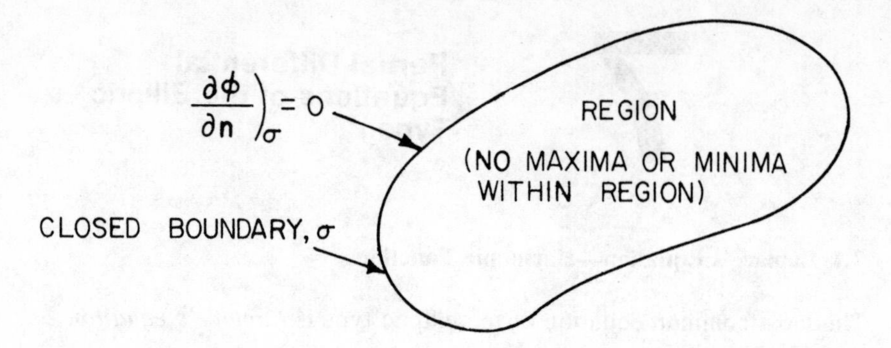

Figure 7-1. Region with Insulated Boundary

$$\alpha\left(\frac{\partial^2 T}{\partial x^2} + \frac{\partial^2 T}{\partial y^2} + \frac{\partial^2 T}{\partial z^2}\right) + g(x,y,z,t) = \frac{\partial T}{\partial t} \tag{7.3}$$

For a steady state situation and in the absence of heat sources, Equation (7.3) becomes

$$\frac{\partial^2 T}{\partial x^2} + \frac{\partial^2 T}{\partial y^2} + \frac{\partial^2 T}{\partial z^2} = 0 \tag{7.4}$$

or

$$\nabla^2 T = 0$$

When heat sources are present, Equation (7.3) yields Poisson's equation, that is,

$$\left(\frac{\partial^2 T}{\partial x^2} + \frac{\partial^2 T}{\partial y^2} + \frac{\partial^2 T}{\partial z^2}\right) = -\frac{g}{\alpha} \tag{7.5}$$

7.3 Potential Flow—Continuity Equation

In Figure 7-2 we show a differential control volume through which a steady flow of an incompressible fluid takes place.

The principle of conservation of mass requires that under steady state conditions the inflow of mass to the volume should equal the outflow of mass from the volume. Taking the coordinates at the center of the cubical volume with u, v, and w representing the fluid velocity in the x-, y-, and z-directions respectively, we can write:

Mass flow into volume per unit time

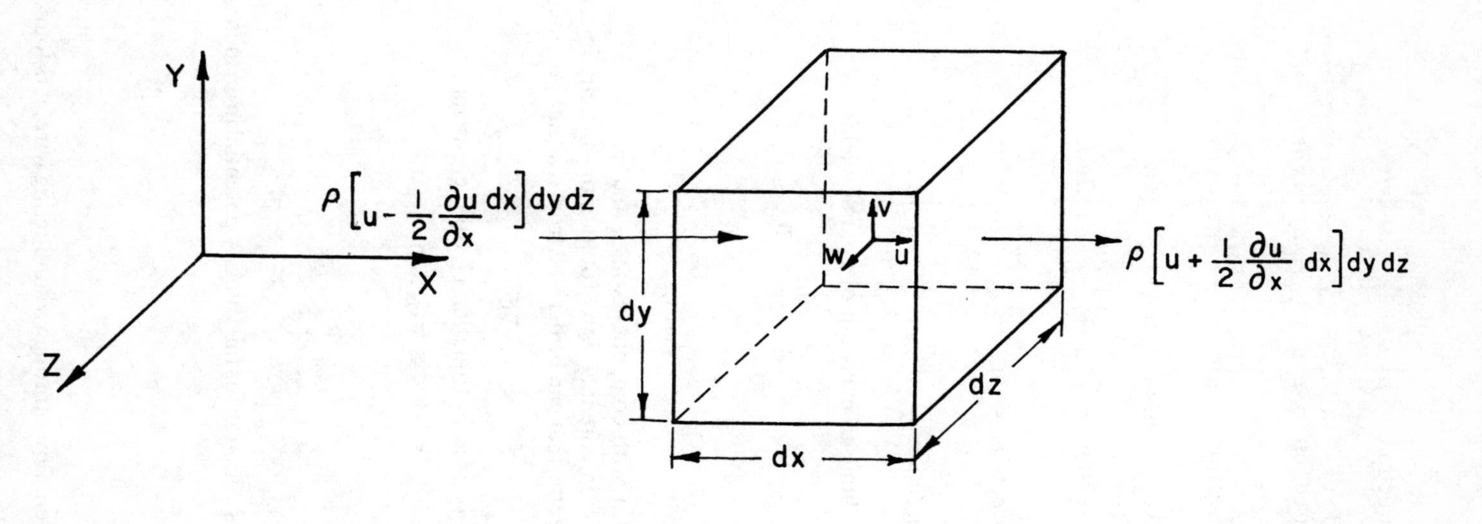

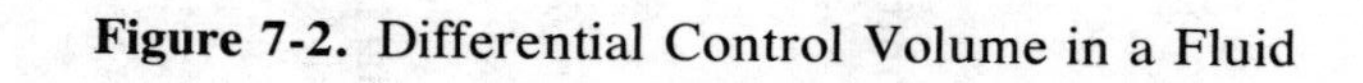

Figure 7-2. Differential Control Volume in a Fluid

$$x\text{-direction} = \rho\left[u - \frac{1}{2}\frac{\partial u}{\partial x}dx\right]dy\,dz \tag{7.6a}$$

$$y\text{-direction} = \rho\left[v - \frac{1}{2}\frac{\partial v}{\partial y}dy\right]dx\,dz \tag{7.6b}$$

$$z\text{-direction} = \rho\left[w - \frac{1}{2}\frac{\partial w}{\partial z}dz\right]dx\,dy \tag{7.6c}$$

Mass flow out of volume per unit time

$$x\text{-direction} = \rho\left[u + \frac{1}{2}\frac{\partial u}{\partial x}dx\right]dy\,dz \tag{7.7a}$$

$$y\text{-direction} = \rho\left[v + \frac{1}{2}\frac{\partial v}{\partial y}dy\right]dx\,dz \tag{7.7b}$$

$$z\text{-direction} = \rho\left[w + \frac{1}{2}\frac{\partial w}{\partial z}dz\right]dx\,dy \tag{7.7c}$$

Equating mass into volume to mass out of volume yields

$$\frac{\partial u}{\partial x} + \frac{\partial v}{\partial y} + \frac{\partial w}{\partial z} = 0 \tag{7.8}$$

or

$$\nabla \cdot \mathbf{V} = 0 \tag{7.9}$$

$\mathbf{V}$ is the velocity vector with components u, v, and w.

In irrotational flow (in potential flow, rotation or vorticity is zero) the fluid velocity $\mathbf{V}$ is a vector that can be derived from a scalar potential ϕ as

$$\mathbf{V} = -\nabla\phi \tag{7.10}$$

When Equation (7.9) is used in Equation (7.10) there results

$$\nabla \cdot \nabla\phi = \nabla^2\phi = 0$$

or

$$\frac{\partial^2\phi}{\partial x^2} + \frac{\partial^2\phi}{\partial y^2} + \frac{\partial^2\phi}{\partial z^2} = 0 \tag{7.11}$$

This is again Laplace's equation for the velocity potential in an incompressible fluid.

7.4 Electric Potential

When an electric current flows through a conducting wire, it is found that a magnetic field is created in the neighborhood of the wire. The electric current I is related to the total charge passing through a given surface σ per unit time as follows

$$I = \int_\sigma \rho \mathbf{V} \cdot \mathbf{d\sigma} = \int_\sigma \mathbf{j} \cdot \mathbf{d\sigma} \tag{7.12}$$

where $\mathbf{V}$ is the velocity of the charge along the wire and ρ is the charge density. When the surface σ encloses a certain region v in space, and when the charge density remains constant within the region, the total charge Q enclosed by the region remains constant, that is,

$$Q = \int_v \rho \, dv = \text{constant} \tag{7.13}$$

Correspondingly,

$$\frac{\partial Q}{\partial t} = \int_v \frac{\partial}{\partial t} \rho \, dv = 0 \tag{7.14}$$

Therefore, the equation of continuity of charge requires that

$$\int_\sigma \mathbf{j} \cdot \mathbf{d\sigma} = \int_v \nabla \cdot \mathbf{j} \, dv = -\int_v \frac{\partial}{\partial t} \rho \, dv = 0 \tag{7.15}$$

or

$$\nabla \cdot \mathbf{j} = 0 \tag{7.16}$$

The electric field E is related to the current j by Ohm's law as

$$E = \frac{j}{\lambda} \tag{7.17}$$

where λ is the electric conductivity of the medium. Accordingly, using Equation (7.17) in Equation (7.16) yields

$$\nabla \cdot E\lambda = 0 \tag{7.18}$$

Because the process is steady, the electric field is irrotational. Therefore, there exists a scalar function $\phi(x,y,z)$ such that

$$E = -\nabla \phi \tag{7.19}$$

Considering λ to be constant, and introducing Equation (7.19) in Equation (7.18) yields

$$\nabla^2 \phi = 0 \tag{7.20}$$

7.5 Solution of the Equations of Elliptic Type

In this section, the method of separation of variables will be used to solve

Laplace's equation in rectangular geometry. A two-dimensional problem is chosen for analysis. It is important to note that this method is applicable directly to the steady state problem shown in Figure 7.3, provided that:

1. The differential equation is homogeneous.
2. There is only one nonhomogeneous boundary condition.
3. The separation constant is chosen so that the homogeneous direction (x-axis in Figure 7-3) results in an eigenvalue problem.

Consider Figure 7.3 to represent a rectangular bar having a side at $y = w$ subjected to an arbitrary temperature distribution $f(x)$ while the rest of the sides are maintained at zero temperature. The objective is to find the steady state temperature distribution in the bar. The bar is very long in the z-direction so that the problem is two-dimensional.

The differential equation and boundary conditions for this problem can be stated as

$$\frac{\partial^2 T}{\partial x^2} + \frac{\partial^2 T}{\partial y^2} = 0 \tag{7.21}$$

$$T(0,y) = 0\,, \qquad T(l,y) = 0 \tag{7.22a,b}$$

$$T(x,0) = 0\,, \qquad T(x,w) = f(x) \tag{7.23a,b}$$

Let

$$T(x,y) = X(x)\,Y(y) \tag{7.24}$$

Introducing this product solution into Equation (7.21) gives

$$\frac{1}{X}\frac{d^2X}{dx^2} = -\frac{1}{Y}\frac{d^2Y}{dy^2} \tag{7.25}$$

Again here, we see that the left-hand side of Equation (7.25) is a function of x while the right-hand side is a function of y. Accordingly, each side should be equal to the same constant. The constant of separation, as indicated earlier in this section, is chosen so that the homogeneous direction results in an eigenvalue problem. The constant of separation then is chosen to be negative, $-\lambda^2$. Hence, there results

$$\frac{d^2X}{dx^2} + \lambda^2 X = 0 \tag{7.26}$$

$$X(0) = 0\,, \qquad X(l) = 0 \tag{7.27a,b}$$

and

$$\frac{d^2Y}{dy^2} - \lambda^2 Y = 0 \tag{7.28}$$

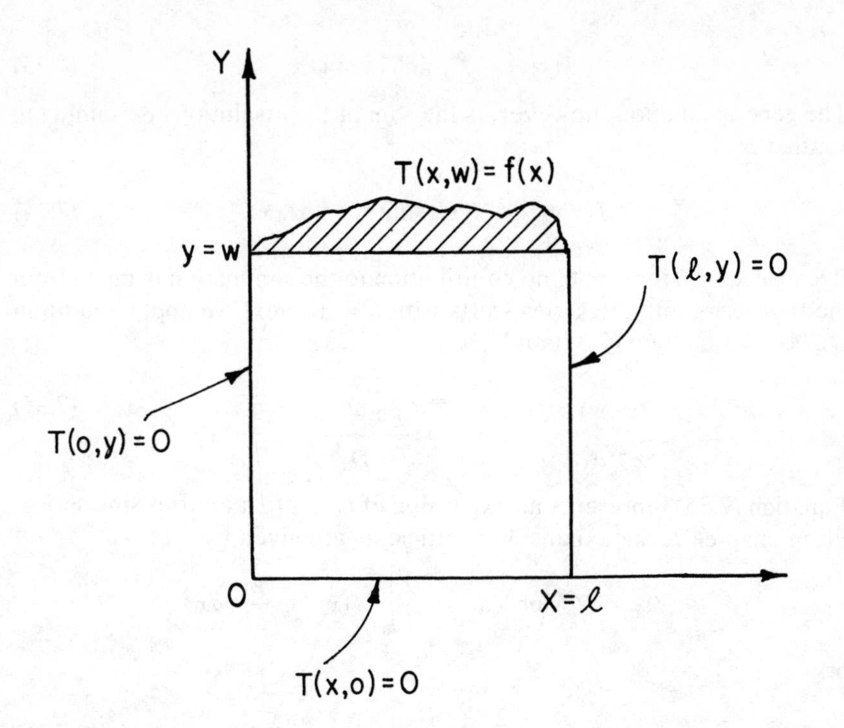

Figure 7-3. Rectangular Rod with Temperature Prescribed on One Side

$$Y(0) = 0\,, \qquad Y(w) = f(x) \tag{7.29a,b}$$

Equation (7.26) has the following solution

$$X(x) = A \sin \lambda x + B \cos \lambda x \tag{7.30}$$

Boundary condition (7.27a) yields $B = 0$. Therefore,

$$X(x) = A \sin \lambda x \tag{7.31}$$

Equation (7.27b) gives

$$X(l) = 0 = A \sin \lambda l$$

Hence,

$$\lambda l = n\pi\,; \qquad \lambda_n = n\pi/l\,, \qquad n = 1, 2, 3, \ldots$$

Equation (7.28) has the solution

$$Y(y) = C \sinh \lambda y + D \cosh \lambda y \tag{7.32}$$

Boundary condition (7.29a) gives $D = 0$. Accordingly, the form of the solution can now be written as

$$T(x,y) = C_1 \sinh \lambda y \sin \lambda x \tag{7.33}$$

The general solution, however, is the sum of the n solutions pertaining to λ_n, that is,

$$T(x,y) = \sum_{n=1}^{\infty} C_n \sinh \lambda_n y \sin \lambda_n x \tag{7.34}$$

Because $\lambda_n = 0$ for $n = 0$, no contribution to the summation is made from the first term, and the series starts with $n = 1$. Next we apply condition (7.29b) to Equation (7.34) and get

$$T(x,w) = f(x) = \sum_{n=1}^{\infty} \underbrace{C_n \sinh \lambda_n w}_{D_n} \sin \lambda_n x \tag{7.35}$$

Equation (7.35) represents an expansion of $f(x)$ into a Fourier sine series. From chapter 2, the expansion coefficients are given by

$$D_n = C_n \sinh \lambda_n w = \frac{2}{l}\int_0^l f(x) \sin \frac{n\pi x}{l} dx$$

or

$$C_n = \frac{2}{l \sinh \lambda_n w}\int_0^l f(x) \sin \frac{n\pi x}{l} dx \tag{7.36}$$

Finally, the solution becomes

$$T(x,y) = \frac{2}{l}\sum_{n=1}^{\infty} \frac{\sinh \dfrac{n\pi y}{l}}{\sinh \dfrac{n\pi w}{l}} \sin \frac{n\pi x}{l} \int_0^l f(x) \sin \frac{n\pi x}{l} dx \tag{7.37}$$

Example 7.1 Find the temperature distribution in the rod of Figure 7-3 if (a) $f(x)$ is a constant $= T_0$, (b) $f(x) = T_0 \sin \pi x/l$.
Solution: (a) Replacing $f(x)$ in Equation (7.37) by T_0 and integrating yields

$$\frac{T(x,y)}{T_0} = \frac{2}{\pi}\sum_{n=1,2,\ldots}^{\infty} \frac{[1-(-1)^n]}{n} \left[\frac{\sinh \dfrac{n\pi y}{l}}{\sinh \dfrac{n\pi w}{l}}\right] \sin \frac{n\pi x}{l} \tag{7.38}$$

(b) If $f(x) = T_0 \sin \pi x/l$, the integral in Equation (7.37) becomes

$$\int_0^l T_0 \sin \frac{\pi x}{l} \sin \frac{n\pi x}{l} dx \tag{7.39}$$

In this case, the integral is zero if $n \neq 1$ (because of the orthogonality of $\sin n\pi x/l$). Hence, the only contribution to the solution is from $n = 1$. The integral, therefore, yields

$$T_0 \int_0^l \sin^2 \frac{\pi x}{l} dx = \frac{lT_0}{2} \tag{7.40}$$

Accordingly, the solution in this case becomes one term only, and Equation (7.37) is reduced to

$$T(x,y) = T_0 \frac{\sinh \frac{\pi y}{l}}{\sinh \frac{\pi w}{l}} \sin \frac{\pi x}{l} \tag{7.41}$$

7.6 Superposition of Solutions

There are cases where a rod like the one analyzed in Article 7.5 is exposed to nonhomogeneous boundary conditions on more than one surface such as different temperatures or different potentials on every surface. Because of the linearity of the problem, cases of this sort can be solved using the principle of superposition. This means that the temperature (or potential) at any point within the medium is equal to the sum of the various temperatures resulting from the various distributions on the boundary.

Consider the solid shown in Figure 7-4. The temperature at any point (x,y) within the solid can be written as

$$T(x,y) = T_1(x,y) + T_2(x,y) + T_3(x,y) + T_4(x,y)$$

Each of T_1, T_2, T_3, and T_4 represents the solution to the problem with only one side being nonhomogeneous and the rest of the sides being maintained at zero. Figure 7-5 schematically illustrates the solution.

The solutions to each of these subproblems are:

$T_1(x,y)$ is given by Equation (7.37) as

$$T_1(x,y) = \frac{2}{l} \sum_{n=1}^{\infty} \frac{\sinh \frac{n\pi y}{l}}{\sinh \frac{n\pi w}{l}} \sin \frac{n\pi x}{l} \int_0^l f_2(x) \sin \frac{n\pi x}{l} dx$$

$T_2(x,y)$ is given by analogy to T_1 as

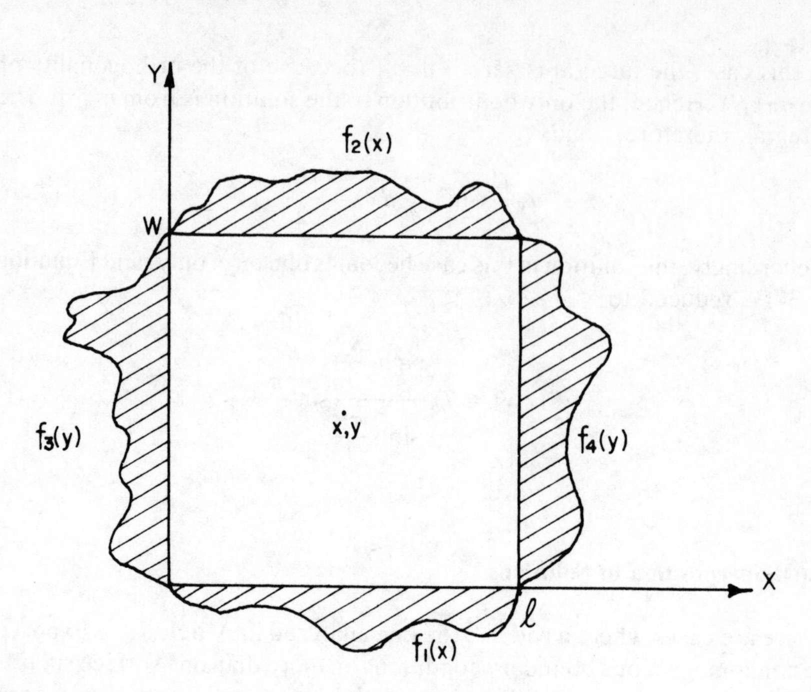

Figure 7-4. Rectangular Rod with a Different Temperature on Each Side

$$T_2(x,y) = \frac{2}{l}\sum_{n=1}^{\infty} \frac{\sinh \dfrac{n\pi(w-y)}{l}}{\sinh \dfrac{n\pi w}{l}} \sin \frac{n\pi x}{l} \int_0^l f_1(x) \sin \frac{n\pi x}{l} dx \quad (7.42)$$

$T_3(x,y)$ is similar to $T_2(x,y)$ with x replaced by y and w replaced by l, that is,

$$T_3(x,y) = \frac{2}{l}\sum_{n=1}^{\infty} \frac{\sinh \dfrac{n\pi(l-x)}{w}}{\sinh \dfrac{n\pi l}{w}} \sin \frac{n\pi y}{w} \int_0^w f_3(y) \sin \frac{n\pi y}{w} dy \quad (7.43)$$

$T_4(x,y)$ is similar to $T_1(x,y)$ with x replaced by y and w replaced by l, that is,

$$T_4(x,y) = \frac{2}{l}\sum_{n=1}^{\infty} \frac{\sinh \dfrac{n\pi x}{w}}{\sinh \dfrac{n\pi l}{w}} \sin \frac{n\pi y}{w} \int_0^w f_4(y) \sin \frac{n\pi y}{w} dy \quad (7.44)$$

Correspondingly, as was mentioned earlier, $T(x,y)$ can be written as

$$T(x,y) = T_1(x,y) + T_2(x,y) + T_3(x,y) + T_4(x,y) \quad (7.45)$$

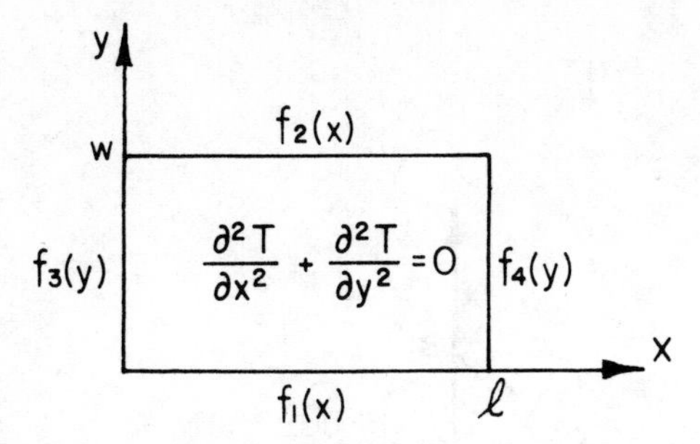

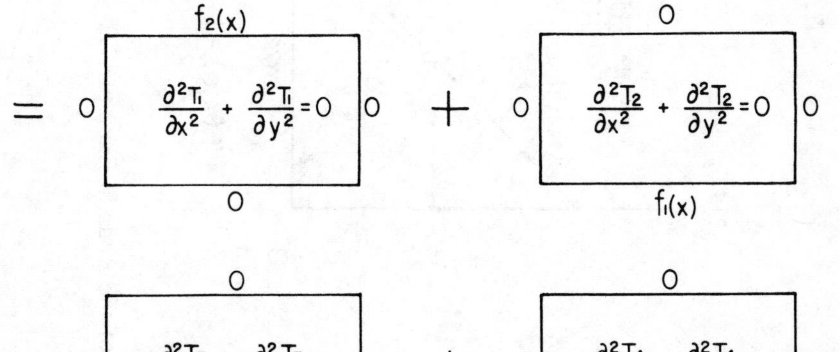

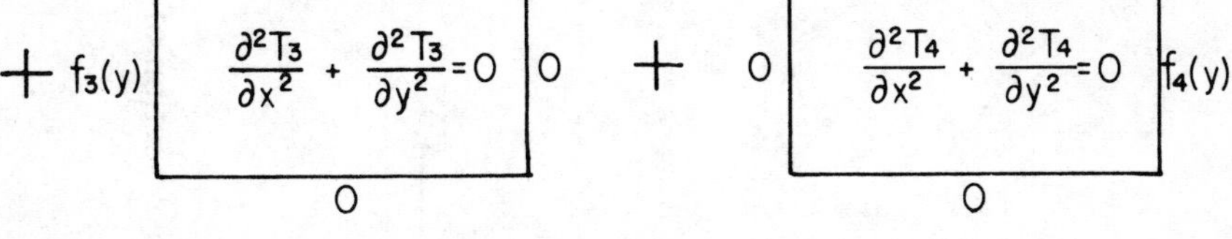

Figure 7-5. Superposition of Solutions

Example 7.2 Find the value of the steady state temperature at the point P in Figure 7-6a and b.
Solution: (a) In Figure 7-6a, the solution can be schematically represented as follows (Figure 7-7). When all the sides have a temperature equal to 1, the temperature at point P will therefore be equal to 1. Accordingly, the contribution from each of the subproblems shown is 1/4. Hence, the temperature at P for part a is 1/4. (b) In Figure 7-6b, the problem is reconstructed as shown in Figure 7-8. A mirror image of the solid is constructed symmetric along the insulated surface. Hence, the temperature at P equals 1/2.

Example 7.3 Suppose that the rod of Figure 7-3 undergoes a constant heat generation G per unit volume. Determine the temperature distribution in the rod.

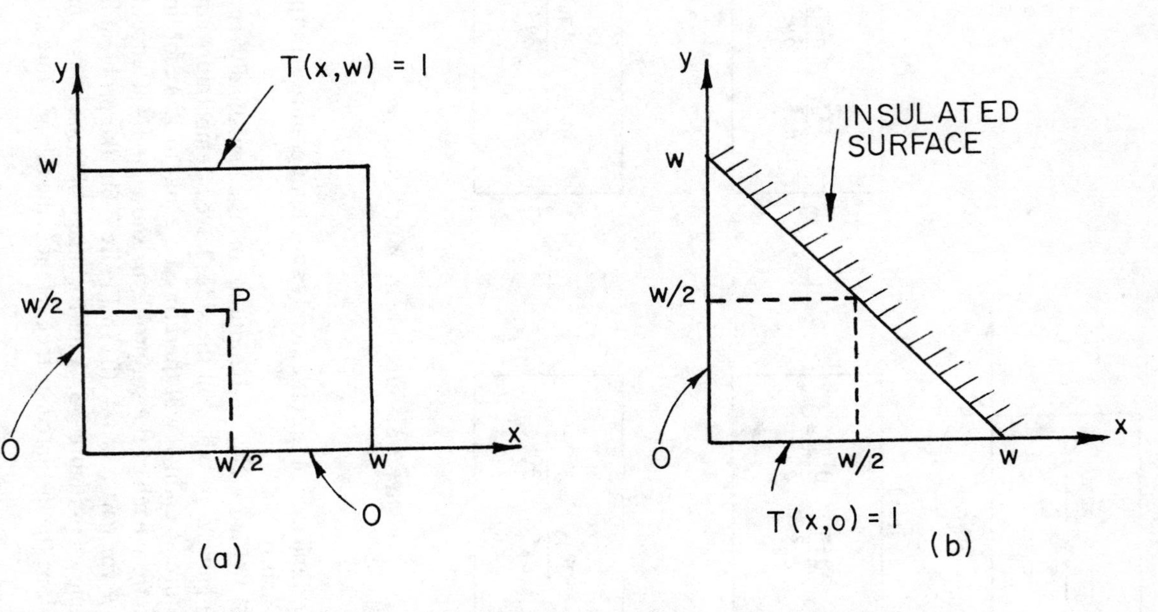

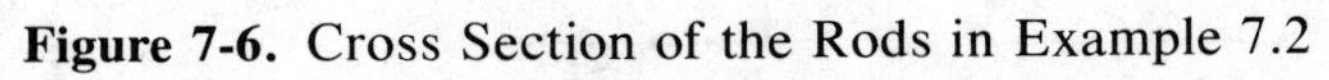

Figure 7-6. Cross Section of the Rods in Example 7.2

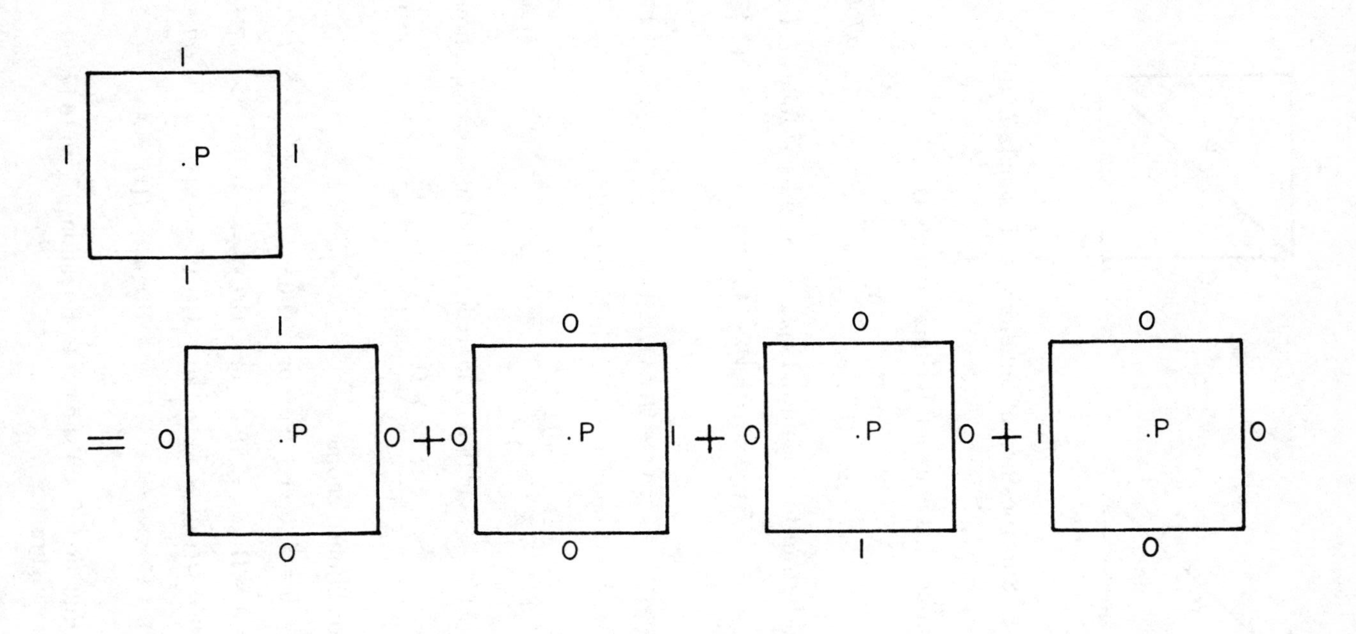

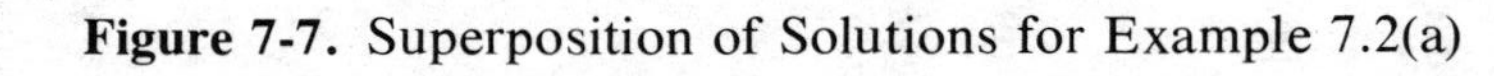

Figure 7-7. Superposition of Solutions for Example 7.2(a)

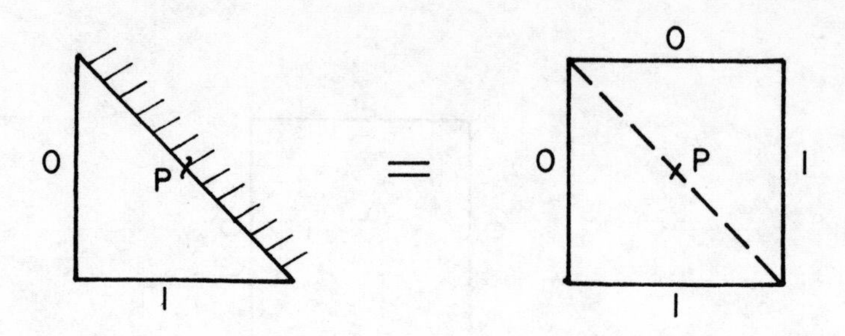

Figure 7-8. Superposition of Solutions for Example 7.2(b)

Solution: The differential equation for this example is

$$\frac{\partial^2 T}{\partial x^2} + \frac{\partial^2 T}{\partial y^2} = -\frac{G}{K} \tag{7.46}$$

This equation can be transformed into a Laplace's equation as follows. Let

$$T(x,y) = \Omega(x,y) + Ax^2 \tag{7.47a}$$

or

$$T(x,y) = \Omega(x,y) + Ay^2 \tag{7.47b}$$

where A is arbitrary. Introducing (7.47a) in (7.46) yields

$$\frac{\partial^2 \Omega}{\partial x^2} + \frac{\partial^2 \Omega}{\partial y^2} + 2A = -\frac{G}{K} \tag{7.48}$$

If A is chosen $= -G/2k$, Equation (7.48) reduces to Laplace's equation,

$$\frac{\partial^2 \Omega}{\partial x^2} + \frac{\partial^2 \Omega}{\partial y^2} = 0 \tag{7.49}$$

The boundary conditions become

$$T(0,y) = 0 = \Omega(0,y) + 0, \quad \text{or} \quad \Omega(0,y) = 0 \tag{7.50}$$

$$T(l,y) = 0 = \Omega(l,y) + Al^2, \quad \text{or} \quad \Omega(l,y) = -Al^2 \tag{7.51}$$

$$T(x,0) = 0 = \Omega(x,0) + Ax^2, \quad \text{or} \quad \Omega(x,0) = -Ax^2 \tag{7.52}$$

$$T(x,w) = f(x) = \Omega(x,w) + Ax^2, \quad \text{or} \quad \Omega(x,w) = f(x) - Ax^2 \tag{7.53}$$

The solution to Equation (7.49) subject to Equations (7.50) to (7.53) has been illustrated in Figure 7-5.

Example 7.4 Consider the conductor shown in Figure 7-9 subjected to the

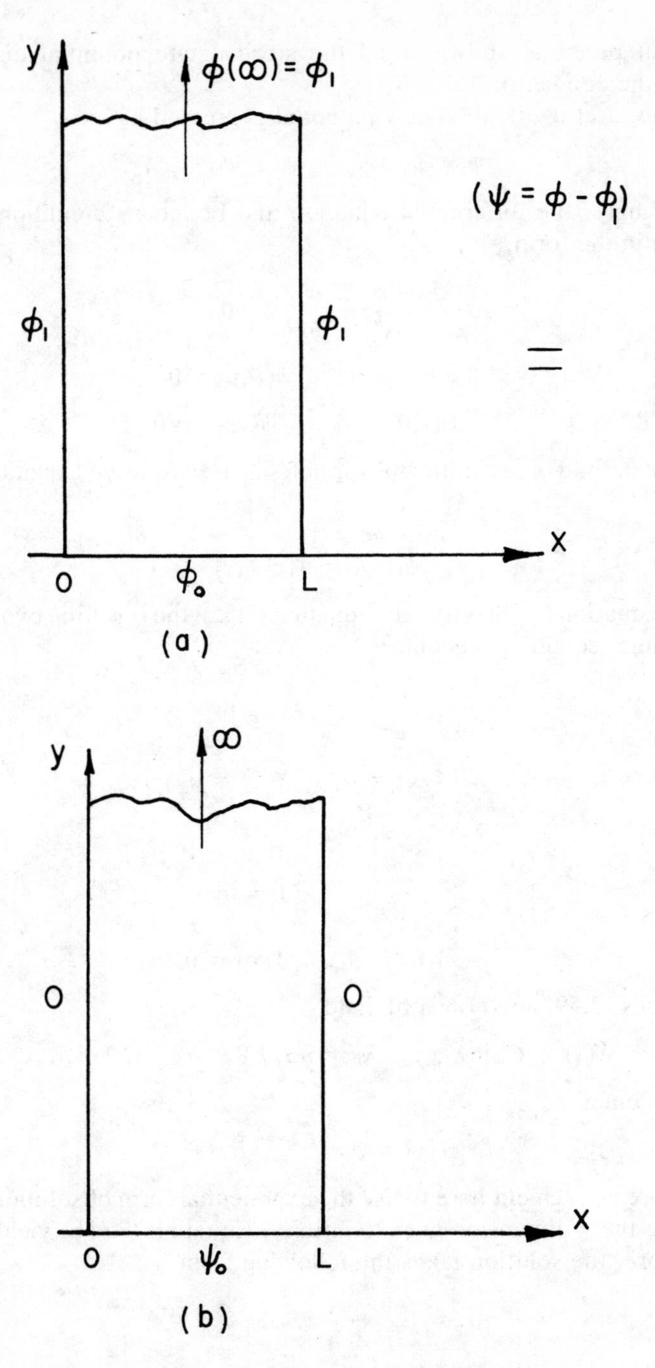

Figure 7-9. The Conductor in Example 7.4

constant potential shown. Find the steady state potential distribution within the conductor.

Solution: Let us define a new potential expressed as

$$\psi = \phi - \phi_1 , \quad \psi_0 = \phi_0 - \phi_1 \tag{7.54}$$

Accordingly, the differential equation and boundary conditions can be written in the form

$$\frac{\partial^2 \psi}{\partial x^2} + \frac{\partial^2 \psi}{\partial y^2} = 0 \tag{7.55}$$

$$\psi(0,y) = 0 , \quad \psi(L,y) = 0 \tag{7.56a,b}$$

$$\psi(x,0) = \psi_0 , \quad \psi(x,\infty) = 0 \tag{7.57a,b}$$

The method of separation of variables is used to solve Equation (7.55).

Let

$$\psi(x,y) = X(x)\, Y(y) \tag{7.58}$$

When Equation (7.58) is used in Equation (7.55), the resulting two ordinary differential equations become

$$\frac{d^2X}{dx^2} + \lambda^2 X = 0 \tag{7.59}$$

$$X(0) = 0 , \quad X(L) = 0 \tag{7.60a,b}$$

and

$$\frac{d^2Y}{dy^2} - \lambda^2 Y = 0 \tag{7.61}$$

$$Y(0) = \psi_0 , \quad Y(\infty) = 0 \tag{7.62a,b}$$

Equations (7.59) and (7.60a,b) yield

$$X(x) = C \sin \lambda_n x ; \quad \lambda_n = n\pi/L , \quad n = 1, 2, 3, \ldots \tag{7.63}$$

$Y(y)$ becomes

$$Y(y) = A\, e^{-\lambda_n y} + B\, e^{\lambda_n y} \tag{7.64}$$

It is more convenient here to use the exponential form of solution for $Y(y)$ because the y dimension goes to infinity. Equation (7.62b) yields $B = 0$. Therefore, the solution takes the following form

$$\psi(x,y) = \sum_{n=1}^{\infty} C_n\, e^{-(n\pi/L)y} \sin \frac{n\pi x}{L} \tag{7.65}$$

The nonhomogeneous condition at $y = 0$ is now applied

$$\psi(x,0) = \psi_0 = \sum_{n=1}^{\infty} C_n \sin \frac{n\pi x}{L} dx \tag{7.66}$$

with

$$C_n = \frac{2}{L}\int_0^L \psi_0 \sin\frac{n\pi x}{L} dx = \frac{4\psi_0}{n\pi}; \quad n = 1, 3, 5, \ldots \tag{7.67}$$

Finally, the solution becomes

$$\frac{\psi(x,y)}{\psi_0} = \frac{\phi - \phi_1}{\phi_0 - \phi_1} = \frac{4}{\pi}\sum_{n=1}^{\infty}\frac{1}{n}e^{-(n\pi/L)y}\sin\frac{n\pi}{L}x \tag{7.68}$$

$$= \frac{4}{\pi}\left[e^{-(\pi/L)y}\sin\frac{\pi}{L}x + \frac{1}{3}e^{-(3\pi/L)y}\sin\frac{3\pi}{L}x + \ldots\right] \tag{7.69}$$

The theory of complex variables yields a closed form solution to this problem in the form

$$\frac{\phi - \phi_1}{\phi_0 - \phi_1} = \frac{2}{\pi}\tan^{-1}\left[\frac{\sin\left(\frac{\pi}{L}\right)x}{\sinh\left(\frac{\pi}{L}\right)y}\right] \tag{7.70}$$

Example 7.5 Obtain the expression for the steady temperature distribution in the cube shown in Figure 7-10. The temperature is nonzero only on the face shown.

Solution: The differential equation for the problem is

$$\frac{\partial^2 T}{\partial x^2} + \frac{\partial^2 T}{\partial y^2} + \frac{\partial^2 T}{\partial z^2} = 0 \tag{7.71}$$

Let

$$T(x,y,z) = X(x)\,Y(y)\,Z(z) \tag{7.72}$$

When Equation (7.72) is used in Equation (7.71) we get

$$\frac{1}{X}\frac{d^2X}{dx^2} + \frac{1}{Y}\frac{d^2Y}{dy^2} + \frac{1}{Z}\frac{d^2Z}{dz^2} = 0 \tag{7.73}$$

Equation (7.73) shows that:

a function of x + a function of y + a function of $z = 0$

Accordingly, each of the terms in Equation (7.73) should be a constant. Let

$$\frac{1}{X}\frac{d^2X}{dx^2} = -\lambda^2, \quad \frac{1}{Y}\frac{d^2Y}{dy^2} = -\beta^2 \tag{7.74a,b}$$

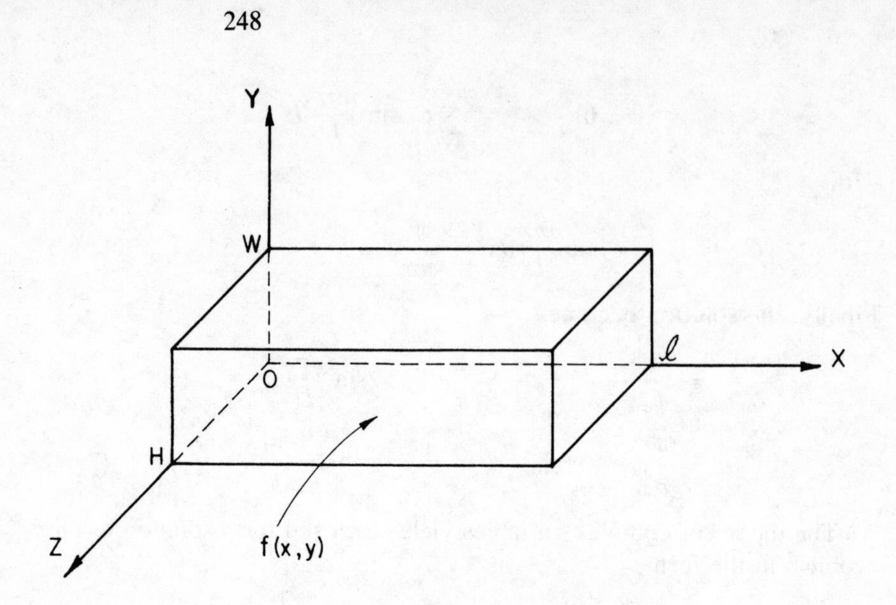

Figure 7-10. Cubic Solid in Example 7.5

Hence,

$$\frac{1}{Z}\frac{d^2Z}{dz^2} = \lambda^2 + \beta^2 = p^2 \tag{7.75}$$

Equations (7.74a,b) and (7.75) yield

$$X(x) = A\cos\lambda x + B\sin\lambda x \tag{7.76}$$

$$X(0) = 0\,, \qquad X(l) = 0 \tag{7.77a,b}$$

$$Y(y) = C\cos\beta y + D\sin\beta y \tag{7.78}$$

$$Y(0) = 0\,, \qquad Y(w) = 0 \tag{7.79a,b}$$

$$Z(z) = E\cosh pz + F\sinh pz \tag{7.80}$$

$$Z(0) = 0\,, \qquad Z(H) = f(x,y) \tag{7.81a,b}$$

Equations (7.76) to (7.81) require that

$$X = B\sin\lambda_n x\,, \qquad \lambda_n = n\pi/l \tag{7.82}$$

$$Y = D\sin\beta_m y\,, \qquad \beta_m = m\pi/w \tag{7.83}$$

$$Z = F\sinh p_{m,n}z\,, \qquad p_{m,n} = \sqrt{\left(\frac{n\pi}{l}\right)^2 + \left(\frac{m\pi}{w}\right)^2} \tag{7.84}$$

Therefore, the general form of the solution becomes

$$T(x,y,z) = \sum_{n=1}^{\infty}\sum_{m=1}^{\infty} A_{mn} \sinh p_{m,n}z \sin \lambda_n x \sin \beta_m y \tag{7.85}$$

Applying now the condition at $z = H$ gives

$$\begin{aligned} T(x,y,H) &= f(x,y) \\ &= \sum_{n=1}^{\infty}\sum_{m=1}^{\infty} A_{mn} \sinh p_{m,n}H \sin \lambda_n x \sin \beta_m y \end{aligned} \tag{7.86}$$

This equation is a double Fourier series representation of $f(x,y)$; see chapter 5. Accordingly, A_{mn} is given by

$$A_{mn} = \left(\frac{1}{\sinh p_{m,n}H}\right)\frac{4}{lw}\int_0^l\int_0^w f(x,y) \sin \frac{n\pi x}{l} \sin \frac{m\pi y}{w} \, dx\,dy \tag{7.87}$$

7.7 Laplace's Equation in Spherical and Cylindrical Coordinates

Laplace's equation can be written in cylindrical and in spherical coordinates using the following transformations:

Cylindrical coordinates, r, θ, z; (Figure 7-11a)

$$x = r\cos\theta\,, \qquad y = r\sin\theta\,, \qquad z = z$$

Hence,

$$\nabla^2 u = \frac{\partial^2 u}{\partial r^2} + \frac{1}{r}\frac{\partial u}{\partial r} + \frac{1}{r^2}\frac{\partial^2 u}{\partial \theta^2} + \frac{\partial^2 u}{\partial z^2} = 0 \tag{7.88}$$

Spherical coordinates, r, θ, ψ; (Figure 7-11b)

$$x = r\cos\psi\sin\theta\,, \qquad y = r\sin\psi\sin\theta\,, \qquad z = r\cos\theta$$

Hence,

$$\begin{aligned} \nabla^2 u &= \frac{1}{r^2}\frac{\partial}{\partial r}\left(r^2\frac{\partial u}{\partial r}\right) + \frac{1}{r^2\sin\theta}\frac{\partial}{\partial\theta}\left(\sin\theta\frac{\partial u}{\partial\theta}\right) \\ &\quad + \frac{1}{r^2\sin^2\theta}\frac{\partial^2 u}{\partial\psi^2} = 0 \end{aligned} \tag{7.89}$$

7.8 Solution of Laplace's Equation in Cylindrical Coordinates

In this section, we develop the solution to Laplace's equation in cylindrical

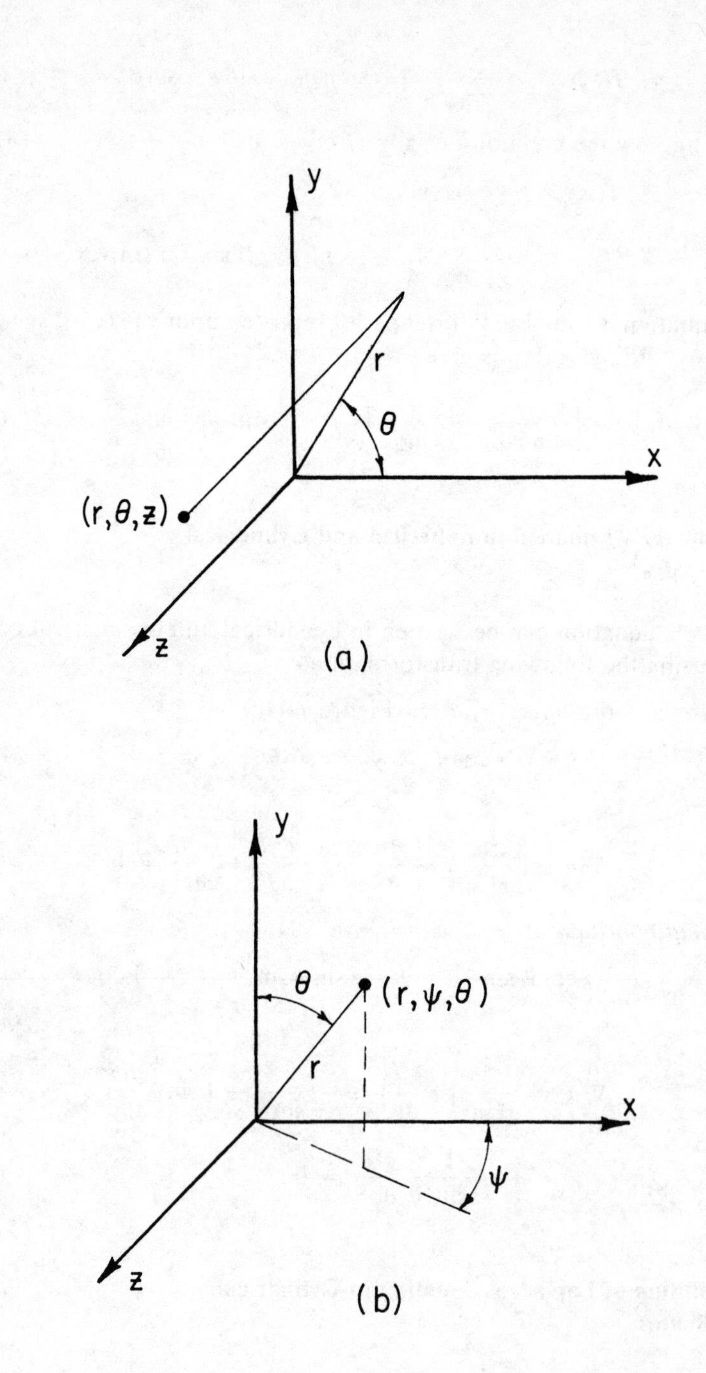

Figure 7-11. Cylindrical and Spherical Coordinates

coordinates with no z dependency. Among the problems that behave according to this solution are a long conducting cylinder in a uniform electric field, a long cylinder in a uniform fluid stream, and a long cylinder subjected to an arbitrary time steady temperature on its surface. In such problems, the electric potential, velocity potential, or temperature distribution may be required within the cylinder (interior problem) or external to the cylinder (exterior problem). In these cases, the differential equation to be solved is Laplace's equation of the form

$$\frac{\partial^2 u}{\partial r^2} + \frac{1}{r}\frac{\partial u}{\partial r} + \frac{1}{r^2}\frac{\partial^2 u}{\partial \theta^2} = 0 \tag{7.90}$$

We like to note that solutions of Laplace's equation that are independent of z are called *circular harmonics*.

In using the method of separation of variables to solve Equation (7.90), we let

$$u(r,\theta) = R(r)\,F(\theta) \tag{7.91}$$

and substitute this product solution into Equation (7.90). There results

$$\frac{r^2}{R}\left(\frac{d^2R}{dr^2} + \frac{1}{r}\frac{dR}{dr}\right) = -\frac{1}{F}\frac{d^2F}{d\theta^2} \tag{7.92}$$

In this equation, the left-hand side is a function of r only, while the right-hand side is a function of θ only. Accordingly, each term must be a constant. We set this constant equal to λ^2 in order to obtain a periodic solution for $F(\theta)$. Hence,

$$\frac{d^2F}{d\theta^2} + \lambda^2\theta = 0 \tag{7.93}$$

and

$$r^2\frac{d^2R}{dr^2} + r\frac{dR}{dr} - \lambda^2 R = 0 \tag{7.94}$$

In Equations (7.93) and (7.94), a nontrivial solution exists for $\lambda = 0$, and therefore, zero and non-zero values for λ should be considered. The solution to Equation (7.93) becomes

$$F(\theta) = A\cos\lambda\theta + B\sin\lambda\theta\,, \quad \lambda \neq 0 \tag{7.95}$$

$$F(\theta) = A_0\theta + B_0\,, \quad \lambda = 0 \tag{7.96}$$

The solution to Equation (7.94) becomes

$$R(r) = Cr^{\lambda} + Dr^{-\lambda}, \quad \lambda \neq 0 \tag{7.97}$$

$$R(r) = C_0\ln r + D_0\,, \quad \lambda = 0 \tag{7.98}$$

The physics of the problem requires that the solution $u(r,\theta)$ be single-valued in θ in the sense that

$$u(r,\theta) = u(r,\theta + 2\pi) = u(r,\theta + 2n\pi), \quad n = 1, 2, 3, \ldots \tag{7.99}$$

To satisfy this condition, A_0 in Equation (7.96) should be zero, and λ should be an integer n. Therefore, the general solution for Equation (7.90) in which $u(r,\theta)$ is single-valued can be written as

$$\begin{aligned} u(r,\theta) &= R(r)\,F(\theta) \\ &= \sum_{n=1}^{\infty} r^n[a_n \cos n\theta + b_n \sin n\theta] \\ &\quad + \sum_{n=1}^{\infty} r^{-n}[c_n \cos n\theta + d_n \sin n\theta] \\ &\quad + a_0 + b_0 \ln r \end{aligned} \tag{7.100}$$

We consider now a few physical problems and apply this equation.

Example 7.6 Consider a very long hollow circular cylinder of radius R made of a conducting material. The upper half of the cylinder is maintained at a potential ϕ_0 while the lower half is grounded (Figure 7-12). Determine the potential distribution within the cylinder.

Solution: The potential should be finite everywhere within the cylinder, including the points along its axis. Therefore, in Equation (7.100) the following constants should vanish: c_n, d_n, and b_0. Hence, Equation (7.100) is reduced to the following expression

$$\phi(r,\theta) = a_0 + \sum_{n=1}^{\infty} r^n[a_n \cos n\theta + b_n \sin n\theta] \tag{7.101}$$

Applying the boundary condition at $r = R$ which is

$$\phi(R,\theta) = f(\theta) = \begin{cases} \phi_0, & 0 < \theta < \pi \\ 0, & \pi < \theta < 2\pi \end{cases} \tag{7.102}$$

we get

$$f(\theta) = a_0 + \sum_{n=1}^{\infty} R^n[a_n \cos n\theta + b_n \sin n\theta] \tag{7.103}$$

This equation is a complete Fourier series representation for $f(\theta)$. From chapter 3, the coefficients of expansion are given by

$$a_0 = \frac{1}{2\pi}\int_0^{2\pi} f(\theta)\,d\theta = \frac{1}{2\pi}\int_0^{\pi} \phi_0\,d\theta + \int_{\pi}^{2\pi} 0\,d\theta = \frac{\phi_0}{2} \tag{7.104}$$

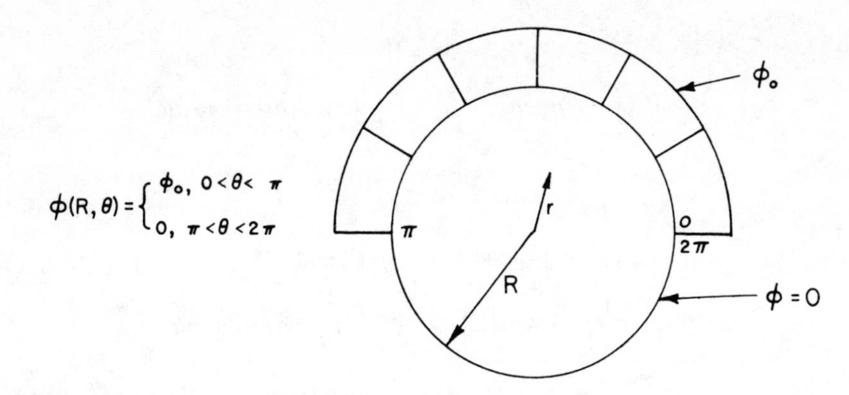

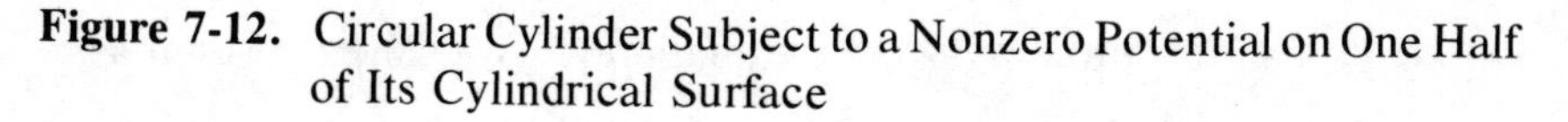

Figure 7-12. Circular Cylinder Subject to a Nonzero Potential on One Half of Its Cylindrical Surface

$$a_n R^n = \frac{1}{\pi}\int_0^{\pi} \phi_0 \cos n\theta \, d\theta = 0 \tag{7.105}$$

$$b_n R^n = \frac{1}{\pi}\int_0^{\pi} \phi_0 \sin n\theta \, d\theta = \begin{cases} \dfrac{2\phi_0}{n\pi} & n \text{ odd} \\ 0 & n \text{ even} \end{cases} \tag{7.106}$$

Accordingly, the solution given by Equation (7.101) is reduced to

$$\phi(r,\theta) = \frac{\phi_0}{2} + \frac{2\phi_0}{\pi}\sum_{n=1}^{\infty}\left(\frac{r}{R}\right)^n \frac{\sin n\theta}{n}, \qquad n = 1, 3, 5, \ldots \tag{7.107}$$

Suppose that in this problem it is required also to find the potential exterior to the cylinder. We must have then the constraints that the potential must be finite as $r \to \infty$. Therefore, in Equation (7.100), the following constants should vanish: a_n, b_n, and b_0. The solution in this case takes the form

$$\phi(r,\theta) = a_0 + \sum_{n=1}^{\infty} r^{-n}[c_n \cos n\theta + d_n \sin n\theta] \tag{7.108}$$

The solution for the internal problem (potential or temperature within a cylinder) as given by Equation (7.101) or the solution for the external problem (potential or temperature exterior to a cylindrical surface) as given by Equation (7.108) can be written in a closed form. If we let the potential or temperature at the surface of a cylinder be prescribed in an arbitrary manner, the solution for the internal problem becomes

$$\phi(r,\theta) = a_0 + \sum_{n=1}^{\infty} r^n[a_n \cos n\theta + b_n \sin n\theta] \tag{7.109}$$

with

$$a_0 = \frac{1}{2\pi}\int_0^{2\pi} f(\theta)\, d\theta\,; \qquad a_n = \frac{1}{R^n\pi}\int_0^{\pi} f(\theta) \cos n\theta\, d\theta$$

$$b_n = \frac{1}{R^n\pi}\int_0^{\pi} f(\theta) \sin n\theta\, d\theta$$

Using these expressions in Equation (7.109) yields

$$\phi(r,\theta) = \frac{1}{\pi}\int_0^{2\pi} f(\theta')\left\{\frac{1}{2} + \sum_{n=1}^{\infty}\left(\frac{r}{R}\right)^n (\cos n\theta \cos n\theta' + \sin n\theta \sin n\theta')\right\} d\theta' \tag{7.110}$$

where θ' is an integration variable. Equation (7.110) can be written as

$$\phi(r,\theta) = \frac{1}{\pi}\int_0^{2\pi} f(\theta')\left\{\frac{1}{2} + \sum_{n=1}^{\infty}\left(\frac{r}{R}\right)^n \cos n(\theta - \theta')\right\} d\theta' \tag{7.111}$$

Let us consider now the following identity

$$\begin{aligned}
\frac{1}{2} + \sum_{n=1}^{\infty} z^n \cos n(\theta - \theta') &= \frac{1}{2} + \frac{1}{2}\sum_{n=1}^{\infty} z^n[e^{in(\theta-\theta')} + e^{-in(\theta-\theta')}] \\
&= \frac{1}{2}\left\{1 + \sum_{n=1}^{\infty} [ze^{i(\theta-\theta')}]^n + [ze^{-i(\theta-\theta')}]^n\right\} \\
&= \frac{1}{2}\left[1 + \frac{ze^{i(\theta-\theta')}}{1 - ze^{i(\theta-\theta')}} + \frac{ze^{-i(\theta-\theta')}}{1 - ze^{-i(\theta-\theta')}}\right] \\
&= \frac{1}{2}\,\frac{1 - z^2}{1 - 2z\cos(\theta - \theta') + z^2}
\end{aligned}$$

Accordingly, in Equation (7.11) $z = r/R$, and the solution becomes

$$\phi(r,\theta) = \frac{1}{2\pi}\int_0^{2\pi} f(\theta')\frac{\left[1 - \left(\frac{r}{R}\right)^2\right] d\theta'}{1 - 2\left(\frac{r}{R}\right)\cos(\theta - \theta') + \left(\frac{r}{R}\right)^2} \qquad r < R \tag{7.112}$$

The relation given by this expression is referred to as *Poisson's integral*.

Similarly, Poisson's integral for the exterior problem can be obtained in the form

$$\phi(r,\theta) = \frac{1}{2\pi}\int_0^{2\pi} f(\theta')\frac{\left[1 - \left(\frac{R}{r}\right)^2\right] d\theta'}{1 - 2\left(\frac{R}{r}\right)\cos(\theta - \theta') + \left(\frac{R}{r}\right)^2} \qquad r > R \tag{7.113}$$

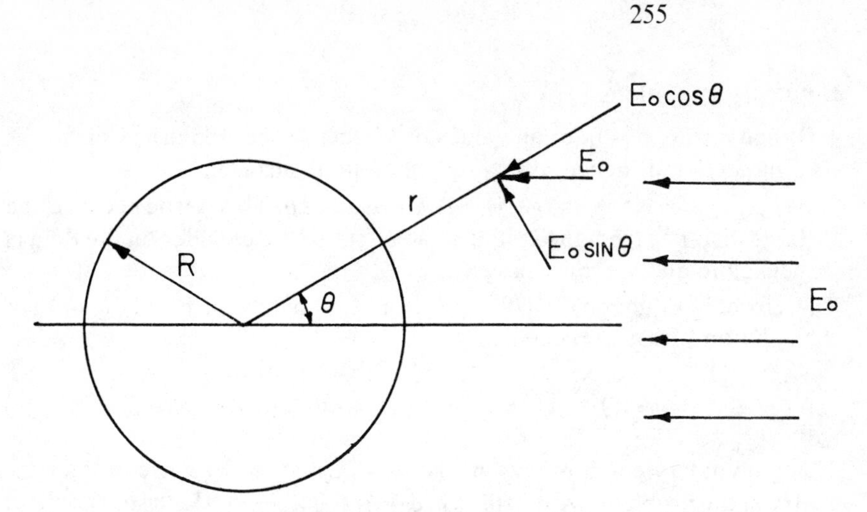

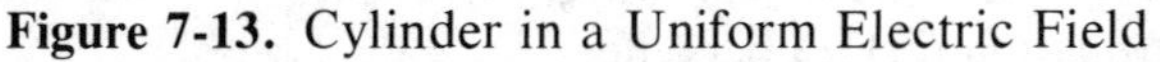
Figure 7-13. Cylinder in a Uniform Electric Field

Equations (7.112) and (7.113) reduce to

$$\phi(r,\theta) = f(\theta') \quad \text{when} \quad r = R$$

Example 7.7 Consider an infinitely long, uncharged, grounded, conducting cylinder of radius R placed in a uniform electric field having a strength E_0 and whose direction is perpendicular to the axis of the cylinder (Figure 7-13). Determine the electric potential and electric field at any point r exterior to the cylinder.
Solution: In Article 7.4 it was found that the electric field E can be represented as the gradient of a scalar electric potential ϕ as

$$E = -\nabla\phi \tag{7.114}$$

and the electric potential satisfies Laplace's equation

$$\nabla^2\phi = 0 \tag{7.115}$$

Accordingly, from Equation (7.114), the two components of the electric potential, namely E_r and E_θ, are given by

$$E_r = -\frac{\partial\phi}{\partial r}, \quad E_\theta = -\frac{1}{r}\frac{\partial\phi}{\partial\theta} \tag{7.116}$$

and the potential ϕ satisfies Laplace's equation

$$\frac{\partial^2\phi}{\partial r^2} + \frac{1}{r}\frac{\partial\phi}{\partial r} + \frac{1}{r^2}\frac{\partial^2\phi}{\partial\theta^2} = 0 \tag{7.117}$$

Boundary conditions:

1. When $r = R$, $\phi = 0$ for all values of θ because the cylindrical surface is grounded and, hence, it is an equipotential surface.
2. When x (or r) is very large, $\phi = E_0 x = E_0 r \cos\theta$. This is true because at a large distance from the cylinder the effect of the cylinder on the field is negligible and we must have: $E_0 = \partial\phi/\partial x$, $\partial\phi/\partial y = 0$.

Therefore, in Equation (7.100), a_n, b_n, and b_0 should vanish and the form of the solution for $\phi(r,\theta)$ becomes

$$\phi(r,\theta) = a_0 + \sum_{n=1}^{\infty} r^{-n}[c_n \cos n\theta + d_n \sin n\theta] \tag{7.118}$$

Using boundary condition (2) yields: $a_0 = E_0 r \cos\theta$. Because of the symmetry of the problem, we must have: $\phi(r,\theta) = \phi(r, -\theta)$. Hence d_n should be zero, and Equation (7.118) becomes

$$\phi(r,\theta) = E_0 r \cos\theta + \sum_{n=1}^{\infty} c_n r^{-n} \cos n\theta \tag{7.119}$$

Using boundary condition (1) in Equation (7.119) we get

$$0 = E_0 R \cos\theta + \sum_{n=1}^{\infty} c_n R^{-n} \cos n\theta \tag{7.120}$$

or

$$-E_0 R \cos\theta = \sum_{n=1}^{\infty} c_n R^{-n} \cos n\theta \tag{7.121}$$

This equation is an identity; furthermore, from the orthogonality properties of the cosine function, we must have $c_n = 0$ for $n > 1$. Accordingly, Equation (7.121) yields

$$-E_0 R = \frac{c_1}{R} \quad \text{or} \quad c_1 = -E_0 R^2 \tag{7.122}$$

Correspondingly, the solution to the problem as given by Equation (7.119) is reduced to

$$\phi(r,\theta) = E_0 r \cos\theta - \left(\frac{E_0 R^2}{r}\right) \cos\theta$$

or

$$\phi(r,\theta) = \left(1 - \frac{R^2}{r^2}\right) E_0 r \cos\theta \tag{7.123}$$

Hence, the radial E_r and the transverse E_θ components of the field become

$$E_r = -\frac{\partial \phi}{\partial r} = -\left(1 + \frac{R^2}{r^2}\right)E_0 \cos\theta \tag{7.124}$$

and

$$E_\theta = -\frac{1}{r}\frac{\partial \phi}{\partial \theta} = \left(1 - \frac{R^2}{r^2}\right)E_0 \sin\theta \tag{7.125}$$

7.9 Heat Transfer in a Finite Cylinder

When considering the temperature or potential distribution in a finite cylinder, it may be then necessary to solve Laplace's equation in cylindrical polar coordinates of the form

$$\frac{\partial^2 T}{\partial r^2} + \frac{1}{r}\frac{\partial T}{\partial r} + \frac{1}{r^2}\frac{\partial^2 T}{\partial \theta^2} + \frac{\partial^2 T}{\partial z^2} = 0 \tag{7.126}$$

Solutions of this equation are referred to as *cylindrical harmonics*. We consider first the case of a temperature distribution within a finite cylinder which is independent of θ. The problem is described by Figure 7-14 and mathematically is represented by

$$\frac{\partial^2 T}{\partial r^2} + \frac{1}{r}\frac{\partial T}{\partial r} + \frac{\partial^2 T}{\partial z^2} = 0 \tag{7.127}$$

For symmetry with respect to θ, the imposed conditions must be independent of θ. Accordingly, the sides and bottom of the cylinder are chosen to be at a constant temperature that may be taken as zero; the temperature at $z = L$ is taken as $f(r)$, that is,

$$\text{at } z = 0, \quad T = 0 \tag{7.128}$$

$$\text{at } z = L, \quad T = f(r) \tag{7.129}$$

$$\text{at } r = a, \quad T = 0 \tag{7.130}$$

$$\text{at } r = 0, \quad T \text{ is finite} \tag{7.131}$$

Let $T(r,z) = R(r)\,Z(z)$. Equation (7.127) then yields

$$\frac{1}{R}\frac{d^2R}{dr^2} + \frac{1}{Rr}\frac{dR}{dr} = -\frac{1}{Z}\frac{d^2Z}{dz^2} \tag{7.132}$$

The left-hand side of this equation is a function of r only, while the right-hand side is a function of z only and so each side must be a constant, $-\lambda^2$. Therefore, there results

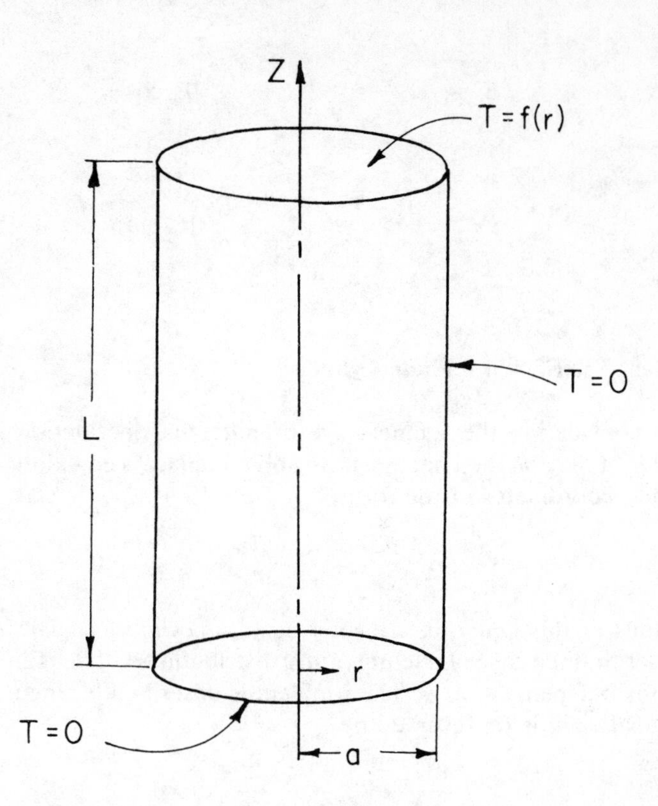

Figure 7-14. Finite Cylinder Exposed to a Prescribed Temperature on One Side

$$\frac{d^2Z}{dz^2} - \lambda^2 Z = 0 \tag{7.133}$$

and

$$\frac{d^2R}{dr^2} + \frac{1}{r}\frac{dR}{dr} + \lambda^2 R = 0 \tag{7.134}$$

The solutions to Equations (7.133) and (7.134), respectively, are

$$Z = A_1 \sinh \lambda z + A_2 \cosh \lambda z \tag{7.135}$$

$$R = A_2 J_0(\lambda r) + A_4 Y_0(\lambda r) \tag{7.136}$$

Hence,

$$\begin{aligned} T(r,z) &= ZR \\ &= (A_1 \sinh \lambda z + A_2 \cosh \lambda z)[A_2 J_0(\lambda r) + A_4 Y_0(\lambda r)] \end{aligned} \tag{7.137}$$

Because $Y_0(\lambda r) \rightarrow \infty$ as $\lambda r \rightarrow 0$, the condition given by Equation (7.131) requires that A_4 be zero. Equation (7.128) makes A_2 also vanish so one obtains (we set $A_1A_2 = A$)

$$T(r,z) = A \sinh(\lambda z) J_0(\lambda r) \tag{7.138}$$

The application of Equation (7.130) gives

$$0 = A \sinh(\lambda z) J_0(\lambda a) \tag{7.139}$$

In order not to have a trivial solution ($T = 0$ for all r and z), Equation (7.139) requires that

$$J_0(\lambda a) = 0 \tag{7.140}$$

or

$$\lambda_n a = \alpha_n \qquad n = 1, 2, 3, \ldots$$

where α_n are the roots of $J_0(x) = 0$. Therefore, a general solution including all values of λ_n can be written as

$$T(r,z) = \sum_{n=1}^{\infty} A_n \sinh(\lambda_n z) J_0(\lambda_n r) \tag{7.141}$$

Now, the application of the nonhomogeneous condition at $z = L$ determines the unknown coefficients A_n, that is,

$$f(r) = \sum_{n=1}^{\infty} A_n \sinh(\lambda_n L) J_0(\lambda_n r) \tag{7.142}$$

From Equation (3.68) of chapter 3, A_n is given by

$$A_n \sinh(\lambda_n L) = \frac{\int_0^a r f(r) J_0(\lambda_n r)\, dr}{\int_0^a r J_0^2(\lambda_n r)\, dr}$$

or

$$A_n = \frac{1}{\sinh(\lambda_n L)} \frac{\int_0^a r f(r) J_0(\lambda_n r)\, dr}{\frac{a^2}{2} J_1^2(\lambda_n a)} \tag{7.143}$$

Finally, the solution becomes

$$T(r,z) = \frac{2}{a^2} \sum_{n=1}^{\infty} \frac{\sinh(\lambda_n z) J_0(\lambda_n r)}{\sinh(\lambda_n L) J_1^2(\lambda_n a)} \times \int_0^a r f(r) J_0(\lambda_n r)\, dr \tag{7.144}$$

When $f(r)$ is a constant $= T_0$, the integral in Equation (7.144) becomes

$$\int_0^a T_0 r J_0(\lambda_n r)\, dr = \frac{T_0 a}{\lambda_n} J_1(\lambda_n a)$$

and Equation (7.144) is reduced to

$$T(r,z) = 2T_0 \sum_{n=1}^{\infty} \frac{1}{\lambda_n a} \frac{\sinh(\lambda_n z) J_0(\lambda_n r)}{\sinh(\lambda_n L) J_1(\lambda_n a)} \tag{7.145}$$

7.10 Nonhomogeneity on the Cylindrical Surface

In this section, the problem analyzed in Article 7.9 is modified to have a nonhomogeneity on the cylindrical surface ($r = a$) while the top and bottom surfaces are maintained at $T = 0$. The steady state temperature distribution is again here obtained by solving Laplace's equation

$$\frac{\partial^2 T}{\partial r^2} + \frac{1}{r}\frac{\partial T}{\partial r} + \frac{\partial^2 T}{\partial z^2} = 0 \tag{7.146}$$

subject to the conditions:

$$T = 0 \quad \text{when } z = 0 \tag{7.147}$$

$$T = 0 \quad \text{when } z = L \tag{7.148}$$

$$T = f(z) \quad \text{when } r = a \tag{7.149}$$

The method of separation of variables yields here equations similar to Equations (7.133) and (7.134). However, because the hyperbolic functions do not form an orthogonal set and cannot be used to express an arbitrary function $f(z)$, the separation constant is chosen as λ^2 (as compared with $-\lambda^2$ in Article 7.9). Accordingly, the two resulting equations become

$$\frac{d^2 Z}{dz^2} + \lambda^2 Z = 0 \tag{7.150}$$

with

$$Z(0) = 0\,; \quad Z(L) = 0 \tag{7.151a,b}$$

and

$$\frac{d^2 R}{dr^2} + \frac{1}{r}\frac{dR}{dr} - \lambda^2 R = 0 \tag{7.152}$$

with

$$R(0) = \text{finite}; \quad R(a) = f(z) \tag{7.153a,b}$$

Equation (7.150) along with Equation (7.151) yields

$$Z = B_1 \sin \lambda_n z \qquad (7.154)$$

$$\lambda_n = n\pi/L, \qquad n = 1, 2, 3, \ldots$$

Equation (7.152) gives

$$R = G\, I_0(\lambda r) + D\, K_0(\lambda r) \qquad (7.155)$$

where I_0 and K_0 are the modified Bessel functions of the first and second kind, respectively, of order zero. Condition (7.153a) requires that $D = 0$. Therefore, the solution for $T(r,z)$ can now be written as

$$T(r,z) = \sum_{n=1}^{\infty} A_n I_0\left(\frac{n\pi r}{L}\right) \sin \frac{n\pi z}{L} \qquad n = 1, 2, 3, \ldots \qquad (7.156)$$

Employing Equation (7.153b) gives

$$T(a,z) = f(z) = \sum_{n=1}^{\infty} A_n I_0\left(\frac{n\pi a}{L}\right) \sin \frac{n\pi z}{L} \qquad (7.157)$$

with

$$A_n I_0\left(\frac{n\pi a}{L}\right) = \frac{2}{L} \int_0^L f(z) \sin \frac{n\pi z}{L} dz \qquad (7.158)$$

Finally, using Equation (7.158) in Equation (7.156) yields the solution as

$$T(r,z) = \frac{2}{L} \sum_{n=1}^{\infty} \frac{I_0\left(\frac{n\pi r}{L}\right)}{I_0\left(\frac{n\pi a}{L}\right)} \sin \frac{n\pi z}{L} \int_0^L f(z) \sin \frac{n\pi z}{L} dz$$

$$n = 1, 2, 3, \ldots \qquad (7.159)$$

If $f(z)$ is a constant $= T_0$, Equation (7.159) becomes

$$T(r,z) = \frac{4T_0}{\pi} \sum_{n=1}^{\infty} \frac{1}{n} \frac{I_0\left(\frac{n\pi r}{L}\right)}{I_0\left(\frac{n\pi a}{L}\right)} \sin \frac{n\pi z}{L} \qquad n = 1, 3, 5, \ldots \qquad (7.160)$$

7.11 Superposition of Solutions

As was done in Example 7.2, the principle of superposition applies here. The method is schematically shown in Figure 7-15. Each of the tempera-

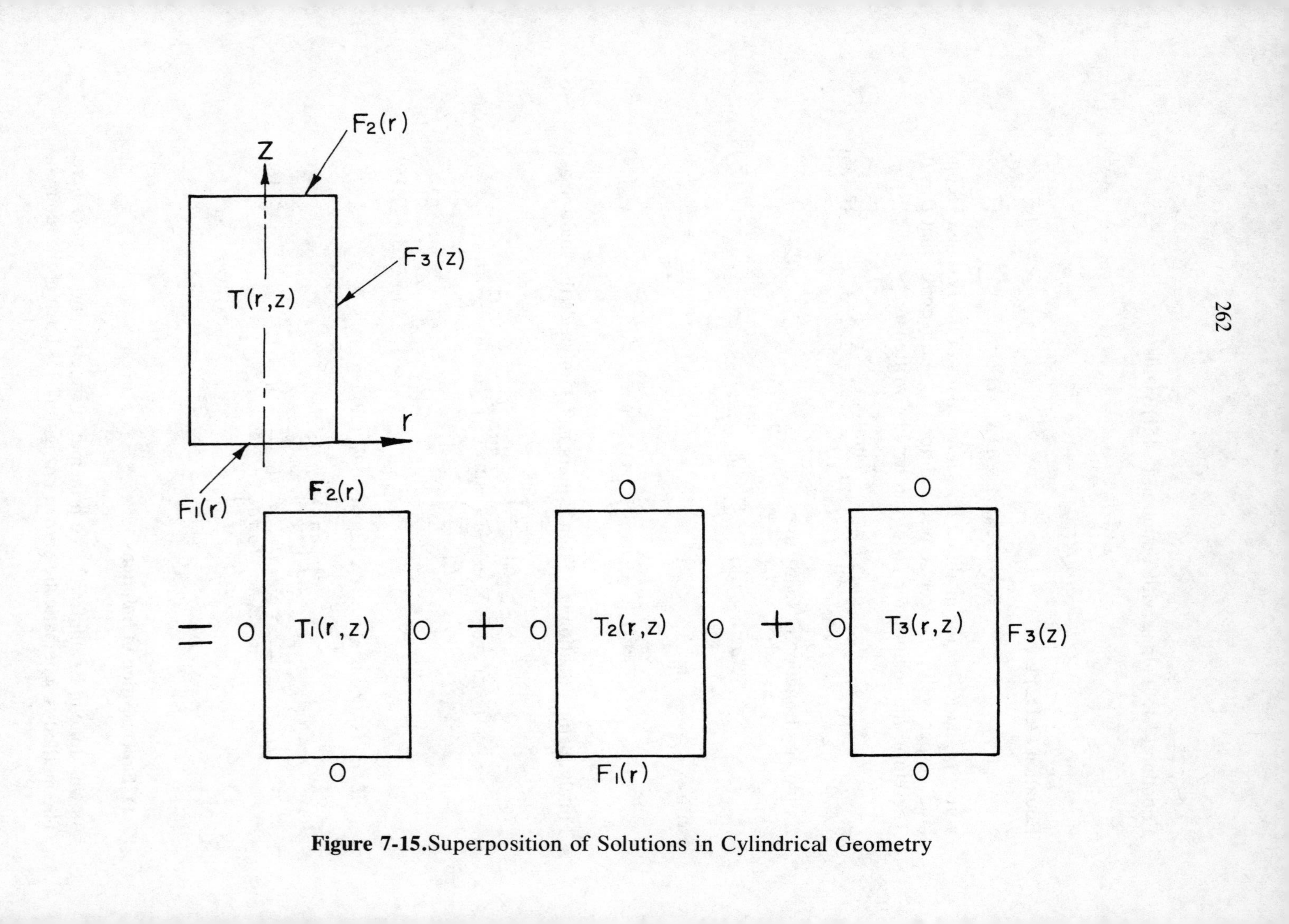

Figure 7-15. Superposition of Solutions in Cylindrical Geometry

tures T_1, T_2, and T_3 satisfies Laplace's equation with one nonhomogeneous boundary condition.

7.12 Solutions of Laplace's Equation in Spherical Coordinates

In many physical problems in heat transfer or field theory involving spherical geometry, there is symmetry about one of the axes (θ) so that the solution to the problem is independent of the angle ψ. In this case, Laplace's equation for the temperature distribution in a sphere becomes (see Figure 7-11)

$$\frac{\partial}{\partial r}\left(r^2\frac{\partial T}{\partial r}\right) + \frac{1}{\sin\theta}\frac{\partial}{\partial\theta}\left(\sin\theta\,\frac{\partial T}{\partial\theta}\right) = 0 \tag{7.161}$$

Let us consider a sphere of radius a subjected at its surface to a temperature distribution $f(\theta)$, while the temperature at the center of the sphere remains finite, that is,

$$T(a,\theta) = f(\theta) \tag{7.162}$$

$$T(0,\theta) = \text{finite} \tag{7.163}$$

The solution to this problem can be obtained by the method of separation of variables by setting $T(r,\theta) = R(r)\,G(\theta)$ and substituting into Equation (7.161). In this case, the only possible orthogonal direction is θ and, therefore, the separation constant λ is chosen to yield orthogonal functions in that direction. Accordingly, we obtain

$$\frac{1}{\sin\theta}\frac{d}{d\theta}\left(\sin\theta\frac{dG}{d\theta}\right) + \lambda G = 0 \tag{7.164}$$

and

$$r^2\frac{d^2R}{dr^2} + 2r\frac{dG}{dr} - \lambda R = 0 \tag{7.165}$$

In Equation (7.165), if we let $r = e^z$, the equation becomes

$$\frac{d^2R}{dz^2} + \frac{dR}{dz} - \lambda R = 0 \tag{7.166}$$

which has the auxiliary equation $p^2 + p - \lambda = 0$. Hence, the sum of the roots of this auxiliary equation is -1 and usually chosen as $p_1 = n$ and $p_2 = -(n + 1)$. This means that $\lambda = p(p + 1) = n(n + 1)$. Therefore, Equation (7.166) has the solution

$$R(z) = A_1 e^{nz} + A_2 e^{-(n+1)z}$$

or

$$R(r) = A_1 r^n + A_2 r^{-(n+1)} \tag{7.167}$$

Equation (7.164) now can be written as

$$\frac{1}{\sin\theta}\frac{d}{d\theta}\left[\frac{1-\cos^2\theta}{\sin\theta}\frac{dG}{d\theta}\right] + n(n+1)G = 0 \tag{7.168}$$

If we introduce the transformation $x = \cos\theta$, this equation becomes

$$\frac{d}{dx}\left[(1-x^2)\frac{dG}{dx}\right] + n(n+1)G = 0 \tag{7.169}$$

This is Legendre equation of degree n. Its solution is

$$G = B_1 P_n(x) + B_2 Q_n(x) \tag{7.170}$$

$Q_n(x)$ is finite only for $x < 1$; therefore, for a solution that is finite at $x = \pm 1$ ($\cos\theta = \pm 1$ or $\theta = 0,\pi$), B_2 should vanish. The resulting form of the solution for $T(r,\theta)$ becomes

$$T(r,\theta) = \left(Ar^n + \frac{B}{r^{n+1}}\right)P_n(\cos\theta)$$

More generally, the solution should include all values for n, and correspondingly, we write

$$T(r,\theta) = \sum_{n=0}^{\infty}\left(A_n r^n + \frac{B_n}{r^{n+1}}\right)P_n(\cos\theta) \tag{7.171}$$

The constants A_n and B_n are determined by the conditions imposed on the problem and they are given by Equations (7.162) and (7.163). Hence, from Equation (7.163) we get $B_n = 0$ and the solution is reduced to

$$T(r,\theta) = \sum_{n=0}^{\infty} A_n r^n P_n(\cos\theta) \tag{7.172}$$

Applying Equation (7.162) yields

$$T(a,\theta) = f(\theta) = \sum_{n=0}^{\infty} A_n a^n P_n(\cos\theta) \tag{7.173}$$

This equation represents the expansion of $f(\theta)$ in terms of Legendre polynomials. The coefficients A_n are obtained as (see chapter 3)

$$A_n = \frac{2n+1}{2a^n}\int_0^{\pi} f(\theta)\,P_n(\cos\theta)\sin\theta\,d\theta \tag{7.174}$$

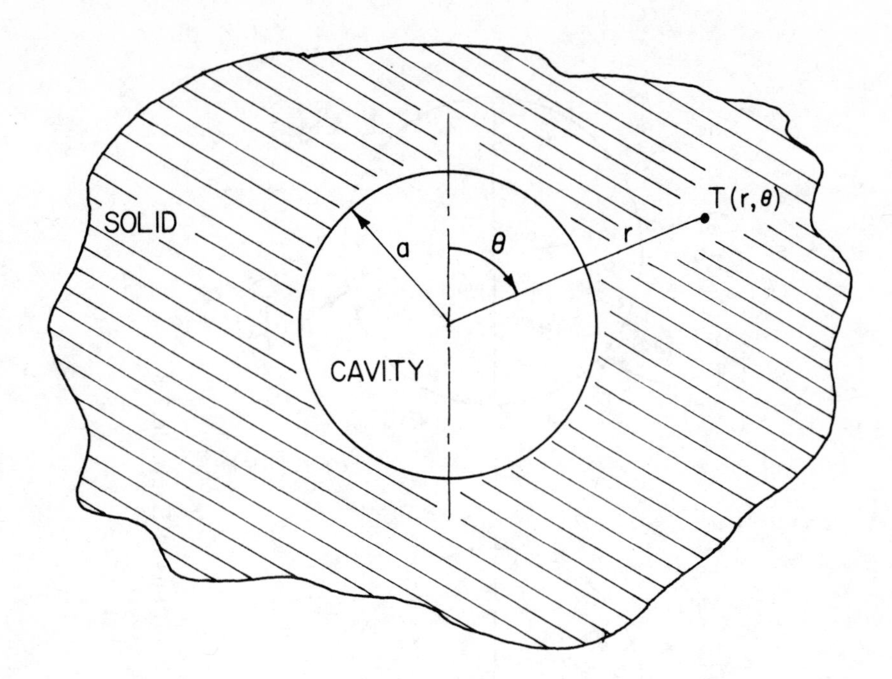

Figure 7-16. Solid Medium Exterior to a Spherical Cavity

When the temperature distribution is desired in a solid exterior to a spherical cavity (Figure 7-16), and for a finite temperature at very large r, Equation (7.171) is reduced to

$$T(r,\theta) = \sum_{n=0}^{\infty} \frac{B_n}{r^{n+1}} P_n(\cos\theta) \tag{7.175}$$

Obviously, the solution presented by Equation (7.173) with coefficients given by Equation (7.174) is applicable to the electric conduction problem for a sphere.

If a spherical case is taken such that $f(\theta)$ is given by

$$f(\theta) = \begin{cases} T_0 & 0 \le \theta < \dfrac{\pi}{2} \\ 0 & \dfrac{\pi}{2} < \theta \le \pi \end{cases}$$

The coefficients A_n are obtained from Equation (7.174) as

$$A_0 = \frac{T_0}{2}, \quad A_1 = \frac{3T_0}{4}, \quad A_2 = 0, \quad A_3 = -\frac{7T_0}{16}, \quad A_4 = 0, \quad \text{etc.}$$

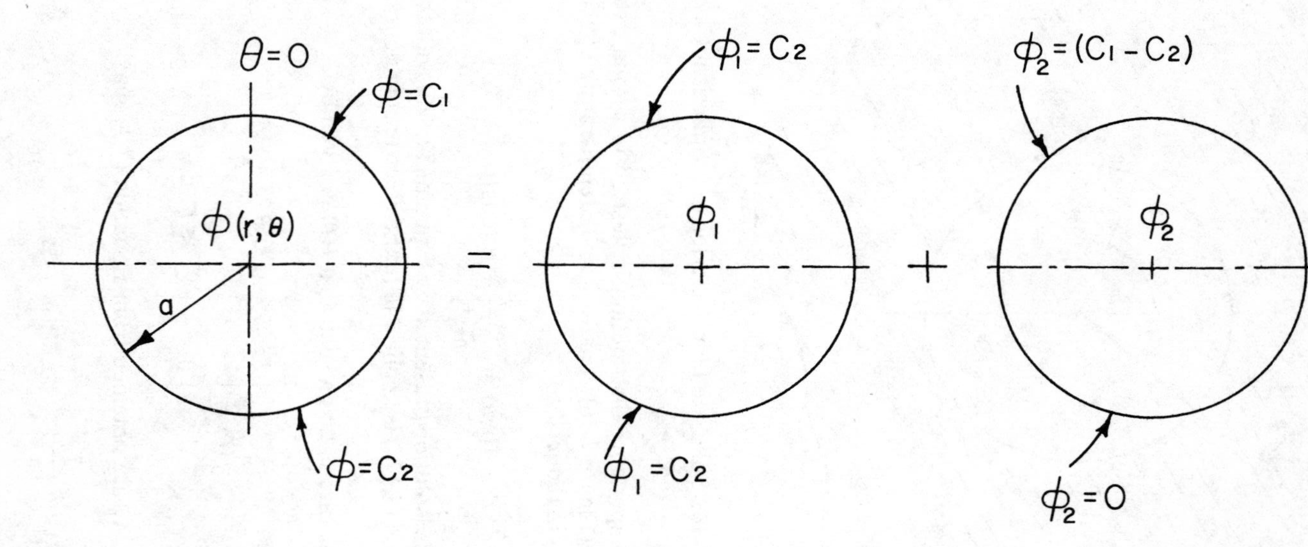

Figure 7-17. Superposition of Solutions in Example 7.8

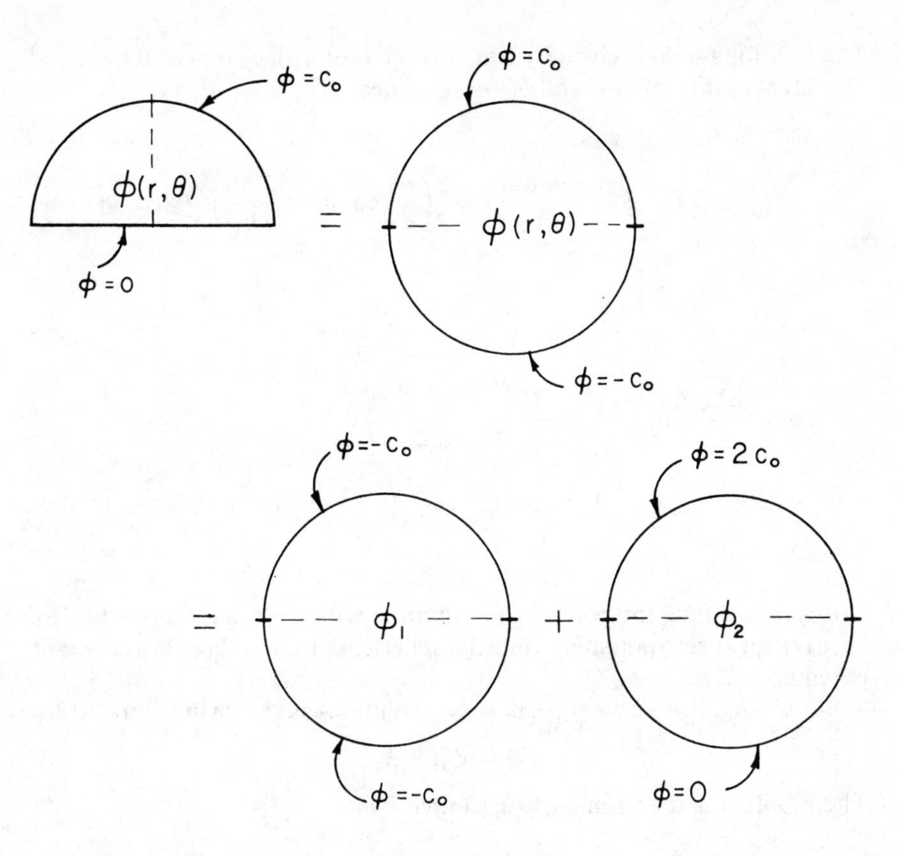

Figure 7-18. Superposition of Solutions in Example 7.9

Hence, the solution represented by Equation (7.173) becomes

$$T(r,\theta) = \frac{T_0}{2}\left[1 + \frac{3}{2}\left(\frac{r}{a}\right)\cos\theta - \frac{7}{8}\left(\frac{r}{a}\right)^3 P_3(\cos\theta) + \frac{11}{16}\left(\frac{r}{a}\right)^5 P_5(\cos\theta) - \ldots\right] \quad (7.176)$$

Example 7.8 Consider a sphere of radius a made up of two hemispherical shells separated by a thin non-conducting layer. One shell is set at a potential C_1 while the other is set at C_2. Determine the value of the potential anywhere within the sphere.
Solution: This problem can easily be solved by superposition of solutions as shown schematically in Figure 7-17. Obviously, the solution for ϕ_1 is

$$\phi_1 = \text{constant everywhere within the sphere} \quad (7.177)$$

The solution for ϕ_2 is given by Equation (7.176) with T_0 replaced by $(C_1 - C_2)$. Hence, the solution for $\phi(r,\theta)$ becomes

$$\begin{aligned}\phi(r,\theta) &= \phi_1 + \phi_2 \\ &= C_2 + \frac{C_1 - C_2}{2}\left[1 + \frac{3}{2}\left(\frac{r}{a}\right)\cos\theta - \frac{7}{8}\left(\frac{r}{a}\right)^3 P_3(\cos\theta)\right. \\ &\left. \quad + \frac{11}{16}\left(\frac{r}{a}\right)^5 P_5(\cos\theta) - \ldots\right] \qquad (7.178)\end{aligned}$$

or

$$\begin{aligned}\phi(r,\theta) = \frac{C_1 + C_2}{2} + \frac{C_1 - C_2}{2}&\left[\frac{3}{2}\left(\frac{r}{a}\right)\cos\theta\right. \\ &\left. - \frac{7}{8}\left(\frac{r}{a}\right)^3 P_3(\cos\theta) + \frac{11}{16}\left(\frac{r}{a}\right)^5 P_5(\cos\theta) - \ldots\right] \qquad (7.179)\end{aligned}$$

Example 7.9 Find the potential distribution within a hemisphere when its base is kept at zero potential while the spherical surface is kept at a constant potential C_0.
Solution: Again here we superpose two solutions as shown in Figure 7-18.

$$\phi = \phi_1 + \phi_2$$

The problem is then similar to Example 7-8.

Integral Equations and Classifications

8.1 Introduction

An equation in which the unknown function appears only under one or more integral signs is designated as an integral equation, e.g.

$$u(x) = f(x) + \lambda \int_a^b K(x,t)\, u(t)\, dt \qquad (8.1)$$

where $a \le x \le b$, $a \le t \le b$

$u(x)$ is the unknown function to be determined

$f(x)$ is a known function of x

$K(x,t)$ is a known function of x and t and is referred to as the kernel of the equation

λ is a known parameter

a, b are known limits of integration; however, in certain equations the upper limit b is a variable

A typical example of an integral equation is

$$u(x) = \sin x + 4 \int_0^{10} (xt + e^{x+t})\, u(t)\, dt \qquad (8.2)$$

We will present in the following chapter areas in which the integral equations play important roles, such as the fields of mechanics, heat transfer, and mathematical physics. We will also relate differential equations and integral equations, as many problems in engineering and physics can be solved more effectively by integral equations than by differential equation methods.

When the unknown function appears both as a derivative and under an integral sign, the equation is classified as an integro-differential equation.

8.2 Classifications of Integral Equations

A linear integral equation in its general form can be written as

$$g(x)\,u(x) = f(x) + \lambda \int_a^s K(x,t)\,u(t)\,dt \tag{8.3}$$

where the upper limits can be a fixed number or a variable x. When the upper limit is a fixed number, b, the equation is called *Fredholm Integral Equation*. When it is a variable, x, the equation is called *Volterra Integral Equation*.

Fredholm Integral Equation

1. When $g(x) = 0$ in Equation (8.3), the equation is referred to as *Fredholm Integral Equation of the First Kind,* that is,

$$f(x) + \lambda \int_a^b K(x,t)\,u(t)\,dt = 0 \tag{8.4}$$

2. When $g(x) = 1$ and $f(x) = 0$ in Equation (8.3), the equation is called *homogeneous*, that is

$$u(x) = \lambda \int_a^b K(x,t)\,u(t)\,dt \tag{8.5}$$

3. When $g(x) = 1$ in Equation (8.3), the equation is called *Fredholm Equation of the Second Kind*, that is

$$u(x) = f(x) + \lambda \int_a^b K(x,t)\,u(t)\,dt \tag{8.6}$$

Volterra Integral Equation

The same types that were listed under Fredholm equations exist under Volterra equations, but of course the upper limit of integration is a variable. We have

1. First Kind $\quad g(x) = 0 \quad$ that is,

$$f(x) + \lambda \int_a^x K(x,t)\,u(t)\,dt = 0 \tag{8.7}$$

2. Homogeneous $\quad g(x) = 1,\ f(x) = 0 \quad$ that is,

$$u(x) = \lambda \int_a^x K(x,t)\,u(t)\,dt \tag{8.8}$$

3. Second Kind $\quad g(x) = 1 \quad$ that is,

$$u(x) = f(x) + \lambda \int_a^x K(x,t)\,u(t)\,dt \tag{8.9}$$

Equation (8.3) sometimes is referred to as an integral equation of the third kind. It can be reduced to an equation of the second kind as follows:

Let us divide both sides of the equation by $\sqrt{g(x)}$; this yields

$$\sqrt{g(x)}\,u(x) = \frac{f(x)}{\sqrt{g(x)}} + \lambda \int_a^s \frac{K(x,t)}{\sqrt{g(x)}} u(t)\,dt$$

Multiply and divide the kernel $K(x,t)$ by $\sqrt{g(t)}$ to give

$$\sqrt{g(x)}\,u(x) = \frac{f(x)}{\sqrt{g(x)}} + \lambda \int_a^s \frac{K(x,t)}{\sqrt{g(x)\,g(t)}} \sqrt{g(t)}\,u(t)\,dt \tag{8.10}$$

Equation (8.10) is an integral equation of the second kind for the unknown $\sqrt{g(x)}\,u(x)$. Let

$$\sqrt{g(x)}\,u(x) = \phi(x)$$

$$\frac{f(x)}{\sqrt{g(x)}} = F(x)$$

$$\frac{K(x,t)}{\sqrt{g(x)\,g(t)}} = \overline{K}(x,t)$$

Equation (8.10) becomes

$$\phi(x) = F(x) + \lambda \int_a^s \overline{K}(x,t)\,\phi(t)\,dt \tag{8.11}$$

which is an equation of the second type. In the above transformation we should emphasize that the function $g(x)$ should not change sign within the interval of integration.

Singular Integral Equations

An integral equation is called singular if one or both of the limits of integration become infinite or if the kernel becomes infinite at any point within the interval of integration. The following integral equations are singular:

$$u(x) = \tan x + \int_0^\infty [\sin xt + \cos xt]\,u(t)\,dt \tag{8.12a}$$

$$u(x) = e^x + 4\int_{-\infty}^{\infty} \exp[-|x-t|]\,u(t)\,dt \tag{8.12b}$$

$$u(x) = 3x + 2\int_0^x \frac{x(e^{x+t})}{(x-t)^2} u(t)\,dt \tag{8.12c}$$

In Equation (8.12c) $K(x,t)$ becomes infinite when $t = x$;

$$u(x) = 2x^2 + x + 10 \int_0^1 \frac{xt + \cos xt}{(t - 1)} u(t)\, dt \tag{8.12d}$$

In Equation (8.12d) $K(x,t)$ becomes infinite when $t = 1$

Nonlinear Integral Equations

When the unknown function $u(t)$ appears under the integral to a power other than unity, the equation is called nonlinear, that is,

$$u(x) = \sinh x + \int_0^{10} 3x(t + e^{-|x-t|})\,[u(t)]^2\, dt \tag{8.13a}$$

The unknown function, however, may appear in a more general way, such as

$$u(x) = x + e^x + \lambda \int_a^b F\{x,t,u(t)\}\, dt \tag{8.13b}$$

where F is any nonlinear function of $u(t)$.

8.3 Relations Between Integral Equations and Differential Equations

In this section we relate the ordinary differential equations with prescribed initial conditions to Volterra integral equations, and the ordinary differential equations with prescribed boundary conditions to Fredholm integral equations. We start with the first case and illustrate the method first by a simple example.

1. Initial Value Problem

Example 8.1 Change the following oscillator equation into a Volterra integral equation.

$$y'' + \omega^2 y = 0 \tag{8.14}$$

$$y(0) = 0, \qquad y'(0) = 1 \tag{8.15a,b}$$

Solution: Let us integrate Equation (8.14) from $x = 0$ to x. The result becomes

$$y'(x) - y'(0) + \omega^2 \int_0^x y(t)\,dt = 0 \tag{8.16}$$

Using condition (8.15b) gives

$$y'(x) - 1 + \omega^2 \int_0^x y(t)\,dt = 0 \tag{8.17}$$

Integrating Equation (8.17) once more we get

$$y(x) - y(0) - \int_0^x (1)\,dt + \omega^2 \int_0^x \int_0^\xi y(t)\,dt\,d\xi = 0 \tag{8.18}$$

Using condition (8.15a) gives

$$y(x) - x + \omega^2 \int_0^x \int_0^\xi y(t)\,dt\,d\xi = 0 \tag{8.19}$$

Let us take now the double integral

$$\int_0^x \int_0^\xi y(t)\,dt\,d\xi \tag{8.20}$$

and write it as

$$\int_0^x d\xi \int_0^\xi y(t)\,dt \tag{8.21}$$

Using integration by parts gives

$$\int_0^x d\xi \int_0^\xi y(t)\,dt = \left[\xi \int_0^\xi y(t)\,dt\right]_{\xi=0}^{\xi=x} - \int_0^x \xi \left\{\frac{d}{d\xi}\int_0^\xi y(t)\,dt\right\} d\xi \tag{8.22}$$

$$= x\int_0^x y(t)\,dt - \int_0^x \xi y(\xi)\,d\xi \tag{8.23}$$

$$= \int_0^x (x - t)\,y(t)\,dt \tag{8.24}$$

where we replaced ξ by t in the second integral of Equation (8.23) because t and ξ are dummy variables. Therefore, Equation (8.19) now becomes

$$y(x) - x + \omega^2 \int_0^x (x - t)\,y(t)\,dt = 0 \tag{8.25}$$

Equation (8.25) is a Volterra integral equation. It can be verified by direct substitution that

$$y = 1/\omega \sin \omega x$$

satisfies both Equation (8.25) and Equation (8.14).

We consider next a more general initial value problem and present two methods to change it into a Volterra integral equation.

General Initial Value Problem

Method 1 — Direct Integration. Let us consider the following ordinary differential equation

$$y''(x) + A(x)\,y'(x) + B(x)\,y(x) = F(x) \tag{8.26}$$

$$y(a) = C_0, \quad y'(a) = C_1 \tag{8.27}$$

Integrating Equation (8.26) from a to x gives

$$y'(x) - y'(a) + A(x)\,y(x) - A(a)\,y(a) - \int_a^x y(t)\,A'(t)\,dt + \int_a^x B(t)\,y(t)\,dt = \int_a^x F(t)\,dt \tag{8.28}$$

Using the initial conditions given by Equation (8.27), we obtain

$$y'(x) - C_1 + A(x)\,y(x) + \int_a^x [B(t) - A'(t)]\,y(t)\,dt - C_0A(a) = \int_a^x F(t)\,dt \tag{8.29}$$

Integrating now Equation (8.29) using the initial conditions, Equation (8.27), yields

$$y(x) - C_0 + \int_a^x A(t)\,y(t)\,dt + \int_a^x \int_a^\xi [B(t) - A'(t)]y(t)\,dt\,d\xi = \int_a^x \int_a^\xi F(t)\,dt\,d\xi + [C_1 + C_0A(a)](x - a) \tag{8.30}$$

The double integral

$$\int_a^x \int_a^\xi F(t)\,dt\,d\xi$$

was found in Example 8.1, following integration by parts, to be equal to

$$\int_a^x (x - t)\,F(t)\,dt \tag{8.31}$$

Repeated application of this procedure results in the following formula:

$$\left\{\int_a^x\right\}^n F(t)\,(dt)^n$$

$$\equiv \int_a^x \int_a^{t_n} \cdots \int_a^{t_3} \int_a^{t_2} F(t_1)\,dt_1\,dt_2 \ldots dt_{n-1}\,dt_n$$

$$= \frac{1}{(n-1)!}\int_a^x (x - t)^{n-1} F(t)\,dt \tag{8.32}$$

Applying now Equation (8.31) or (8.32) to Equation (8.30) results in the following expression:

$$y(x) = C_0 + [C_1 + C_0A(a)](x - a) + \int_a^x (x - t)\,F(t)\,dt$$

$$- \int_a^x \{A(t) + (x - t)[B(t) - A'(t)]\}y(t)\,dt \tag{8.33}$$

Let us set

$$F(x) = C_0 + [C_1 + C_0A(a)](x - a) + \int_a^x (x - t)\,F(t)\,dt \tag{8.34}$$

$$K(x,t) = -\{A(t) + (x - t)[B(t) - A'(t)]\} \tag{8.35}$$

We can write then Equation (8.33) as

$$y(x) = F(x) + \int_a^x K(x,t)\,y(t)\,dt \tag{8.36}$$

which is a Volterra integral equation of the second kind for Equations (8.26) and (8.27).

Method 2 — Introducing an Unknown Function. Let

$$y''(x) = \phi(x) \tag{8.37}$$

where $\phi(x)$ is an unknown function.
Integrating Equation (8.37) from a to x yields

$$y'(x) = \int_a^x \phi(t)\,dt + K_1 \tag{8.38}$$

and

$$y(x) = \int_a^x (x - t)\phi(t)\,dt + K_1 x + K_2 \tag{8.39}$$

where K_1 and K_2 are constants to be determined from the initial conditions of the problem. Using the conditions given by Equation (8.27) to evaluate K_1 and K_2, we obtain

$$K_1 = C_1 \tag{8.40}$$

and

$$K_2 = C_0 - C_1 a \tag{8.41}$$

Substituting Equations (8.37), (8.40) and (8.41) in Equation (8.26) yields

$$\phi(x) + A(x)\left\{\int_a^x \phi(t)\,dt + C_1\right\}$$

$$+ B(x)\left\{\int_a^x (x - t)\phi(t)\,dt + C_1 x + (C_0 - C_1 a)\right\} = F(x) \tag{8.42}$$

or

$$\phi(x) = F(x) - C_1 A(x) - B(x)[C_1 x + C_0 - C_1 a]$$

$$- \int_a^x [A(x) + B(x)(x - t)]\phi(t)\,dt \tag{8.43}$$

Setting

$$F_1(x) = F(x) - C_1 A(x) - B(x)(C_1 x + C_0 - C_1 a) \tag{8.44}$$

and

$$K(x,t) = -[A(x) + B(x)(x - t)] \tag{8.45}$$

Equation (8.43) can be written as

$$\phi(x) = F_1(x) + \int_a^x K(x,t)\phi(t)\,dt \tag{8.46}$$

Equation (8.46) is a Volterra integral equation for $\phi(x)$. Once $\phi(x)$ is obtained, we can use, then Equations (8.39) to (8.41) to obtain $y(x)$.

Example 8.2 Applying method 2, change the linear oscillator equation of Example 8.1 to an integral equation.
Solution: We have from Example 8.1

$$A = 0, \quad B = \omega^2, \quad F = 0, \quad a = 0, \quad C_0 = 0, \quad C_1 = 1$$

Using Equation (8.43) we get

$$\phi(x) = -\omega^2 x - \omega^2 \int_0^x (x - t)\phi(t)\,dt$$

where $y''(x) = \phi(x)$

Example 8.3 Change the following initial value problem to a Volterra integral equation using method 2.

$$y'' - 5y' + 6y = 0$$

$$y(0) = 0, \quad y'(0) = -1$$

Applying Equation (8.43) we have

$$A = -5, \quad B = 6, \quad F = 0, \quad a = 0, \quad C_0 = 0, \quad C_1 = -1$$

Therefore, Equation (8.43) is reduced to the following expression

$$\phi(x) = 5 - 6x - \int_0^x [-5 + 6(x - t)]\phi(t)\,dt$$

where $\phi(x) = y''(x)$

2. Boundary Value Problem

Here we illustrate the equivalence between the boundary value problems and the Fredholm-type integral equations.

Let us take again the equation

$$y''(x) + A(x)\,y'(x) + B(x)\,y(x) = F(x) \tag{8.47}$$

with the boundary conditions

$$y(a) = C_0, \quad y(b) = B_0 \tag{8.48}$$

Integrating Equation (8.47) from a to x yields

$$y'(x) - C + A(x)\,y(x) - A(a)\,C_0 - \int_a^x [A'(t) - B(t)]\,y(t)\,dt = \int_a^x F(t)\,dt \tag{8.49}$$

where we used $y(a) = C_0$, and C is a constant of integration. Another integration reduces Equation (8.49) to the following

$$y(x) - C_0 - [A(a)C_0 + C](x-a) + \int_a^x \{A(t) - (x-t)[A'(t) - B(t)]\}y(t)\,dt = \int_a^x (x-t)F(t)\,dt \quad (8.50)$$

Setting $x = b$ where $y(b) = B_0$ yields the value of C as follows:

$$B_0 - C_0 + \int_a^b \{A(t) - (b-t)[A'(t) - B(t)]\}y(t)\,dt - [A(a)C_0 + C](b-a) = \int_a^b (b-t)F(t)\,dt \quad (8.51)$$

or

$$C = \left(\frac{1}{b-a}\right)\left\{(B_0 - C_0) + \int_a^b \{A(t) - (b-t)[A'(t) - B(t)]\}y(t)\,dt - \int_a^b (b-t)F(t)\,dt\right\} - A(0)C_0 \quad (8.52)$$

Using Equation (8.52) in Equation (8.50) we get

$$y(x) = C_0 + \frac{(x-a)}{(b-a)}\left\{(B_0 - C_0) + \int_a^b \{A(t) - (b-t)[A'(t) - B(t)]\}y(t)\,dt - \int_a^b (b-t)F(t)\,dt\right\} + \int_a^x (x-t)\,F(t)\,dt - \int_a^x \{A(t) - (x-t)[A'(t) - B(t)]\}y(t)\,dt \quad (8.53)$$

Example 8.4 Let us now look at a simplified form of Equation (8.53) by considering the following differential equation

$$y'' + \omega^2 y = 0 \quad (8.54)$$

$$y(0) = 0, \quad y(b) = 0 \quad (8.55)$$

Comparing Equation (8.54) with Equation (8.53) we obtain

$$A = 0, \quad a = 0, \quad C_0 = 0, \quad b = b, \quad B_0 = 0, \quad F(x) = 0, \quad B = \omega^2$$

Equation (8.53) becomes

$$y(x) = \frac{x}{b}\int_0^b -(b - t)(-\omega^2)\, y(t)\, dt - \int_0^x -(x - t)\,(-\omega^2)\,(y(t)\, dt$$

or

$$y(x) = \frac{\omega^2 x}{b}\int_0^b (b - t)\, y(t)\, dt - \omega^2\int_0^x (x - t)\, y(t)\, dt \qquad (8.56)$$

As

$$\frac{x}{b}(b - t) - (x - t) = \frac{t}{b}(b - x)$$

we can write Equation (8.56) as

$$y(x) = \omega^2\int_0^x \frac{t}{b}(b - x)\, y(t)\, dt + \omega^2\int_x^b \frac{x}{b}(b - t)\, y(t)\, dt \qquad (8.57)$$

Define a kernel as (Figure 8-1)

$$K(x,t) = \begin{cases} \dfrac{t}{b}\,(b - x) & t < x \qquad (8.58a) \\ \\ \dfrac{x}{b}\,(b - t) & x < t \qquad (8.58b) \end{cases}$$

we can write then

$$y(x) = \omega^2\int_0^b K(x,t)\, y(t)\, dt \qquad (8.59)$$

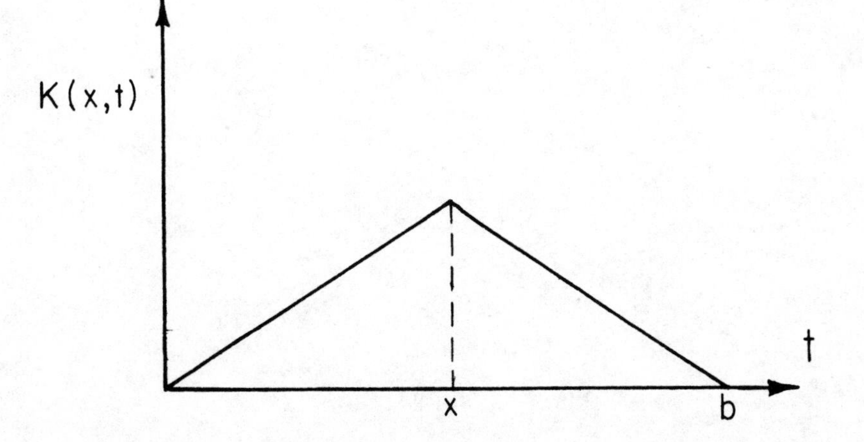

Figure 8-1. The Kernal Defined by Equation (8.58) in Example 8.4

which is a Fredholm-type integral equation.

The kernel defined by Equation (8.58) is referred to as a symmetric kernel, that is,

$$K(x,t) = K(t,x)$$

Methods of Solution of Integral Equations

9.1 Introduction

In this chapter we present a number of methods for the solution of the linear integral equations. Solutions to some integral equations can be expressed in closed form. Others have to be solved by methods that are approximate in some sense, the solution being obtained in the form of analytical approximation such as polynomials, power series, or as numerical approximation. We like to emphasize that approximate methods need not be designated as inferior relative to the exact method if the degree of approximation has been well-defined and is not exceeded. Furthermore, approximate techniques might be the only approach to solve a large class of problems.

9.2 Algebraic Method for the Integral Equations of Fredholm Type

Integral equations of the Fredholm type can be replaced by simultaneous algebraic equations whenever the kernel $K(x,t)$ is separable in the sense that

$$K(x,t) = \sum_{j=1}^{n} M_j(x)\, N_j(t) \tag{9.1}$$

where n is finite. Such kernels are called also degenerate. Examples of such kernels are all polynomials and many transcendental functions. The following kernels are degenerate:

$$K(x,t) = x + t \tag{9.2}$$

$$K(x,t) = xt + x^2 t^2 + xt^3 + x^3 t^4 \tag{9.3}$$

$$K(x,t) = \cos x \sin t \tag{9.4}$$

$$K(x,t) = \cos(x - t) = \cos x \cos t + \sin x \sin t \tag{9.5}$$

In Equation (9.2):

$$n = 2 \quad M_1 = x \quad N_1 = 1 \quad M_2 = 1 \quad N_2 = t$$

In Equation (9.5):

$$n = 2 \quad M_1 = \cos x \quad N_1 = \cos t \quad M_2 = \sin x \quad N_2 = \sin t$$

Before we discuss the method in a general form, we present an example to illustrate the approach.

Example 9.1 Solve the following Fredholm integral equation

$$u(x) = 2x + \lambda \int_0^1 e^x e^t u(t) \, dt \tag{9.6}$$

Solution: Let us write the equation as

$$u(x) = 2x + \lambda e^x \int_0^1 e^t u(t) \, dt \tag{9.7}$$

Since the integration is with respect to t, the expression

$$\int_0^1 e^t u(t) \, dt \tag{9.8}$$

is simply a constant after the integration is carried out. Let us denote the result by a number, say C. Hence, we can write

$$u(x) = 2x + \lambda C e^x \tag{9.9}$$

This is the form of the solution to Equation (9.6). In order to determine C, let us substitute the expression for $u(x)$ given by Equation (9.9) back into Equation (9.6) in both sides of the equation and obtain

$$2x + \lambda C e^x = 2x + \lambda e^x \int_0^1 e^t (2t + \lambda C e^t) \, dt \tag{9.10}$$

Performing the integration in Equation (9.10) and solving for C yields

$$C = \frac{2}{2 - \lambda(e^2 - 1)} \tag{9.11}$$

Therefore, the solution of the integral equation, Equation (9.6), becomes

$$u(x) = 2x + \frac{2\lambda e^x}{2 - \lambda(e^2 - 1)} \tag{9.12}$$

The equation has a solution as long as the denominator in Equation (9.12) is not equal to zero. When the denominator is equal to zero, λ, then, takes the following value

$$\lambda = \frac{2}{e^2 - 1} \tag{9.13}$$

For this value of λ the integral equation has no solution.

We proceed now with the general presentation of the method. Let us take now the Fredholm equation of the second kind

$$u(x) = f(x) + \lambda \int_a^b K(x,t)\, u(t)\, dt \tag{9.14}$$

Substituting Equation (9.1) in Equation (9.14) results in the following equation

$$u(x) = f(x) + \lambda \sum_{j=1}^{n} M_j(x) \int_a^b N_j(t)\, u(t)\, dt \tag{9.15}$$

where we have interchanged the processes of integration and summation. Looking at the integral

$$\int_a^b N_j(t)\, u(t)\, dt \tag{9.16}$$

we see that after the integration is carried out, the result is a number, a constant, say C_j. Hence, Equation (9.15) becomes

$$u(x) = f(x) + \lambda \sum_{j=1}^{n} C_j M_j(x) \tag{9.17}$$

Equation (9.17) suggests that the solution for the unknown $u(x)$ depends upon the parameter λ and the constants C_j; C_j is still an unknown quantity.

Let us multiply Equation (9.17) by $N_i(x)$ and then integrate over (a,b). The results become

$$\underbrace{\int_a^b N_i(x)\, u(x)\, dx}_{\equiv C_i} = \underbrace{\int_a^b N_i(x) f(x)\, dx}_{\equiv f_i} + \lambda \sum_{j=1}^{n} C_j \underbrace{\int_a^b N_i(x)\, M_j(x)\, dx}_{\equiv a_{ij}} \tag{9.18}$$

or

$$C_i = f_i + \lambda \sum_{j=1}^{n} a_{ij} C_j \quad i = 1, \ldots, n \tag{9.19}$$

where

$$C_i \equiv \int_a^b N_i(x)\, u(x)\, dx \tag{9.20}$$

$$f_i \equiv \int_a^b N_i(x) f(x)\, dx \tag{9.21}$$

$$a_{ij} \equiv \int_a^b N_i(x)\, M_j(x)\, dx \tag{9.22}$$

and

$$f_i \text{ and } a_{ij} \text{ are known quantities.}$$

Equation (9.19) represents a system of n algebraic equations for the unknown C_j given by

$$C_i - \lambda \sum_{j=1}^{n} a_{ij} C_j = f_i \quad i = 1, \ldots, n \tag{9.23}$$

When expanded, Equation (9.23) becomes

$$\begin{aligned}
(1 - \lambda a_{11}) C_1 - \lambda a_{12} C_2 - \lambda a_{13} C_3 - \ldots &= f_1 \\
-\lambda a_{21} C_1 + (1 - \lambda a_{22}) C_2 - \lambda a_{23} C_3 - \ldots &= f_2 \\
-\lambda a_{31} C_1 - \lambda a_{32} C_2 + (1 - \lambda a_{33}) C_3 - \ldots &= f_3 \\
&\vdots \\
-\lambda a_{nl} C_1 - \lambda a_{n2} C_2 - \ldots + (1 - \lambda a_{nn}) C_n &= f_n
\end{aligned} \tag{9.24}$$

The determinant $D(\lambda)$ of the system given by Equation (9.24) is

$$D(\lambda) = \begin{vmatrix} 1 - \lambda a_{11} & -\lambda a_{12} & \ldots & -\lambda a_{1n} \\ -\lambda a_{21} & 1 - \lambda a_{22} & \ldots & -\lambda a_{2n} \\ \vdots & & & \\ -\lambda a_{n1} & -\lambda a_{n2} & \ldots & 1 - \lambda a_{nn} \end{vmatrix} \tag{9.25}$$

The determinant $D(\lambda)$ is a polynomial in λ, and it is not identically zero, because when $\lambda = 0$, $D(\lambda) = 1$.

The system of Equation (9.24) will have a unique solution when

$$D(\lambda) \neq 0 \tag{9.26}$$

This solution, of course, results in the solution of the integral equation Equation (9.14) with its new form stated by Equation (9.17). When $D(\lambda) = 0$, the nonhomogeneous integral equation Equation (9.14) and, hence, the system given by Equation (9.24) in general will not have a solution.

Let us consider now the case when our integral equation is homogeneous, that is, $f(x) = 0$. The equation is reduced to

$$u(x) = \lambda \int_a^b K(x,t)\, u(t)\, dt \tag{9.27}$$

and from Equation (9.21) $f_i \equiv 0$. This reduces the system given by Equation

(9.24) to a homogeneous set of equations. This set will have a solution only if the determinant of the coefficients of C_i is equal to zero, that is

$$D(\lambda) = 0 \tag{9.28}$$

Hence, setting $D(\lambda)$ as given by Equation (9.25) equal to zero yields a set of values for the parameter λ. This set is the roots of

$$D(\lambda) = 0$$

The values of λ for which $D(\lambda) = 0$ are the eigenvalues of our problem and only for these values of λ's the homogeneous integral equation has a solution. Therefore, we can outline these findings as follows:

1. When $D(\lambda) \neq 0$, the nonhomogeneous Fredholm integral equation

$$u(x) = f(x) + \lambda \int_a^b K(x,t)\, u(t)\, dt$$

does have a solution. The solution exists for all values of λ's except generally for the values of λ's that make $D(\lambda) = 0$, "that is, λ's which are the roots of $D(\lambda) = 0$; an exceptional case is discussed later."

2. When $D(\lambda) \neq 0$, the homogeneous integral equation

$$u(x) = \lambda \int_a^b K(x,t)\, u(t)\, dt$$

has only the trivial solution $u(x) \equiv 0$; this is deduced directly from Equation (9.23) when $f_i \equiv 0$.

3. When $D(\lambda) = 0$, the homogeneous equation

$$u(x) = \lambda \int_a^b K(x,t)\, u(t)\, dt$$

has a number of solutions which is equal to the number of the eigenvalues as obtained by setting $D(\lambda) = 0$. In this case the solution is written as

$$u(x) = A_0\, \phi_0(x) + A_1\, \phi_1(x) + A_2\, \phi_2(x) + \ldots \tag{9.29}$$

where A_0, A_1, and A_2 are arbitrary constants and ϕ_0, ϕ_1, and ϕ_2 are the eigenfunctions associated with the eigenvalues $\lambda_0, \lambda_1, \lambda_2, \ldots$, which are the roots of $D(\lambda) = 0$. The eigenfunctions are obtained from Equation (9.17) when $f(x) = 0$, that is,

$$u(x) = \lambda \sum_{j=1}^{n} C_j\, M_j(x) \tag{9.30}$$

once the values of C_j are determined. The eigenfunctions, ϕ_α, are essentially the solution of the homogeneous equation

$$\phi_\alpha(x) = \lambda \int_a^b K(x,t)\,\phi_\alpha(t)\,dt \qquad \alpha = 0, 1, 2, \ldots, n \tag{9.31}$$

4. When $D(\lambda) = 0$, the nonhomogeneous equation has no solution in general. However, if certain conditions pertaining to the nonhomogeneous term $f(x)$ are met, there will be solutions. To obtain these conditions, let us multiply both sides of the nonhomogeneous equation

$$u(x) = f(x) + \lambda_0 \int_a^b K(x,t)\,u(t)\,dt \tag{9.32}$$

by $\phi_\alpha(x)$ where $\phi_\alpha(x)$ is the eigenfunction satisfying

$$\phi_\alpha(x) = \lambda_0 \int_a^b K(x,t)\,\phi_\alpha(t)\,dt$$

and λ_0 is one of the eigenvalues. As shown in Equation (9.32), the integral equation is nonhomogeneous in which $\lambda = \lambda_0$ is an eigenvalue. Following this multiplication, we obtain

$$\phi_\alpha(x)\,u(x) = \phi_\alpha(x) f(x) + \phi_\alpha(x)\,\lambda_0 \int_a^b K(x,t)\,u(t)\,dt \tag{9.33}$$

Integrating Equation (9.33) over (a,b) yields

$$\int_a^b \phi_\alpha(x) f(x)\,dx = \int_a^b \phi_\alpha(x)\,u(x)\,dx - \lambda_0 \int_a^b \phi_\alpha(x) \left\{ \int_a^b K(x,t)\,u(t)\,dt \right\} dx \tag{9.34}$$

In the last integral on the right-hand side of Equation (9.34), $\phi_\alpha(x)$ is constant with respect to t and, therefore, can be placed under the second integral. Doing that and changing the order of integration, the last term can, then, be written as

$$\int_a^b u(t) \left\{ \lambda_0 \int_a^b \phi_\alpha(x)\,K(x,t)\,dx \right\} dt \tag{9.35}$$

The quantity in the braces in Equation (9.35) is simply $\phi_\alpha(x)$ as can be seen from Equation (9.31). Therefore, Equation (9.34) becomes

$$\int_a^b \phi_\alpha(x) f(x)\,dx = \int_a^b \phi_\alpha(x)\,u(x)\,dx - \int_a^b \phi_\alpha(x)\,u(x)\,dx$$

or

$$\int_a^b \phi_\alpha(x) f(x)\, dx = 0 \tag{9.36}$$

This is the condition on $f(x)$ to be met so that the nonhomogeneous equation has a solution when $\lambda = \lambda_0$, an eigenvalue. Equation (9.36) implies that $\phi_\alpha(x)$ and $f(x)$ are orthogonal.

Example 9.2 Solve the following integral equation

$$u(x) = 1 + \lambda \int_0^{\pi/2} \cos(x - t)\, u(t)\, dt \tag{9.37}$$

and find the values of λ for which a solution exists.
Solution: Let us write the kernel $K(x,t)$ as

$$\cos(x - t) = \cos x \cos t + \sin x \sin t \tag{9.38}$$

Therefore, we have from Equations (9.19) to (9.22) the following:

$$M_1 = \cos x, \quad N_1 = \cos t; \quad M_2 = \sin x, \quad N_2 = \sin t \tag{9.39}$$

$$\begin{aligned} a_{11} &= \int_0^{\pi/2} \cos^2 t\, dt = \frac{\pi}{4} \\ a_{22} &= \int_0^{\pi/2} \sin^2 t\, dt = \frac{\pi}{4} \end{aligned} \tag{9.40}$$

$$\begin{aligned} a_{12} &= \int_0^{\pi/2} \cos t \sin t\, dt = \frac{1}{2} \\ a_{21} &= \int_0^{\pi/2} \sin t \cos t\, dt = \frac{1}{2} \end{aligned} \tag{9.41}$$

$$\begin{aligned} f_1 &= \int_0^{\pi/2} \cos t \cdot 1\, dt = 1 \\ f_2 &= \int_0^{\pi/2} \sin t \cdot 1\, dt = 1 \end{aligned} \tag{9.42}$$

Therefore, from Equation (9.19) we have

$$C_1 = 1 + \lambda \left[\frac{\pi}{4} C_1 + \frac{1}{2} C_2\right] \tag{9.43}$$

$$C_2 = 1 + \lambda \left[\frac{1}{2} C_1 + \frac{\pi}{4} C_2\right] \tag{9.44}$$

Equations (9.43) and (9.44), when solved simultaneously, yield

$$C_1 = C_2 = \left[1 - \frac{\lambda}{4}(\pi + 2)\right]^{-1} \tag{9.45}$$

Hence, the solution to our equation as represented by Equation (9.17), which is

$$u(x) = f(x) + \lambda \sum_{j=1}^{n} C_j M_j(x) \tag{9.46}$$

becomes

$$u(x) = 1 + \frac{\lambda}{1 - \lambda/4\,(\pi + 2)}[\cos x + \sin x] \tag{9.47}$$

We notice that in Equation (9.47) the denominator becomes zero when

$$\lambda = \frac{4}{\pi + 2} \tag{9.48}$$

For this value of λ no solution exists for this integral equation. Hence, we may further conclude now that the value of λ given by Equation (9.48) is an eigenvalue for the homogeneous part of Equation (9.37), which is solved in the following example.

Example 9.3 Solve the homogeneous integral equation

$$u(x) = \lambda \int_0^{\pi/2} \cos(x - t)\, u(t)\, dt \tag{9.49}$$

Solution: From Example 9.2 we have

$$a_{11} = \pi/4 = a_{22} \qquad a_{12} = 1/2 = a_{21}$$

$$M_1 = \cos x, \quad N_1 = \cos t; \quad M_2 = \sin x, \quad N_2 = \sin t$$

Here $f = 0$. Therefore, from Equation (9.19) or Equation (9.25) we have

$$\begin{vmatrix} 1 - \lambda a_{11} & -\lambda a_{12} \\ -\lambda a_{21} & 1 - \lambda a_{22} \end{vmatrix} = 0 \tag{9.50}$$

Determinant (9.50) yields

$$\lambda_1 = \frac{4}{\pi + 2} \qquad \lambda_2 = \frac{4}{\pi - 2} \tag{9.51}$$

These are the two eigenvalues for which solution to the homogeneous equation exists. Therefore, the two solutions for the equation are readily obtained from Equation (9.17) in which now $f(x) = 0$, $n = 2$, and $\lambda = \lambda_j$, $j = 1, 2$, that is,

$$u(x) = \lambda_j \sum_{j=1}^{2} C_j \, M_j(x) \tag{9.52}$$

Hence,

$$u_1(x) = C_1 \frac{4}{\pi + 2} (\cos x) \tag{9.53}$$

$$u_2(x) = C_2 \frac{4}{\pi - 2} (\sin x) \tag{9.54}$$

where

C_1 and C_2 are arbitrary constants.

Cos x and sin x are the eigenfunctions corresponding to λ_1 and λ_2 respectively.

9.3 Fredholm Integral Equations in Engineering Problems

In the formulation and analysis of radiative heat transfer problems, integral equations constitute a major part of the analysis. We present in this section an introductory view as to the occurrence and methods of solutions for such equations.

Integral Equations for the Radiosity in an Enclosure

In an enclosure of N gray[a] surfaces, the radiosity (quantity of radiation leaving) for surface i can be written as (Figure 9-1)

$$B_i(\mathbf{r}_i) = \epsilon_i \sigma T_i^4(\mathbf{r}_i) + (1 - \epsilon_i) H_i(\mathbf{r}_i) \tag{9.55}$$

The radiosity B_i is equal to:

(1) Emitted radiation, $\epsilon_i \sigma T_i^4$, where ϵ_i is emissivity of surface i, σ is Stefan Boltzmann constant, and T_1 is the temperature of surface i, and

(2) Reflected radiation, $(1 - \epsilon_i) H_i$; $(1 - \epsilon_i)$ is the reflectivity of the surface; H_i is the incident radiation on i from all surrounding surfaces.

The incident radiation H_i is represented as

$$H_i(\mathbf{r}_i) = \sum_{j=1}^{N} \int_{A_j} B_j(\mathbf{r}_i) \, dF_{dA_i - dA_j} \tag{9.56}$$

[a] Gray implies frequency independent properties.

where $dF_{dA_i-dA_j}$ is the shape factor representing the fraction of radiation arriving at i from j and it can be easily shown to be given by[b] (Figure 9-2)

$$dF_{dA_i-dA_j} = \frac{\cos\theta_i \cos\theta_j \, dA_j}{\pi r^2} \tag{9.57}$$

Substituting Equations (9.56) and (9.57) in Equation (9.55) we obtain

$$B_i = \epsilon_i \sigma T_i^4 + (1 - \epsilon_i) \sum_{j=1}^{N} \int_{A_j} B_j \frac{\cos\theta_i \cos\theta_j \, dA_j}{\pi r^2} \tag{9.58}$$

Equation (9.58) is a Fredholm integral equation of the second type for the radiosity B_i. Because the quantity of interest in such problems is the net radiative heat flux to or from a surface, we relate now B_i to the net flux q_i from surface i.

From the defining relation for the net flux we have

$$q_i = \text{emitted flux} - \text{absorbed flux from all incident radiation}$$

or

$$q_i = \underbrace{\epsilon_i \sigma T_i^4}_{\text{emitted flux}} - \underbrace{\alpha_i H_i}_{\text{absorbed flux}} \tag{9.59}$$

α_i — Absorptivity of surface i; H_i — Incident radiation

where in this problem ϵ_i and α_i are taken to be equal. From Equation (9.55) we can solve for H_i and get

$$H_i = \frac{B_i - \epsilon_i \sigma T_i^4}{1 - \epsilon_i} \tag{9.60}$$

Substituting the expression for H_i from Equation (9.60) in Equation (9.59) yields

$$q_i = \frac{\epsilon_i}{1-\epsilon_i}[\sigma T_i^4 - B_i] \tag{9.61}$$

which is the required expression for the net flux q_i. Usually, for a certain surface, ϵ_i and either T_i or q_i are known. If T_i is known then q_i is the quantity to be determined. However, before q_i can be determined, the radiosity B_i has to be found. We illustrate the method through the following examples.

Example 9.4 Derive the integral equation for the radiosity in a spherical

---

[b] See Robert Siegel and J.R. Howell, *Thermal Radiation Heat Transfer* (New York: McGraw-Hill, 1972).

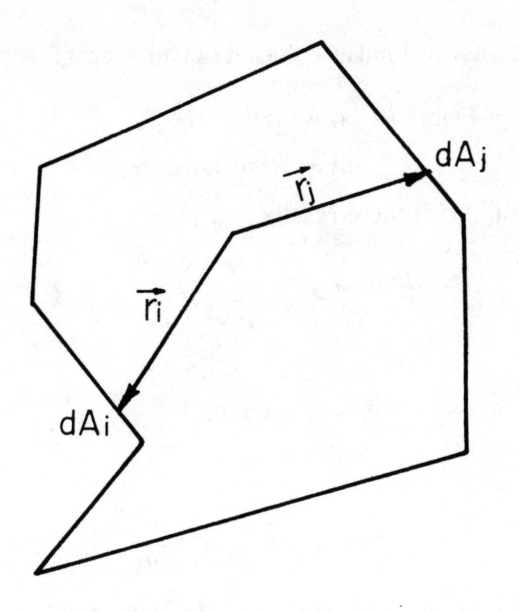

Figure 9-1. Enclosure of *N* Surfaces

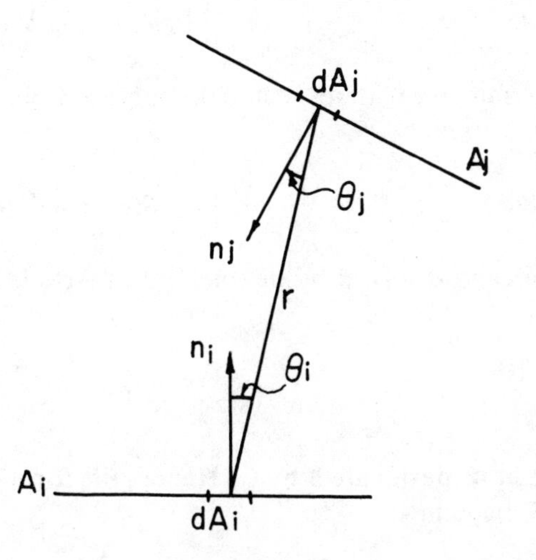

Figure 9-2. Radiative Interchange Between Two Differential Surfaces

enclosure and solve it. Find also the net radiative heat flux for the spherical surface.

Solution: From Figure (9-3a) we have

$$dA = R^2 \sin\phi \, d\psi \, d\phi \tag{9.62}$$

Using Equation (9.57) there results

$$dF_{dA_1-dA_2} = \frac{\cos\theta_2 \cos\theta_1}{\pi r^2} dA_2 \tag{9.63}$$

and

$$\cos\theta_2 = \cos\theta_1 = \frac{r}{2R} \tag{9.64}$$

so that

$$\cos\theta_2 \cos\theta_1 = \frac{r^2}{4R^2} \tag{9.65}$$

Therefore, using Equations (9.62) and (9.65) in Equation (9.63) yields

$$dF_{dA_1-dA_2} = \frac{r^2R^2 \sin\phi \, d\psi \, d\phi}{4R^2\pi r^2} = \frac{\sin\phi \, d\psi \, d\phi}{4\pi} \tag{9.66}$$

The integral equation for radiosity becomes "from Equation (9.58) or Equation (9.55)"

$$B(\psi, \phi) = \epsilon\sigma T^4 + \frac{(1-\epsilon)}{4\pi}\int_{\phi'=0}^{\pi}\int_{\psi'=0}^{2\pi} B(\psi', \phi') \sin\phi' \, d\phi' \, d\psi' \tag{9.67}$$

If we consider symmetry with respect to the angle ψ, Equation (9.67) can be integrated over ψ to give

$$B(\phi) = \epsilon\sigma T^4 + \frac{(1-\epsilon)}{2}\int_{\phi'=0} B(\phi') \sin\phi' \, d\phi' \tag{9.68}$$

Equation (9.68) can be solved by the method of Article 9.2 as follows:

The integral

$$\int_{\phi'=0}^{\pi} B(\phi') \sin\phi' \, d\phi'$$

is a constant. Let us designate it by C. Hence, the form of the solution to Equation (9.68) becomes

$$B(\phi) = \epsilon\sigma T^4(\phi) + \frac{(1-\epsilon)}{2} C \tag{9.69}$$

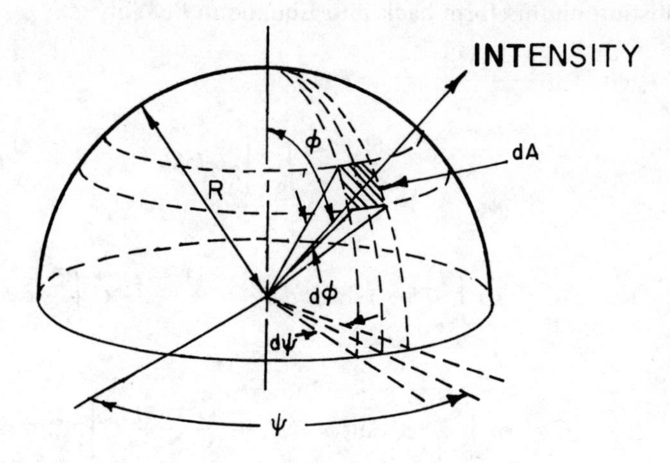

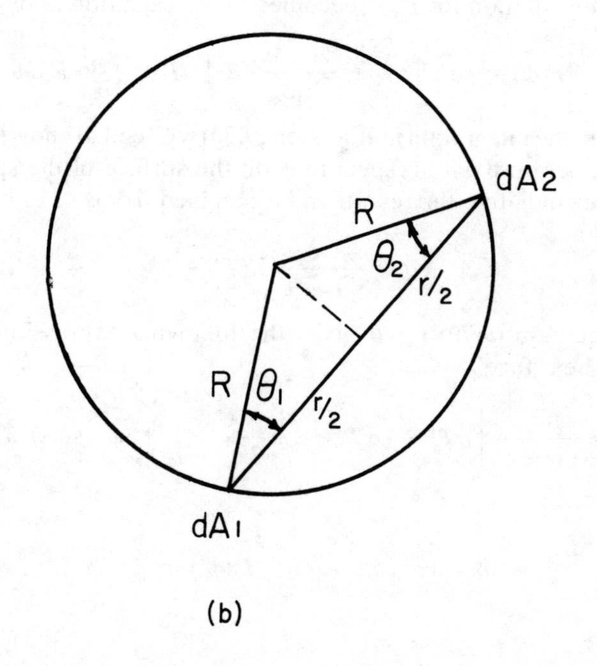

Figure 9-3. Radiosity in a Spherical Enclosure

Substituting this form back into Equation (9.68) gives

$$\epsilon\sigma T^4(\phi) + \frac{(1-\epsilon)}{2} C = \epsilon\sigma T^4(\phi) + \frac{(1-\epsilon)}{2}\int_0^{\pi}\left[\epsilon\sigma T^4(\phi') + \frac{(1-\epsilon)}{2} C\right]\sin\phi'\,d\phi'$$

or

$$C = \epsilon\sigma\int_0^{\pi} T^4(\phi')\sin\phi'\,d\phi' + \frac{(1-\epsilon)}{2} C\int_0^{\pi}\sin\phi'\,d\phi'$$

or

$$C = \epsilon\sigma\int_0^{\pi} T^4(\phi')\sin\phi'\,d\phi' + \frac{[1-\epsilon)}{2} C\Big[-\cos\phi'\Big]_0^{\pi}$$

Solving for C yields

$$C = \sigma\int_0^{\pi} T^4(\phi')\sin\phi'\,d\phi'$$

Finally, the solution for $B(\phi)$ becomes (from Equation (9.69))

$$B(\phi) = \epsilon\sigma T^4(\phi) + \frac{(1-\epsilon)}{2}\sigma\int_0^{\pi} T^4(\phi')\sin\phi'\,d\phi' \tag{9.70}$$

To perform the integration in Equation (9.70) we need to know the variation of the temperature with respect to ϕ on the surface of the sphere.

The net radiative flux can then be obtained from

$$q = \frac{\epsilon}{1-\epsilon}[\sigma T^4 - B]$$

Using Equation (9.70) for B gives the following expression for the net radiative heat flux

$$q = \frac{\epsilon}{1-\epsilon}\left[\sigma T^4 - \epsilon\sigma T^4 - \frac{1-\epsilon}{2}\sigma\int_0^{\pi} T^4(\phi')\sin\phi'\,d\phi'\right] \tag{9.71}$$

or

$$q = \epsilon\sigma\left[T^4 - \frac{1}{2}\int_0^{\pi} T^4(\phi')\sin\phi'\,d\phi'\right] \tag{9.72}$$

Example 9.5 A very large array of thin fins of thickness b, width W, and length L, which is much larger than W, are attached to a base that is held at a constant temperature T_b as shown in Figure 9-4. Find an expression for the radiosity of the surface of one fin. Write an energy balance for an element dx along one fin as shown in Figure 9-4. Assume gray surfaces and all quantities are time independent.

Solution: From the symmetry of the problem we have

$$T(\xi) = T(x) \tag{9.73}$$

The expression for radiosity $B(x)$ becomes (Equation (9.58))

$$B(x) = \underbrace{\epsilon\sigma T^4(x)}_{\text{emission}} + \underbrace{(1-\epsilon)\int_{\xi=0}^{\xi=W} B(\xi)\, dF_{dx-d\xi}}_{\substack{\text{contribution to incident}\\ \text{radiation from adjacent fin}}}$$

$$+ \underbrace{(1-\epsilon)\int_{\eta=0}^{\eta=a} B(\eta)\, dF_{dx-d\eta}}_{\substack{\text{contribution to incident}\\ \text{radiation from the base}}} \tag{9.74}$$

An energy balance on dx yields the following (Figure 9-5)

$$\text{Conduction in, } Q_{c,i} = -k\frac{b}{2}\frac{dT}{dx} \tag{9.75}$$

where k is thermal conductivity of the fin.

$$\begin{aligned} \text{Conduction out, } Q_{c,0} &= Q_{c,i} + dQ_{c,i} \\ &= -k\frac{b}{2}\frac{dT}{dx} + \frac{d}{dx}\left(-k\frac{b}{2}\frac{dT}{dx}\right)dx \end{aligned} \tag{9.76}$$

Therefore, under steady state constraint

$$\text{Energy into element } dx = \text{Energy out of element } dx$$

$$Q_{c,i} + H\,dx = B\,dx + Q_{c,0} \tag{9.77}$$

or

$$H\,dx - B\,dx = Q_{c,0} - Q_{c,i} = -\frac{d}{dx}\left(k\frac{b}{2}\frac{dT}{dx}\right)dx \tag{9.78}$$

If the thermal conductivity k is constant, we can write then

$$B - H = \frac{kb}{2}\frac{d^2T}{dx^2} \tag{9.79}$$

Let $B - H = q_r$; then, the energy balance becomes

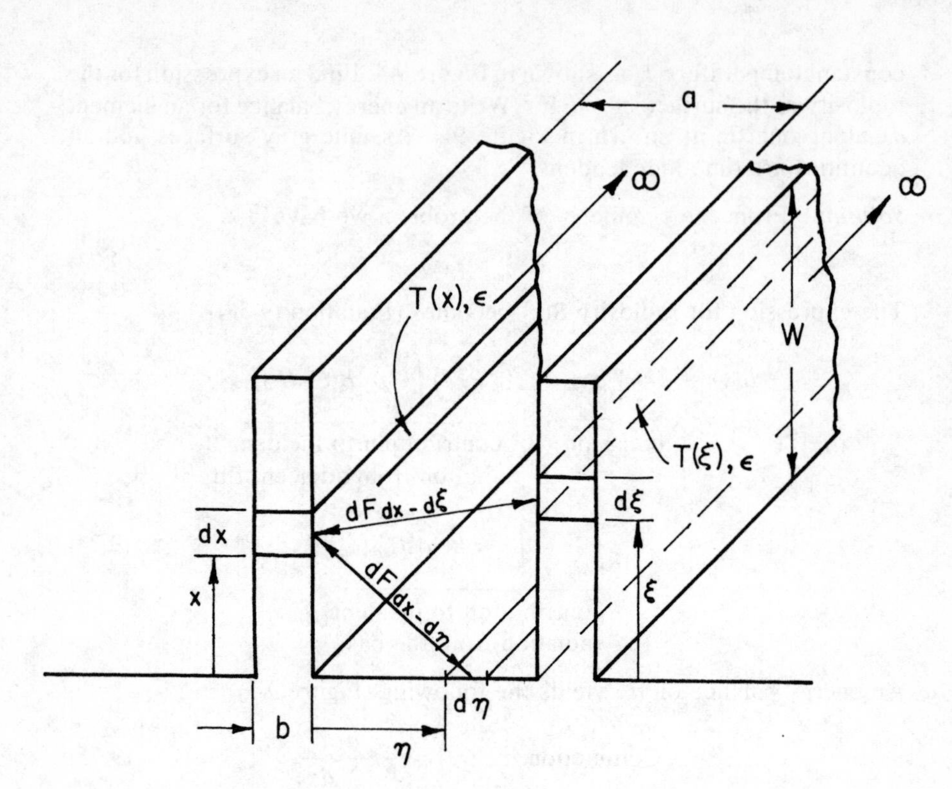

Figure 9-4. Radiation Between Parallel Fins

$$\frac{d^2T}{dx^2} = \frac{2}{k\,b} q_r \tag{9.80}$$

As noted in this section q_r can be written as

$$q_r = \frac{\epsilon}{1 - \epsilon}[\sigma T^4 - B] \tag{9.81}$$

Therefore, Equation (9.80), with the help of Equation (9.74), can be written as an integro-differential equation in the form

$$\frac{d^2T}{dx^2} = \left(\frac{2}{kb}\right)\frac{\epsilon}{1 - \epsilon}\left[\sigma T^4 - \epsilon\sigma T^4 - (1 - \epsilon)\int_{\xi=0}^{\xi=W} B(\xi)\, dF_{dx-d\xi}\right.$$

$$\left. - (1 - \epsilon)\int_{\eta=0}^{\eta=a} B(\eta)\, dF_{dx-d\eta}\right] \tag{9.82}$$

Equation (9.82) requires numerical techniques for its solution.

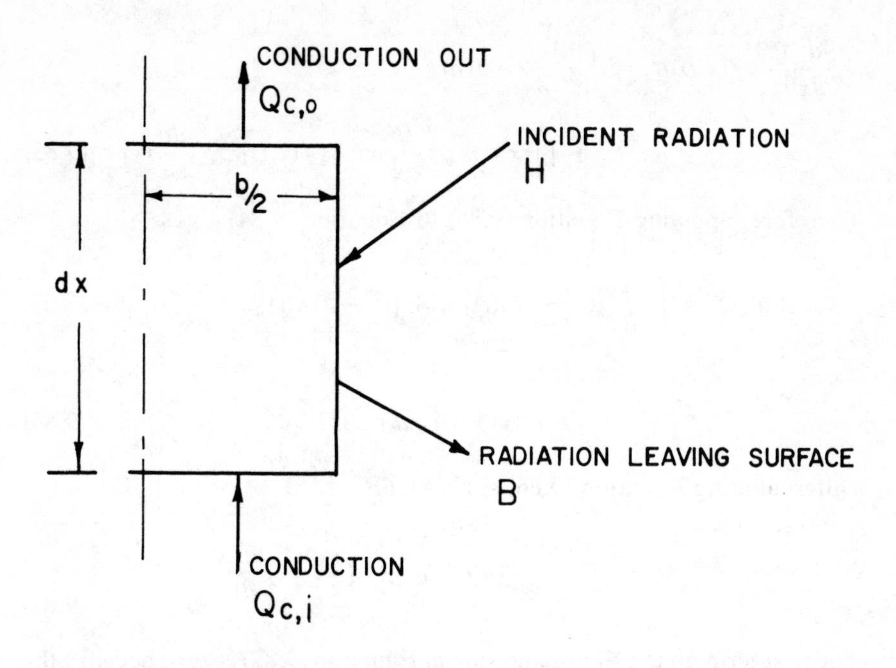

Figure 9-5. Heat Balance on a Section of a Fin

9.4 Volterra Integral Equations Solved by Differentiation

We found in chapter 8 that integral equations and differential equations can be related, and either equation can be a mathematical representation of the same physical problem. In this section, we present a solution to the integral equation of the Volterra type by changing it to a differential equation. We illustrate the method by the following example.

Example 9.6 Solve the following Volterra integral equation

$$u(x) = 10 + \int_0^x (x - t)\, u(t)\, dt \qquad (9.83)$$

Solution: Differentiating Equation (9.83) with respect to x and signifying the differentiation by a prime yields

$$u'(x) = 0 + \frac{d}{dx}\int_0^x (x - t)\, u(t)\, dt \qquad (9.84)$$

In differentiating an integral with a parameter we use the following formula referred to as Liebnitz formula:

$$\frac{d}{dx}\int_{a(x)}^{b(x)} f(x,t)\,dt = \int_{a(x)}^{b(x)} \frac{\partial f}{dx}(x,t)\,dt$$

$$+ [f(x,t)]_{t=b(x)}\frac{db}{dx}(x) - [f(x,t)]_{t=a(x)}\frac{da}{dx}(x) \quad (9.85)$$

Therefore, applying Equation (9.85) to Equation (9.84) gives

$$u'(x) = \int_0^x \frac{\partial}{\partial x}[(x - t)\,u(t)]\,dt + [\overbrace{(x - t)}^{0}u(t)]_{t=x}\,\frac{dx}{dx} - 0$$

or

$$u'(x) = \int_0^x u(t)\,dt \quad (9.86)$$

Differentiating Equation (9.86) again yields

$$u''(x) = \int_0^x \overbrace{\frac{\partial}{\partial x}[u(t)]}^{0}\,dt + [u(t)]_{t=x}\,\frac{dx}{dx} \quad (9.87)$$

The first term on the right-hand side in Equation (9.87) is zero because the integrand is a function of t only. Hence, Equation (9.87) becomes

$$u''(x) - u(x) = 0 \quad (9.88)$$

This equation now is a second order ordinary differential equation with constant coefficients. The solution to Equation (9.88) can be written as

$$u(x) = A\,e^{-x} + B\,e^{+x} \quad (9.89)$$

The constants A and B are determined by satisfying the integral equation as follows. From Equations (9.83) and (9.84) respectively, we have:

$$\text{when } x = 0,\ u(x) = 10 \qquad \text{when } x = 0,\ u'(x) = 0 \quad (9.90)$$

Applying conditions (9.90) to Equation (9.89) we get

$$10 = A + B \quad (9.91)$$

and

$$0 = -A + B \quad (9.92)$$

Therefore,

$$A = B = 5 \quad (9.93)$$

and the solution to our original integral equation becomes

$$u(x) = 5\,[e^{x} + e^{-x}] \quad (9.94)$$

Example 9.7 Solve the following Volterra integral equation of the first kind for $u(x)$

$$x = \int_0^x e^{(x-t)} u(t)\, dt \tag{9.95}$$

Solution: Differentiating Equation (9.95) once with respect to x yields

$$1 = \int_0^x \frac{\partial}{\partial x} [e^{(x-t)} u(t)]\, dt + u(x)$$

or

$$1 = \underbrace{\int_0^x e^{(x-t)} u(t)\, dt}_{= x} + u(x) \tag{9.96}$$

From Equation (9.95) the first term on the right-hand side of Equation (9.96) is equal to x. Therefore, the solution becomes, directly from Equation (9.96)

$$u(x) = 1 - x \tag{9.97}$$

9.5 Abel's Integral Equation—The Tautochrome

Abel posed a very famous problem which can be stated as follows:

A particle of mass m (Figure 9-6) situated at point P, an arbitrary elevation, starts to descend under the action of gravity along a smooth curve. Determine the shape of that curve "path of the particle" so that the time of fall from P to 0 is a given function of time $T(y)$. Let us derive now the equation for the curve of descent.

The velocity, v, of the particle at any location, say ξ, can be obtained by equating the gain in the kinetic energy of the particle to the loss in its potential energy.

$$\text{Kinetic energy gain} = \tfrac{1}{2} mv^2$$

$$\text{Potential energy loss} = mg(y - \xi)$$

Hence,

$$\frac{1}{2} mv^2 = mg(y - \xi)$$

or

$$v = \sqrt{2g(y - \xi)} \tag{9.98}$$

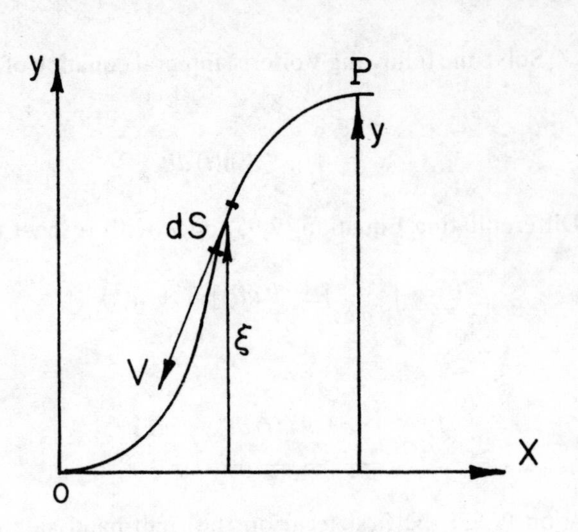

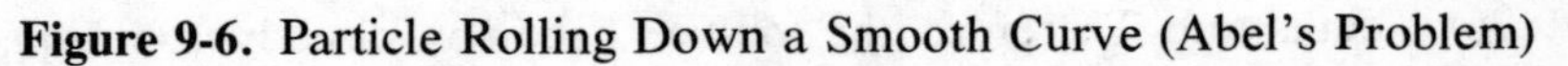

Figure 9-6. Particle Rolling Down a Smooth Curve (Abel's Problem)

The total time of fall of the particle from point P to point 0 can be written as

$$T(y) = \int_0^P \frac{ds}{v} = \int_0^P \frac{ds}{\sqrt{2g(y - \xi)}} \tag{9.99}$$

where ds is a line segment of the required curve. The segment ds can be written as

$$ds = \left(\frac{ds}{d\xi}\right) d\xi \tag{9.100}$$

Therefore, Equation (9.99) becomes

$$T(y) = \frac{1}{\sqrt{2g}} \int_0^y \frac{\left(\frac{ds}{d\xi}\right) d\xi}{\sqrt{(y - \xi)}} \tag{9.101}$$

Equation (9.101) is referred to as Abel's integral equation for the function $(ds/d\xi)$. It is a singular Volterra integral equation of the first kind. Equation (9.101) represents also the classical problem of the tautochrome (the curve of equal descent). It is a special form of the following more general singular integral equation

$$F(y) = \int_0^y \frac{u(\xi)\, d\xi}{(y - \xi)^n} \qquad 0 < n < 1 \tag{9.102}$$

One way to solve Equation (9.102) is presented as follows: let us multiply both sides of Equation (9.102) by $dy/((z - y)^{1-n})$ and integrate the result with respect to y from $y = 0$ to $y = z$. The result becomes

$$\int_0^z \frac{F(y)\,dy}{(z - y)^{1-n}} = \int_0^z \frac{dy}{(z - y)^{1-n}} \int_0^y \frac{u(\xi)\,d\xi}{(y - \xi)^n} \tag{9.103}$$

Reversing the order of integration on the right-hand side of Equation (9.103) yields (Figure 9-7)

$$\int_0^z \frac{F(y)\,dy}{(z - y)^{1-n}} = \int_0^z u(\xi)\,d\xi \int_y^z \frac{dy}{(z - y)^{1-n}(y - \xi)^n} \tag{9 .104}$$

The integral

$$\int_y^z \frac{dy}{(z - y)^{1-n}(y - \xi)^n}$$

can be evaluated by the Beta function as

$$\begin{aligned} \int_y^z \frac{dy}{(z - y)^{1-n}(y - \xi)^n} &= \beta(1 - \alpha, \alpha) \\ &= (-\alpha)!(\alpha - 1)! \\ &= \frac{\pi}{\sin \pi\alpha} \end{aligned} \tag{9.105}$$

Therefore, Equation (9.104) becomes

$$\frac{\sin \pi\alpha}{\pi} \int_0^z \frac{F(y)\,dy}{(z - y)^{1-n}} = \int_0^z u(\xi)\,d\xi \tag{9.106}$$

Differentiating Equation (9.106) with respect to z gives

$$u(z) = \frac{\sin \pi\alpha}{\pi} \frac{d}{dz} \int_0^z \frac{F(y)\,dy}{(z - y)^{1-n}} \tag{9.107}$$

which is the required solution. This solution could have been obtained by the method of Laplace transform.

9.6 The Method of Successive Substitution

In this section, the method of successive substitution is introduced to the reader as a tool that, for certain cases, is quite convenient and powerful to solve integral equations. The method will be applied to both the Fredholm type and the Volterra type integral equations.

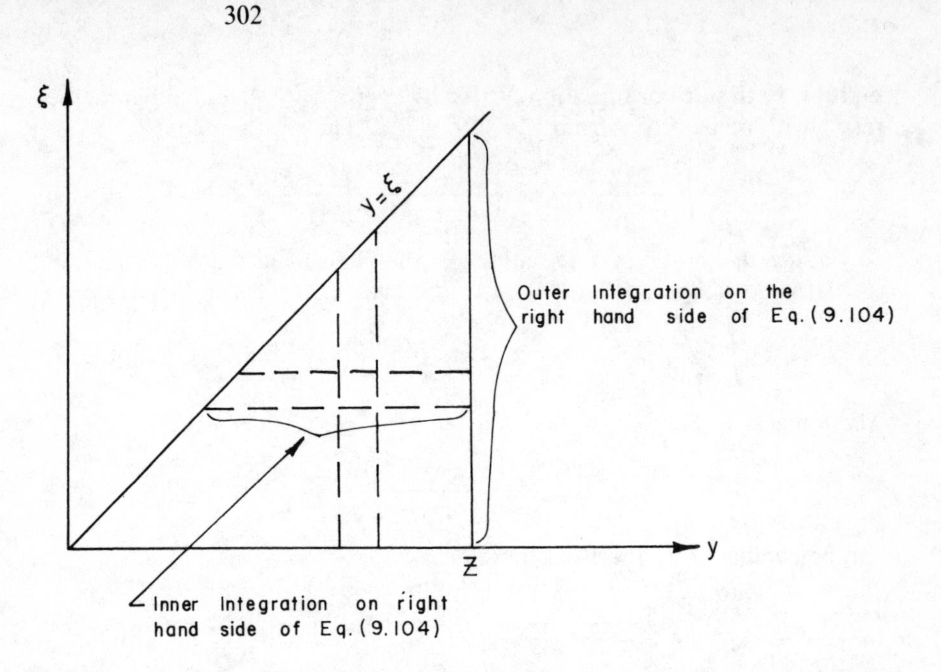

Figure 9-7. Graphical Interpretation of the Integrals in Equation (9.103)

1. Fredholm Integral Equations

As presented earlier, Fredholm integral equation is

$$u(x) = f(x) + \lambda \int_a^b K(x,t)\, u(t)\, dt \tag{9.108}$$

It is assumed in this method that

$K(x,t)$ is real and continuous in $a \leq x \leq b$; $a \leq t \leq b$

$f(x)$ is real, continuous and $\neq 0$ in $a \leq x \leq b$

λ is a constant

The expression for $u(x)$ as given by Equation (9.108) is now substituted for $u(t)$ under the integral in Equation (9.108). The result becomes

$$u(x) = f(x) + \lambda \int_a^b K(x,t) \underbrace{\left[f(t) + \lambda \int_a^b K(t,t_1)\, u(t_1)\, dt_1 \right]}_{u(t) \text{ as given by Equation (9.108)}} dt$$

or

$$u(x) = f(x) + \lambda \int_a^b K(x,t)\, f(t)\, dt$$

$$+ \lambda^2 \int_a^b K(x,t) \int_a^b K(t,t_1)\, u(t_1)\, dt_1\, dt \qquad (9.109)$$

Substituting now for $u(t_1)$ in Equation (9.109) the expression for $u(x)$ as given by Equation (9.108), we obtain

$$u(x) = f(x) + \lambda \int_a^b K(x,t)\, f(t)\, dt \qquad (9.110)$$

$$+ \lambda^2 \int_a^b K(x,t) \int_a^b K(t,t_1)$$

$$\underbrace{\left[f(t_1) + \lambda \int_a^b K(t_1,t_2)\, u(t_2)\, dt_2 \right]}_{u(t_1)} dt_1\, dt$$

When the process of substitution is continued, we get

$$u(x) = f(x) + \lambda \int_a^b K(x,t)\, f(t)\, dt$$

$$+ \lambda^2 \int_a^b K(x,t) \int_a^b K(t,t_1)\, f(t_1)\, dt_1\, dt$$

$$+ \ldots$$

$$+ \lambda^n \int_a^b K(x,t) \int_a^b K(t,t_1) \ldots \int_a^b K(t_{n-2}, t_{n-1})$$

$$f(t_{n-1})\, dt_{n-1} \ldots dt_1\, dt + r_{n+1}(x) \qquad (9.111)$$

where $r_{n+1}(x)$ is the remainder of the solution and can be written as

$$r_{n+1}(x) = \lambda^{n+1} \int_a^b K(x,t) \int_a^b K(t,t_1)$$

$$\ldots \int_a^b K(t_{n-1}, t_n)\, u(t_n)\, dt_n \ldots dt_1\, dt \qquad (9.112)$$

Now we ask ourselves the question: Under what conditions does the above series converge? The nth term of the series is given by

$$T_n(x) = \lambda^n \int_a^b K(x,t) \int_a^b K(t,t_1)$$

$$\ldots \int_a^b K(t_{n-2},t_{n-1})\, f(t_{n-1})\, dt_{n-1} \ldots dt_1\, dt \qquad (9.113)$$

Let the upper bound on the following quantities be as follows:

$|K(x,t)| \leq K_0$ K_0 is the maximum value for K in $a \leq x \leq b$; $a \leq t \leq b$

$|f(x)| \leq f_0$ f_0 is the maximum value for f in $a \leq x \leq b$

$|u(x)| \leq U$ U is the maximum value for u in $a \leq x \leq b$

Therefore, the bound on the nth term becomes

$$|T_n(x)| \leq |\lambda^n| f_0 K_0^n (b - a)^n \tag{9.114}$$

since every integration introduces the following:

$$\lambda, K_0, (b - a)$$

Hence, the series with its general term represented by Equation (9.114) converges only when

$$|\lambda| K_0 (b - a) < 1 \tag{9.115a}$$

or

$$|\lambda| < \frac{1}{K_0 (b - a)} \tag{9.115b}$$

In this case, the remainder r_{n+1} as given by Equation (9.112) has the following limit

$$|r_{n+1}(x)| < |\lambda^{n+1}| U K_0^{n+1}(b - a)^{n+1} \tag{9.116}$$

As Equation (9.115a) holds for convergence, we obtain

$$\lim_{n \to \infty} r_{n+1}(x) = 0$$

Therefore, it can be concluded that the solution to Fredholm equation as given by Equation (9.111) does represent the exact solution as $n \to \infty$, and when the condition for convergence as given by Equation (9.115b) is met. The method is illustrated by the following example.

Example 9.8 Solve the following integral equation using the method of successive substitution

$$u(x) = \frac{5}{6} x + \frac{1}{2} \int_0^1 xt \, u(t) \, dt \tag{9.117}$$

Solution: Let us check first if the method of successive substitution applies to Equation (9.117). We have

$$K(x,t) = xt; \quad \text{limit } a = 0; \quad \text{limit } b = 1 \quad b - a = 1$$

$$f(x) = 5/6\,x; \quad 0 \le x \le 1, \quad 0 \le t \le 1$$

$$|K(x,t)| = |xt| \le 1 = K_0$$

$$\lambda = 1/2$$

Hence,

$$|\lambda|\, K_0\, (b - a) = (1/2)(1)(1) = 1/2 \tag{9.118}$$

Therefore, according to Equation (9.115) the method of successive substitution applies and the series representing the solution converges.

The nth term of the series can be written as

$$T_n(x) = \lambda^n \int_0^1 xt \int_0^1 tt_1 \ldots \int_0^1 t_{n-2}\, t_{n-1} \left(\frac{5}{6} t_{n-1}\right) dt_{n-1}\, dt_{n-2} \ldots dt_1\, dt \tag{9.119}$$

Integrating first t_{n-1}, and taking 5/6 outside the integrals, yields

$$T_n(x) = \frac{5}{6}\lambda^n \int_0^1 xt \int_0^1 tt_1 \ldots \int_0^1 t_{n-3}\, t_{n-2}\, \frac{t_{n-2}}{3} \tag{9.120}$$

Integrating now t_{n-2} yields another 1/3. If we integrate then n times, we obtain

$$T_n(x) = \frac{5}{6}\lambda^n \frac{1}{3^n} x \tag{9.121}$$

The solution, therefore, can be written as the sum of the terms in Equation (9.121) as

$$u(x) = \frac{5}{6}x + T_1(x) + T_2(x) + \ldots \tag{9.122}$$

or

$$u(x) = \frac{5}{6}x \left[1 + \frac{1}{6} + \left(\frac{1}{6}\right)^2 + \ldots \right] \tag{9.123}$$

or

$$u(x) = \frac{5}{6}x \left[\frac{1}{1 - 1/6} \right] \tag{9.124}$$

or

$$u(x) = x \tag{9.125}$$

2. Volterra Integral Equations

When the method of successive substitution is used to solve Volterra's equation, the procedure is quite similar to that of Fredholm's equation; however, the upper limit of integration is a variable. In this case the nth term of the series solution becomes

$$T_n(x) = \lambda^n \int_a^x K(x,t) \int_a^t K(t,t_1)$$

$$\ldots \int_a^{t_{n-2}} K(t_{n-2}, t_{n-1})\, f(t_{n-1})\, dt_{n-1} \ldots dt_1\, dt \qquad (9.126)$$

Equation (9.126) has a different condition for convergence than the one required for Fredholm's equation. Since we have

$$|K(x,t)| \leq K_0; \qquad |f(x)| \leq f_0$$

then

$$|T_n(x)| \leq |\lambda^n|\, f_0\, K_0^n \frac{(x-a)^n}{n!}$$

$$\leq |\lambda^n|\, f_0 \frac{[K_0(b-a)]^n}{n!} \qquad a \leq x \leq b \qquad (9.127)$$

Equation (9.127) represents the nth term of the series solution. Such a series is convergent for all values of

$$\lambda, \quad f_0, \quad K_0, \quad (b-a)$$

Therefore, this method does not put a restriction on the value of λ when applied to equations of the Volterra type.

Example 9.9 Solve the following Volterra's integral equation by the method of successive substitution

$$u(x) = x + \int_0^x (t-x)\, u(t)\, dt \qquad (9.128)$$

Solution: In this problem we have

$$K(x,t) = t - x; \quad \lambda = 1; \quad f(x) = x$$

The nth term of the series becomes

$$T_n(x) = \lambda^n \int_0^x \int_0^t \int_0^{t_1} \ldots \int_0^{t_{n-3}} \int_0^{t_{n-2}} (t-x)(t_1-t)$$

$$(t_2 - t_1) \ldots (t_{n-2} - t_{n-3})\, (t_{n-1} - t_{n-2})$$
$$t_{n-1}\, dt_{n-1} \ldots dt \qquad (9.129)$$

We integrate first t_{n-1} and get

$$\int_0^{t_{n-2}} [t_{n-1}^2 - t_{n-2}\, t_{n-1}]\, dt_{n-1}$$

$$= \left[\frac{t_{n-1}^3}{3} - t_{n-2}\,\frac{t_{n-1}^2}{2}\right]_{t_{n-1}=0}^{t_{n-1}=t_{n-2}}$$

$$= \left(\frac{1}{3} - \frac{1}{2}\right) t_{n-2}^3 = -\frac{1}{(2)(3)}\, t_{n-2}^3$$

$$= -\frac{1}{3!}\, t_{n-2}^3 \tag{9.130}$$

The next integral is

$$\int_0^{t_{n-3}} \left[-(t_{n-2} - t_{n-3})\,\frac{1}{3!}\, t_{n-2}^3\right] dt_{n-2}$$

$$= -\frac{1}{3!}\left(\frac{1}{5} - \frac{1}{4}\right) t_{n-3}^5 = +\frac{1}{3!}\left[\frac{1}{(4)(5)}\right] t_{n-3}^5 \tag{9.131}$$

$$= +\frac{1}{5!}\, t_{n-3}^5 \tag{9.132}$$

Therefore, it can easily be seen that

integrating once yields power 3 for t with $-$ coefficient
integrating twice yields power 5 for t with $+$ coefficient
integrating three times yields power 7 for t with $-$ coefficient

Hence, the nth term of the series can be written as

$$T_n(x) = \frac{(-1)^n\, x^{2n+1}}{(2n + 1)!} \tag{9.133}$$

and the solution of the equation becomes

$$u(x) = \sum_{n=0}^{\infty} \frac{(-1)^n\, x^{2n+1}}{(2n + 1)!}$$

$$= x - \frac{x^3}{3!} + \frac{x^5}{5!} - \frac{x^7}{7!} + \ldots \tag{9.134}$$

This series is the expansion of $\sin x$. Therefore, the solution is

$$u(x) = \sin x \tag{9.135}$$

We note that Equation (9.128) could have been solved in a much simpler way by changing it into a differential equation as follows: Differentiating Equation (9.128) with respect to x yields

$$u'(x) = 1 + \int_0^x -u(t)\,dt \tag{9.136}$$

A second differentiation results in

$$u''(x) = -u(t)$$

or

$$u'' + u = 0 \tag{9.137}$$

The solution to Equation (9.137) is

$$u(x) = A\cos x + B\sin x \tag{9.138}$$

From Equations (9.128) and (9.136) we have

$$u(x) = 0, \quad u'(x) = 1 \quad \text{when } x = 0 \tag{9.139}$$

Therefore,

$$A = 0 \quad B = 1 \tag{9.140}$$

The solution becomes

$$u(x) = \sin x \tag{9.141}$$

Example 9.10 The temperature distribution $T(x,t)$ of a semi-infinite solid, (Figure 9-8), initially at a uniform temperature T_i and with a time dependent heat flux $f(t)$ incident at its surface, can be written as[c]

$$V(x,t) \equiv T(x,t) - T_i$$

$$= \frac{1}{\sqrt{k\rho c}\,\sqrt{\pi}} \int_0^t f(t-\tau)\, e^{-(x^2/4\pi\tau)} \frac{d\tau}{\tau^{1/2}} \tag{9.142}$$

where t = time

k = thermal conductivity

ρ = density

c = specific heat

The surface temperature at $x = 0$ becomes

$$V(0,t) = V_0(t) = \frac{1}{\sqrt{k\rho c}\,\sqrt{\pi}} \int_0^t f(t-\tau) \frac{d\tau}{\tau^{1/2}} \tag{9.143}$$

When a beam of constant intensity I_0 and the surface absorptivity α_i is taken as constant, then

---

[c] H.S. Carslaw and J.C. Jaeger, *Conduction of Heat in Solids* (Oxford: Oxford University Press, 1959).

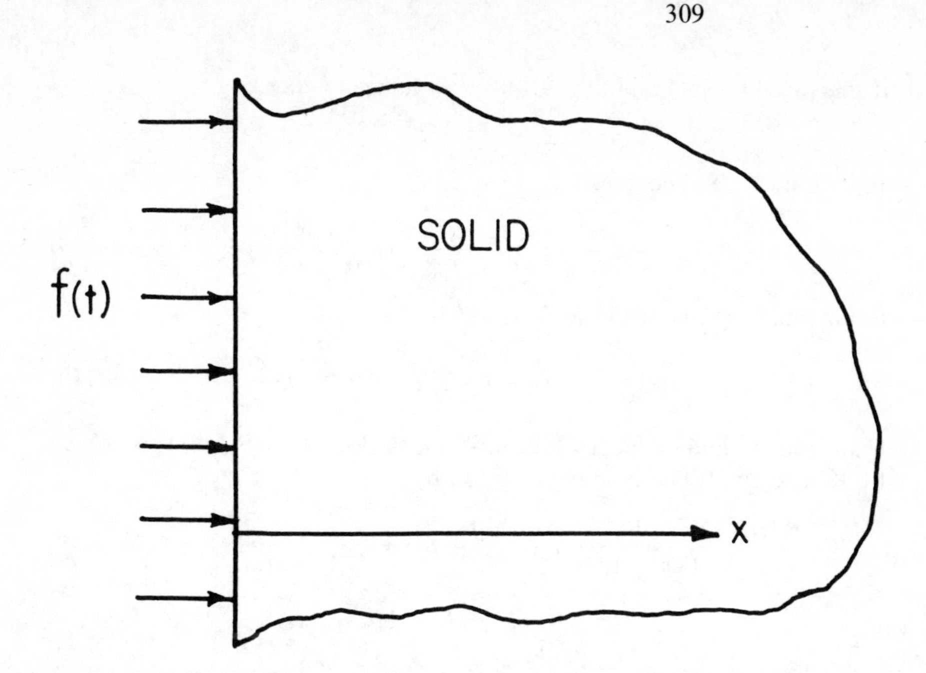

Figure 9-8. Semi-infinite Solid Exposed to a Time Dependent Heat Flux

$$f(t) = I_0 \, \alpha_i = \text{constant} \tag{9.144}$$

and Equation (9.143) can be integrated directly to yield

$$V = \frac{I_0 \alpha_i}{\sqrt{k\rho c}} \frac{2}{\sqrt{\pi}} t^{1/2} \equiv \frac{2}{\sqrt{\pi}} E t^{1/2} \tag{9.145}$$

Find the expression for the temperature distribution and for the surface temperature of the solid when the absorptivity of the solid is a function of temperature given as

$$\alpha = \alpha_i \frac{T}{T_i} \tag{9.146}$$

Solution: When α is given by Equation (9.146), the flux absorbed by the surface becomes

$$f(t) = I_0 \, \alpha = I_0 \, \alpha_i \left[1 + \frac{V_0(t)}{T_i} \right] \tag{9.147}$$

When Equation (9.147) is substituted in Equation (9.143) we obtain the following Volterra's integral equation

$$V_0(t) = \frac{I_0 \, \alpha_i}{\sqrt{k\rho c} \, \sqrt{\pi}} \int_0^t \left[1 + \frac{V_0(t - \tau)}{T_i} \right] \frac{d\tau}{\tau^{1/2}} \tag{9.148}$$

If the variables in Equation (9.148) are changed according to

$$\tau = t - \xi$$

Equation (9.148) becomes

$$V_0(t) = E\frac{2}{\sqrt{\pi}}t^{1/2} + \frac{E}{T_i\sqrt{\pi}}\int_0^t \frac{V_0(\xi)}{(t-\xi)^{1/2}}d\xi \tag{9.149}$$

Equation (9.149) is of the form

$$V_0(t) = f(t) + \lambda\int_0^t K(t, \xi)\, V_0(\xi)\, d\xi \tag{9.150}$$

Using the method of successive substitution, Equation (9.149) was solved by K.R. Chun.[d] The results are given by

$$\begin{aligned}\frac{V_0(t)}{T_i} &= \frac{1}{\Gamma(1+1/2)}Y + \frac{1}{\Gamma(2)}Y^2 + \frac{1}{\Gamma(2+1/2)}Y^3 + \dots \\ &\quad + \frac{1}{\Gamma(1+n/2)}Y^n + \dots \\ &= \sum_{n=1}^{\infty}\frac{1}{\Gamma(1+n/2)}Y^n \end{aligned}\tag{9.151}$$

where

$$Y = (E/T_i)t^{1/2}$$

$$\Gamma(n) = (n-1)!$$

$$\Gamma(n+1/2) = (n-0.5)\dots(2.5)(1.5)(0.5)\sqrt{\pi}$$

When Equation (9.151) is used in Equation (9.147) and then $f(t)$ is introduced in Equation (9.142) the temperature distribution within the solid is then obtained.

9.7 The Method of Successive Approximations

This method is different than the method of successive substitution, and it is illustrated as follows: Consider the Fredholm type equation

$$u(x) = f(x) + \lambda\int_a^b K(x,t)\, u(t)\, dt \tag{9.152}$$

We take a zero-order approximation for $u(x)$ to be $u_0(x)$. $u_0(x)$ can be taken a constant, however, usually it is taken equal to $f(x)$, that is,

[d] K.R. Chun, "Surface Heating of Metallic Mirrors in High Power Laser Cavities," *Journal of Heat Transfer* 96 (1974), pp. 43-47.

$$u_0(x) = f(x) \tag{9.153}$$

Next this expression for $u_0(x)$ is substituted in the right-hand side of Equation (9.152) to give the first-order approximation as

$$u_1(x) = f(x) + \lambda \int_a^b K(x,t)\, u_0(t)\, dt \tag{9.154}$$

When this expression for $u_1(x)$ is substituted into Equation (9.152), a second approximation results, and it is

$$u_2(x) = f(x) + \lambda \int_a^b K(x,t)\, u_1(t)\, dt \tag{9.155}$$

Hence, we can write

$$u_n(x) = f(x) + \lambda \int_a^b K(x,t)\, u_{n-1}\,(t)\, dt \tag{9.156}$$

If we adopt the following notation

$$\begin{aligned} K_2(x,t) &= \int K(x,t_1)\, K(t_1,t)\, dt_1 \\ K_3(x,t) &= \int K(x,t_1)\, K_2(t_1,t)\, dt_1 \\ &\vdots \\ K_m(x,t) &= \int K(x,t_1)\, K_{m-1}(t_1,t)\, dt_1 \end{aligned} \tag{9.157}$$

we can write then the following expressions (with $u_0(x) = f(x)$)

$$\begin{aligned} u_1(x) &= f(x) + \lambda \int_a^b K(x,t) f(t)\, dt \\ u_2(x) &= f(x) + \lambda \int_a^b K(x,t) f(t)\, dt \\ &\quad + \lambda^2 \int_a^b K(x,t) [K(t,t_1) f(t_1)\, dt_1]\, dt \end{aligned}$$

or

$$u_2(x) = f(x) + \lambda \int_a^b K(x,t) f(t)\, dt + \lambda^2 \int_a^b K_2\,(x,t) f(t)\, dt$$

and

$$u_3(x) = f(x) + \lambda\int_a^b K(x,t)f(t)\,dt + \lambda^2\int_a^b K_2(x,t)f(t)\,dt$$

$$+ \lambda^3\int_a^b K_3(x,t)f(t)\,dt$$

Hence, the nth approximation becomes

$$u_n(x) = f(x) + \sum_{i=1}^{n} \lambda^i \int_a^b K_i(x,t)f(t)\,dt \tag{9.158}$$

As n approaches ∞, there results what is referred to as Neumann series, stated as

$$u(x) = \lim_{n\to\infty} u_n(x) = f(x) + \sum_{i=1}^{\infty} \lambda^i \int_a^b K_i(x,t)f(t)\,dt \tag{9.159}$$

The series given by Equation (9.158) or Equation (9.159) converges when

$$|\lambda|\, K_0\,(b - a) < 1 \tag{9.160}$$

which is the same condition for convergence found for the method of successive substitution when applied to Fredholm equation; see Equations (9.113) through (9.115). The following example illustrates further this method.

Example 9.11 Using the method of successive approximation, solve the following integral equation

$$u(x) = e^x + \frac{1}{2}\int_0^1 e^{x-t}\,u(t)\,dt \tag{9.161}$$

Solution: Applying Equation (9.159) gives

$$K_1(x,t) = e^{x-t}$$

$$K_2(x,t) = \int_0^1 e^{x-t_1} e^{t_1-t}\,dt_1 = e^{x-t} = K_1(x,t)$$

Repeated application of the method yields

$$K_1(x,t) = K_2(x,t) = K_3(x,t) = K_i(x,t) = e^{x-t} \tag{9.162}$$

Rewriting Equation (9.158) as

$$u(x) = f(x) + \lambda\int_a^b \sum_{i=1}^{\infty} \lambda^{i-1} K_i(x,t)f(t)\,dt \tag{9.163}$$

and expanding Equation (9.163) gives

$$u(x) = e^x + \lambda\,[1 + \lambda + \lambda^2 + \ldots]\int_0^1 (e^{x-t})\,e^t\,dt \qquad (9.164)$$

$$= e^x + \frac{\lambda}{(1 - \lambda)}\int_0^1 e^x\,dt \qquad (9.165)$$

Performing the integration yields

$$u(x) = e^x\left[1 + \frac{\lambda}{1 - \lambda}\right] = \frac{e^x}{1 - \lambda} \qquad (9.166)$$

With $\lambda = 1/2$, the solution becomes

$$u(x) = 2e^x \qquad (9.167)$$

Note: the series for λ in Equation (9.164) converges only for $\lambda < 1$.

This example can be solved by the method of Article 9.2 as follows because the equation has a degenerate kernel:

$$u(x) = e^x + \frac{1}{2}e^x\int_0^1 e^{-t}\,u(t)\,dt \qquad (9.168)$$

or

$$u(x) = e^x + \frac{1}{2}e^x\,C \qquad \text{where } C = \int_0^1 e^{-t}\,u(t)\,dt \qquad (9.169)$$

Therefore,

$$e^x\left[1 + \frac{C}{2}\right] = e^x + \frac{1}{2}e^x\int_0^1 e^{-t}\left[e^t + e^t\frac{C}{2}\right]dt \qquad (9.170)$$

$$= e^x + \frac{e^x}{2}\int_0^1\left(1 + \frac{C}{2}\right)dt \qquad (9.171)$$

or

$$e^x\left[1 + \frac{C}{2}\right] = e^x\left[1 + \frac{1}{2}\left(1 + \frac{C}{2}\right)\right] \qquad (9.172)$$

Equation (9.172) yields $C = 2$ and the solution becomes

$$u(x) = 2e^x \qquad (9.173)$$

Equation (9.163) can be written as

$$u(x) = f(x) + \lambda\int_a^b \Gamma(x,t;\,\lambda)\,f(t)\,dt \qquad (9.174)$$

where

$$\Gamma(x,t;\lambda) = \sum_{i=1}^{\infty} \lambda^{i-1} K_i(x,t) \tag{9.175}$$

$\Gamma(x,t;\lambda)$ is referred to as the resolvent kernel of the integral equation. It is a known quantity.

Example 9.12 Solve the following integral equation by the method of successive approximation

$$B(\phi) = \epsilon\sigma T^4(\phi) + \frac{(1-\epsilon)}{2}\int_0^{\pi} \sin\phi' \, B(\phi')\, d\phi' \tag{9.176}$$

This is the equation for the radiosity as obtained earlier in Article 9.3 for a spherical geometry.
Solution: The resolvent kernel $\Gamma(x,t;\lambda)$ is

$$\Gamma(x,t;\lambda) = \sum_{i=1}^{\infty} \lambda^{i-1} K_i(x,t) \tag{9.177}$$

$$K_1 = \sin\phi \tag{9.178}$$

$$K_2 = \int_0^{\pi} \sin\phi \sin\phi' \, d\phi' = 2\sin\phi \tag{9.179}$$

$$K_3 = \int_0^{\pi} 2\sin\phi \sin\phi' \, d\phi' = 2^2 \sin\phi \tag{9.180}$$

and

$$K_i = 2^{i-1} \sin\phi \tag{9.181}$$

Therefore, applying Equation (9.174) to Equation (9.176) gives

$$\begin{aligned} B(\phi) &= \epsilon\sigma T^4(\phi) + \frac{(1-\epsilon)}{2}\int_0^{\pi} \Gamma(x,t;\lambda)\, \epsilon\sigma T^4(\phi')\, d\phi' \\ &= \epsilon\sigma T^4(\phi) + \frac{(1-\epsilon)}{2}\int_0^{\pi} \sum_{i=1}^{\infty} \lambda^{i-1} \\ &\quad (2^{i-1}\sin\phi')\, \epsilon\sigma T^4(\phi')\, d\phi' \end{aligned} \tag{9.182}$$

However,

$$\lambda = \frac{1-\epsilon}{2} \tag{9.183}$$

Therefore, Equation (9.182) becomes

$$\begin{aligned} B(\phi) = \epsilon\sigma T^4(\phi) + \sigma\frac{(1-\epsilon)}{2}\int_0^{\pi} \sum_{i=1}^{\infty} \epsilon\left(\frac{1-\epsilon}{2}\right)^{i-1} \\ 2^{i-1}\sin\phi' \; T(\phi')\, d\phi' \end{aligned} \tag{9.184}$$

or

$$B(\phi) = \epsilon\sigma T^4(\phi) + \frac{\sigma(1-\epsilon)}{2}\left\{ \epsilon[1 + (1-\epsilon)^1 + (1-\epsilon)^2 + (1-\epsilon)^3 + \ldots] \int_0^\pi \sin\phi' \, T(\phi')\, d\phi' \right\} \quad (9.185)$$

or

$$B(\phi) = \epsilon\sigma T^4(\phi) + \frac{\sigma(1-\epsilon)}{2} \frac{\epsilon}{1-(1-\epsilon)} \int_0^\pi \sin\phi' \, T(\phi')\, d\phi' \quad (9.186)$$

or

$$B(\phi) = \epsilon\sigma T^4(\phi) + \frac{\sigma(1-\epsilon)}{2} \int_0^\pi \sin\phi' \, T(\phi')\, d\phi' \quad (9.187)$$

This is the same result obtained in Example 9.4.

Example 9.13 Solve the following equation by successive approximation

$$u(x) = 10\,x + \int_0^1 xt\, u(t)\, dt \quad (9.188)$$

Solution: The resolvent kernel in this case yields

$$K_1 = xt \quad (9.189)$$

$$K_2 = \int_0^1 (xt_1)(t_1 t)\, dt_1 = \frac{xt}{3} \quad (9.190)$$

$$K_3 = \int_0^1 \left(\frac{xt_1}{3}\right)(t_1 t)\, dt_1 = \frac{xt}{9} \quad (9.191)$$

Therefore,

$$K_i = \frac{xt}{3^{i-1}} \quad (9.192)$$

$$\Gamma(x,t;\lambda) = \sum_{i=1}^{\infty} (1)^{i-1} \frac{xt}{3^{i-1}} \quad (9.193)$$

Hence, applying Equation (9.174) we obtain

$$u(x) = 10x + \int_0^1 \sum_{i=1}^{\infty} \frac{xt}{3^{i-1}} (10t)\, dt \quad (9.194)$$

$$= 10x + 10x\left[\sum_{i=1}^{\infty} \frac{1}{3^i}\right] \quad (9.195)$$

$$= 10x + 10x\left[\frac{1}{1 - 1/3}\right] \tag{9.196}$$

or

$$u(x) = 10x\left(\frac{5}{2}\right) = 25x \tag{9.197}$$

9.8 Hilbert-Schmidt Theory

Another very useful and powerful method of solving integral equations of the Fredholm type is the Hilbert-Schmidt Method,[e] which is outlined below.

Let us recall first the homogeneous Fredholm equation of the second kind

$$u(x) = \lambda \int_a^b K(x,t)\, u(t)\, dt \tag{9.198}$$

$K(x,t)$ in the present analysis is assumed to be symmetric and real, that is,

$$K(x,t) = K(t,x)$$

The solution to Equation (9.198), as was presented earlier in this chapter, was shown to be a set of eigenfunctions, say $\phi_i(x)$, $i = 1, 2, \ldots$, which correspond to the eigenvalues λ_i, $i = 1, 2, \ldots$. the eigenfunctions are orthogonal (see chapter 3), that is,

$$\int_a^b \phi_i(x)\, \phi_j(x)\, dx = 0 \qquad j \neq i \tag{9.199}$$

and the λ's are real.

If $\phi(x)$ is an eigenfunction belonging to an eigenvalue λ_0, so that

$$\phi(x) = \lambda_0 \int_a^b K(x,t)\, \phi(t)\, dt \tag{9.200}$$

then another function $\psi(x)$ given as

$$\psi(x) = C\, \phi(x) \qquad C \neq 0 \tag{9.201}$$

is also an eigenfunction belonging to the same eigenvalue λ_0. The constant C can be chosen so that

---

[e] We note that Hilbert-Schmidt theory focuses on establishing that the eigenfunctions are orthogonal and that the eigenvalues are real. Solutions of the homogeneous equation are obtained by methods similar to the ones covered earlier in this chapter. However, the method and the form of solution that follow are referred to as the Hilbert-Schmidt method of solution.

$$\int_a^b \psi^2(x)\,dx = 1 \tag{9.202}$$

In this case the function $\psi(x)$ is said to be normalized. Therefore, if we write

$$\int_a^b [C\,\phi(x)]^2\,dx = \int_a^b \psi^2(x)\,dx = 1$$

or

$$C^2 \int_a^b \phi^2(x)\,dx = 1$$

the value of C needed to normalize the eigenfunction $\phi(x)$ becomes

$$C = \pm \frac{1}{\sqrt{\int_a^b \phi^2(x)\,dx}} \tag{9.203}$$

Hence,

$$\psi(x) = \frac{\phi(x)}{\sqrt{\int_a^b \phi^2(x)\,dx}} \tag{9.204}$$

Let the kernel in Equation (9.198) be represented as a series of normalized eigenfunctions pertaining to Equation (9.198) as

$$K(x,t) = \sum_{n=1}^{\infty} a_n(x)\,\psi_n(t) \tag{9.205}$$

Substituting Equation (9.205) in Equation (9.198) yields

$$u(x) = \lambda \int_a^b \left[\sum_{n=1}^{\infty} a_n(x)\,\psi_n(t) \right] u(t)\,dt \tag{9.206}$$

However, $\psi_i(x)$ is a solution to Equation (9.198), that is,

$$u(x) = u_i(x) = \psi_i(x) \tag{9.207}$$

Equation (9.206), then, after changing the order of integration and summation becomes

$$\psi_i(x) = \lambda_i \sum_{n=1}^{\infty} a_n(x) \int_a^b \psi_n(t)\,\psi_i(t)\,dt \tag{9.208}$$

$\psi_n(t)$ and $\psi_i(t)$ are orthogonal; therefore, the integral on the right-hand side of Equation (9.208) is zero when $i \neq n$. Hence, Equation (9.208) becomes

$$\psi_i(x) = \lambda_i\, a_i(x) \tag{9.209}$$

and

$$a_i(x) = \frac{\psi_i(x)}{\lambda_i} \qquad i = n \tag{9.210}$$

When Equation (9.210) is introduced in Equation (9.205), the kernel $K(x,t)$ can then be expressed as

$$K(x,t) = \sum_{n=1}^{\infty} \frac{\psi_i(x)\,\psi_n(t)}{\lambda_n} \tag{9.211}$$

Nonhomogeneous Fredholm Integral Equation of the Second Kind—λ is not an Eigenvalue

Suppose now that it is required to find the solution to the nonhomogeneous Fredholm integral equation of the second kind

$$u(x) = f(x) + \lambda \int_a^b K(x,t)\,u(t)\,dt \tag{9.212}$$

Let first the solutions to the corresponding homogeneous integral equation (that is, $f(x) = 0$ in Equation (9.212)) be known and given in terms of normalized eigenfunctions, that is,

$$\psi_n(x) = \lambda_n \int_a^b K(x,t)\,\psi_n(t)\,dt \tag{9.213}$$

Let $u(x)$ and $f(x)$ in Equation (9.212) be expanded in terms of the eigenfunctions $\psi_n(x)$ as

$$u(x) = \sum_{n=1}^{\infty} A_n\,\psi_n(x) \tag{9.214}$$

and

$$f(x) = \sum_{n=1}^{\infty} B_n\,\psi_n(x) \tag{9.215}$$

Substituting Equations (9.214) and (9.215) in Equation (9.212) yields

$$\sum_{n=1}^{\infty} A_n\,\psi_n(x) = \sum_{n=1}^{\infty} B_n\,\psi_n(x) + \lambda \int_a^b K(x,t) \sum_{n=1}^{\infty} A_n\,\psi_n(t)\,dt \tag{9.216}$$

Interchanging the order of integration and summation in Equation (9.216) and recognizing that from Equation (9.213) we have

$$\frac{\psi_n(x)}{\lambda_n} = \int_a^b K(x,t)\,\psi_n(t)\,dt \tag{9.217}$$

Equation (9.216) then becomes

$$\sum_{n=1}^{\infty} A_n\,\psi_n(x) = \sum_{n=1}^{\infty} B_n\,\psi_n(x) + \lambda \sum_{n=1}^{\infty} \frac{A_n\,\psi_n(x)}{\lambda_n} \tag{9.218}$$

Multiplying now Equation (9.218) by $\psi_i(x)$ and integrating over (a,b) using the orthogonality of the eigenfunctions $\psi_n(x)$ yield (*note:* $\int_a^b \psi_n^2\,dx = 1$)

$$A_i = B_i + \lambda \frac{A_i}{\lambda_i} \tag{9.219}$$

Equation (9.219) can be written as

$$A_i = B_i + \frac{\lambda\,B_i}{\lambda_i - \lambda} \tag{9.220}$$

From Equation (9.215) the expression for B_i (or B_n) can be expressed as (see chapter 3)

$$B_i = \int_a^b f(t)\,\psi_i(t)\,dt \tag{9.221}$$

From Equation (9.219) we have also

$$\frac{A_n}{\lambda_n} = \frac{A_i}{\lambda_i} = \frac{B_i}{\lambda_i - \lambda} = \frac{\int_a^b f(t)\,\psi_i(t)\,dt}{\lambda_i - \lambda} \tag{9.222}$$

Hence, using Equations (9.221) and (9.222) in Equation (9.218) yields the solution to the nonhomogeneous equation in the form

$$\underbrace{\sum_{n=1}^{\infty} A_n\,\psi_n(x)}_{u(x)} = \underbrace{\sum_{n=1}^{\infty} B_n\,\psi_n(x)}_{f(x)} + \lambda \sum_{i=1}^{\infty} \frac{\int_a^b f(t)\,\psi_i(t)\,dt}{\lambda_i - \lambda}\,\psi_i(x) \tag{9.223}$$

or

$$u(x) = f(x) + \lambda \sum_{i=1}^{\infty} \frac{\int_a^b f(t)\,\psi_i(t)\,dt}{\lambda_i - \lambda}\,\psi_i(x) \tag{9.224}$$

Equation (9.224) holds for $\lambda \neq \lambda_i$; in other words, the equation holds for values of λ which are not eigenvalues. However, if $f(x) = 0$, solution exists only for $\lambda = \lambda_i$.

Nonhomogeneous Fredholm Integral Equation of the Second Kind—λ is an Eigenvalue

If λ in the nonhomogeneous equation is an eigenvalue, say λ_q, one might ask as to whether the equation can have a solution even if $\lambda = \lambda_q$. If $\lambda = \lambda_q$ is substituted in Equation (9.224), the last term on the right-hand side becomes infinite. However, to avoid this difficulty, let us return to Equation (9.219) and rewrite it for $\lambda = \lambda_q$. The result becomes

$$A_q = B_q + \lambda_q \frac{A_q}{\lambda_q} = B_q + A_q \tag{9.225}$$

or

$$B_q = 0 \tag{9.226}$$

From Equation (9.221) we have then

$$\int_a^b f(t)\,\psi_q(t)\,dt = 0 \tag{9.227}$$

This means that $f(t)$ and $\psi_q(t)$ are orthogonal. A similar conclusion was reached in Article 9.2, Equation (9.36).

Therefore, for $\lambda = \lambda_q$ (which is one of the λ_i), Equation (9.224) still holds for $i \neq q$. The summation in this case is taken over i excluding $i = q$, and the solution is written then as

$$u(x) = f(x) + A_q \psi_q + \lambda_q \sum_{\substack{i=1 \\ i \neq q}}^{\infty} \frac{\int_a^b f(t)\,\psi_i(t)\,dt}{\lambda_i - \lambda_q} \psi_i(x) \tag{9.228}$$

Equation (9.228) is very similar to the solution of the linear nonhomogeneous differential equation. The interpretation is that to the particular solution of the equation we may add any constant times a solution of the corresponding homogeneous differential equation. This is the case in Equation (9.228).

Example 9.14 Using Hilbert-Schmidt method, solve the following integral equation

$$u(x) = x + \int_0^1 (x + t)\,u(t)\,dt \tag{9.229}$$

Solution: First we check if λ in Equation (9.229) is an eigenvalue for the homogeneous portion of the equation written as

$$u(x) = \lambda \int_0^1 (x + t)\,u(t)\,dt \tag{9.230}$$

To find the eigenvalues for Equation (9.230) we need to find $D(\lambda)$ from Equation (9.25) and then set it equal to zero. Therefore, for $K(x,t) = x + t$ in Equation (9.230), $D(\lambda)$ becomes

$$D(\lambda) = 1 - \lambda - \frac{\lambda^2}{12} \tag{9.231}$$

Setting Equation (9.231) equal to zero yields the two eigenvalues as

$$\lambda_1 = -6 + 4\sqrt{3}; \qquad \lambda_2 = -6 - 4\sqrt{3} \tag{9.232}$$

In Equation (9.229), $\lambda = 1$ and therefore, it is not an eigenvalue. Equation (9.224), then, applies as a solution to this problem.

Next, we need to find the normalized eigenfunctions, $\psi_i(t)$, and, hence, the eigenfunctions for the homogeneous equation should be obtained.

The solution to Equation (9.230) is obtained using the method of Article 9.2, leading to the following two eigenfunctions:

$$\phi_1(x) = 1 + \sqrt{3}\,x \tag{9.233}$$

$$\phi_2(x) = 1 - \sqrt{3}\,x \tag{9.234}$$

The corresponding normalized eigenfunctions become

$$\psi_1(x) = \frac{\phi_1(x)}{\sqrt{\int_0^1 \phi_1^2(x)\,dx}} = \frac{1 + \sqrt{3}x}{\sqrt{\int_0^1 (1 + \sqrt{3}x)^2\,dx}} \tag{9.235}$$

$$\psi_2(x) = \frac{\phi_2(x)}{\sqrt{\int_0^1 \phi_2^2(x)\,dx}} = \frac{1 - \sqrt{3}x}{\sqrt{\int_0^1 (1 - \sqrt{3}x)^2\,dx}} \tag{9.236}$$

or

$$\psi_1(x) = \frac{1 + \sqrt{3}\,x}{(2 + \sqrt{3})^{1/2}} \tag{9.237}$$

$$\psi_2(x) = \frac{1 - \sqrt{3}\,x}{(2 - \sqrt{3})^{1/2}} \tag{9.238}$$

Applying now Equation (9.224) yields

$$u(x) = f(x) + \lambda \sum_{i=1}^{2} \frac{\int_0^1 f(t)\,\psi_i(t)\,dt}{\lambda_i - \lambda}\,\psi_i(x) \tag{9.239}$$

In Equation (9.239) we have

$f(x) = x$; $\lambda = 1$; $\lambda_i = -6 \pm 4\sqrt{3}$ (from Equation (9.232))

$\psi_i(x)$ are given by Equations (9.237) and (9.238). Hence, we can write

$$\begin{aligned} u(x) &= x + (1)\left[\frac{1}{(-6 + 4\sqrt{3}) - 1}\int_0^1 t\,\frac{1 + \sqrt{3}\,t}{(2 + \sqrt{3})^{1/2}}\,dt\right]\frac{1 + \sqrt{3}\,x}{(2 + \sqrt{3})^{1/2}} \\ &\quad + (1)\left[\frac{1}{(-6 - 4\sqrt{3}) - 1}\int_0^1 t\,\frac{1 - \sqrt{3}\,t}{(2 - \sqrt{3})^{1/2}}\,dt\right]\frac{1 - \sqrt{3}\,x}{(2 - \sqrt{3})^{1/2}} \end{aligned} \tag{9.240}$$

Performing the integration in Equation (9.240) gives the solution as

$$u(x) = -6x - 4 \tag{9.241}$$

This example could have been solved in a much easier way using the method of Article 9.2. However, it is important to note that Hilbert-Schmidt method is a more powerful method needed to solve more difficult problems.

Bibliography

Ordinary Differential Equations and Fourier Series

Bowman, Frank. *Introduction to Bessel Functions*. New York: Dover, 1958.

Boyce, W.E. and R.C. Diprima. *Elementary Differential Equations and Boundary Value Problems*. 2nd ed. New York: John Wiley and Sons, 1969.

Brauer, Fred and J.A. Nohel. *Ordinary Differential Equations*, New York: W.A. Benjamin, 1967.

Byerly, W.E. *Fourier Series*. New York: Dover, 1959.

Carslaw, H.S. *Introduction to the Theory of Fourier Series and Integrals*. New York: Dover, 1959.

Churchill, R.V. *Fourier Series and Boundary Value Problems*. New York: McGraw-Hill, 1963.

Davis, H.F. *Fourier Series and Orthogonal Functions*. Boston: Allyn and Bacon, 1963.

Farrel, O.J., and Bertran Ross. *Solved Problems: Gamma and Beta Functions, Legendre Polynomials, Bessel Functions*. New York: Macmillan, 1963.

Hildebrand, F.B. *Advanced Calculus for Applications*. Englewood Cliffs, N.J.: Prentice-Hall, 1962.

Hochstadt, Harry. *Differential Equations*. New York: Holt, Rinehart and Winston, 1964.

Kreyszig, Erwin. *Advanced Engineering Mathematics*. 3rd ed. New York: John Wiley and Sons, 1972.

Tolstov, G.P. *Fourier Series*. Englewood Cliffs, N.J.: Prentice-Hall, 1962.

Applied Mathematics

Arfken, George. *Mathematical Methods for Physicists*. New York: Academic Press, 1970.

Arpaci, Vedat. *Conduction Heat Transfer*. Reading, Mass.: Addison-Wesley, 1966.

Broman, A. *Introduction to Partial Differential Equations: From Fourier Series to Boundary Value Problems*. Reading, Mass.: Addison-Wesley, 1970.

Carslaw, H.S., and J.C. Jaeger. *Conduction of Heat in Solids*. Oxford: Oxford University Press, 1959.

Courant, Richard, and D. Hilbert. *Methods of Mathematical Physics*. vol. 2. New York: Interscience Publishers, 1962.

Dixon, Charles. *Applied Mathematics of Science and Engineering*. London: John Wiley and Sons, 1971.

Hildebrand, F.B. *Methods of Applied Mathematics*. Englewood Cliffs, N.J.: Prentice-Hall, 1965.

Irving, John, and N. Mullineaux. *Mathematics in Physics and Engineering*. New York: Academic Press, 1959.

Jeffreys, Harold, and Bertha Jeffreys. *Methods of Mathematical Physics*. Cambridge: Cambridge University Press, 1956.

Jenson, V.G., and G.V. Jeffreys. *Mathematical Methods in Chemical Engineering*. London: Academic Press, 1963.

Lamb, Horace. *Hydrodynamics*. 6th ed. New York: Dover, 1945.

Lin, C.C., L.A. Segal, and G.H. Handelman. *Mathematics Applied to Deterministic Problems in the Natural Sciences*. New York: Macmillan, 1974.

MacRobert, T.M. *Spherical Harmonics*. Oxford: Pergamon Press, 1967.

Moon, P.H., and D.E. Spencer. *Field Theory for Engineers*. New York: D. Van Nostrand, 1961.

Myint-U, T. *Partial Differential Equations of Mathematical Physics*. New York: American Elsevier, 1973.

Pipes, L.A., and L.R. Harwill. *Applied Mathematics for Engineers and Physicists*. 3rd ed. New York: McGraw-Hill, 1970.

Pollard, Harry. *Applied Mathematics: An Introduction*. Reading, Mass.: Addison-Wesley, 1972.

Raven, F.H. *Mathematics of Engineering Systems*. New York: McGraw-Hill, 1966.

Siegel, Robert, and J.R. Howell. *Thermal Radiation Heat Transfer*. New York: McGraw-Hill, 1972.

Smith, M.G. *Introduction to the Theory of Partial Differential Equations*. Princeton, N.J.: D. Van Nostrand, 1967.

Sobolov, S.L. *Partial Differential Equations of Mathematical Physics*. Reading, Mass.: Addison-Wesley, 1964.

Sokolnikoff, I.S., and R.M. Redheffer. *Mathematics of Physics and Modern Engineering*. New York: McGraw-Hill, 1966.

Tikhonov, A.N., and A.A. Samarskii. *Equations of Mathematical Physics*. New York: Pergamon Press, 1963.

Wylie, C.R., Jr. *Advanced Engineering Mathematics*. 3rd ed. New York: McGraw-Hill, 1966.

Integral Equations

Bôcher, Maxime. *An Introduction to the Study of Integral Equations*. Cambridge: Cambridge University Press, 1913.

Green, C.D. *Integral Equation Methods*. London: Thomas Nelson and Sons, Ltd., 1969.

Hochstadt, Harry. *Integral Equations*. New York: John Wiley and Sons, 1973.

Hoheisel, Guido. *Integral Equations*. New York: Ungar Publishing Co., 1968.

Kanwal, R.P. *Linear Integral Equations*. New York: Academic Press, 1971.

Lovitt, W.V. *Linear Integral Equations*. New York: Dover, 1950.

Mikhlin, S.G. *Integral Equations and their Applications to Certain Problems in Mechanics, Mathematical Physics and Technology*. New York: Pergamon Press, 1957. [Muskhelishvili, N.I. *Singular Integral Equations*. Holland: Noordhoff, 1953.]

Pegorzelski, W. *Integral Equations and their Applications*. vol. 1. New York: Pergamon Press, 1966.

Petrovskii, I.G. *Lectures on the Theory of Integral Equations*. Rochester, N.Y.: Graylock Press, 1957.

Smirnov, V.I. *A Course of Higher Mathematics*. vol. 4. London: Pergamon Press, 1964.

Smithies, F. *Integral Equations*. Cambridge: Cambridge University Press, 1958.

Tricomi, F.G. *Integral Equations*. New York: Interscience Publishers, 1957.

Widom, Harold. *Lectures on Integral Equations*. New York: Van Nostrand-Reinhold, 1969.

Index

About the Author

I. S. Habib is a professor of mechanical engineering at the University of Michigan—Dearborn and was the chairman of the department from 1971 to 1975. He received the MSME from the Virginia Polytechnic Institute and the Ph.D. from the University of California, Berkeley. Widely known for his research in heat and mass transfer, Dr. Habib has published numerous articles on radiation heat transfer, turbulent flow, change of phase, liquid spray cooling, and mathematical methods in thermo-fluid sciences. His industrial experience and consultation includes work on compact heat exchangers, hydraulic network, and high temperature gas discharge.